Atilla Elçi, Mamadou Tadiou Koné, and Mehmet A. Orgun (Eds.)

Semantic Agent Systems

T0205336

Studies in Computational Intelligence, Volume 344

Editor-in-Chief

Prof. Janusz Kacprzyk
Systems Research Institute
Polish Academy of Sciences
ul. Newelska 6
01-447 Warsaw
Poland
E-mail: kacprzyk@ibspan.waw.pl

Further volumes of this series can be found on our
homepage: springer.com

Vol. 320. xxx

Vol. 321. Dimitri Plemenos and Georgios Miaoulis (Eds.)
Intelligent Computer Graphics 2010
ISBN 978-3-642-15689-2

Vol. 322. Bruno Baruque and Emilio Corchado (Eds.)
Fusion Methods for Unsupervised Learning Ensembles, 2010
ISBN 978-3-642-16204-6

Vol. 323. Yingxu Wang, Du Zhang, and Witold Kinsner (Eds.)
Advances in Cognitive Informatics, 2010
ISBN 978-3-642-16082-0

Vol. 324. Alessandro Soro, Vargiu Eloisa, Giuliano Armano,
and Gavino Paddeu (Eds.)
*Information Retrieval and Mining in Distributed
Environments*, 2010
ISBN 978-3-642-16088-2

Vol. 325. Quan Bai and Naoki Fukuta (Eds.)
Advances in Practical Multi-Agent Systems, 2010
ISBN 978-3-642-16097-4

Vol. 326. Sheryl Brahnam and Lakhmi C. Jain (Eds.)
*Advanced Computational Intelligence Paradigms in
Healthcare 5*, 2010
ISBN 978-3-642-16094-3

Vol. 327. Slawomir Wiak and
Ewa Napieralska-Juszczak (Eds.)
*Computational Methods for the Innovative Design of
Electrical Devices*, 2010
ISBN 978-3-642-16224-4

Vol. 328. Raoul Huys and Viktor K. Jirsa (Eds.)
Nonlinear Dynamics in Human Behavior, 2010
ISBN 978-3-642-16261-9

Vol. 329. Santi Caballé, Fatos Xhafa, and Ajith Abraham (Eds.)
*Intelligent Networking, Collaborative Systems and
Applications*, 2010
ISBN 978-3-642-16792-8

Vol. 330. Steffen Rendle
Context-Aware Ranking with Factorization Models, 2010
ISBN 978-3-642-16897-0

Vol. 331. Athena Vakali and Lakhmi C. Jain (Eds.)
New Directions in Web Data Management 1, 2011
ISBN 978-3-642-17550-3

Vol. 332. Jianguo Zhang, Ling Shao, Lei Zhang, and
Graeme A. Jones (Eds.)
Intelligent Video Event Analysis and Understanding, 2011
ISBN 978-3-642-17553-4

Vol. 333. Fedja Hadzic, Henry Tan, and Tharam S. Dillon
Mining of Data with Complex Structures, 2011
ISBN 978-3-642-17556-5

Vol. 334. Álvaro Herrero and Emilio Corchado (Eds.)
Mobile Hybrid Intrusion Detection, 2011
ISBN 978-3-642-18298-3

Vol. 335. Radomir S. Stankovic and Radomir S. Stankovic
From Boolean Logic to Switching Circuits and Automata, 2011
ISBN 978-3-642-11681-0

Vol. 336. Paolo Remagnino, Dorothy N. Monekosso, and
Lakhmi C. Jain (Eds.)
Innovations in Defence Support Systems – 3, 2011
ISBN 978-3-642-18277-8

Vol. 337. Sheryl Brahnam and Lakhmi C. Jain (Eds.)
*Advanced Computational Intelligence Paradigms in
Healthcare 6*, 2011
ISBN 978-3-642-17823-8

Vol. 338. Lakhmi C. Jain, Eugene V. Aidman, and
Canicious Abeynayake (Eds.)
Innovations in Defence Support Systems – 2, 2011
ISBN 978-3-642-17763-7

Vol. 339. Halina Kwasnicka, Lakhmi C. Jain (Eds.)
Innovations in Intelligent Image Analysis, 2010
ISBN 978-3-642-17933-4

Vol. 340. Heinrich Hussmann, Gerrit Meixner, and
Detlef Zuehlke (Eds.)
Model-Driven Development of Advanced User Interfaces, 2011
ISBN 978-3-642-14561-2

Vol. 341. Stéphane Doncieux, Nicolas Bredeche, and
Jean-Baptiste Mouret(Eds.)
New Horizons in Evolutionary Robotics, 2011
ISBN 978-3-642-18271-6

Vol. 342. Federico Montesino Pouzols, Diego R. Lopez, and
Angel Barriga Barros
*Mining and Control of Network Traffic by Computational
Intelligence*, 2011
ISBN 978-3-642-18083-5

Vol. 343. XXX

Vol. 344. Atilla Elçi, Mamadou Tadiou Koné, and
Mehmet A. Orgun (Eds.)
Semantic Agent Systems, 2011
ISBN 978-3-642-18307-2

Atilla Elçi, Mamadou Tadiou Koné,
and Mehmet A. Orgun (Eds.)

Semantic Agent Systems

Foundations and Applications

 Springer

Dr. Atilla Elçi
Software Engineering Program,
Toros University,
Mersin, Turkey
E-mail: atilla.elci@gmail.com

Dr. Mehmet A. Orgun
Department of Computing
Macquarie University
Sydney, NSW 2109
Australia
E-mail: mehmet.orgun@mq.edu.au

Dr. Mamadou Tadiou Koné
Independent Computing Research, Boston,
Roxbury, MA, USA
E-mail: Kone.Mamadou@gmail.com

ISBN 978-3-642-26694-2   ISBN 978-3-642-18308-9 (eBook)

DOI 10.1007/978-3-642-18308-9

Studies in Computational Intelligence   ISSN 1860-949X

© 2011 Springer-Verlag Berlin Heidelberg
Softcover reprint of the hardcover 1st edition 2011

Typeset & Cover Design: Scientific Publishing Services Pvt. Ltd., Chennai, India.

Printed on acid-free paper

9 8 7 6 5 4 3 2 1

springer.com

Preface

To bring our contribution to the advancement of the semantic Web and agent technologies, around November 2005, we initiated the series of *IEEE International Workshops on Engineering Semantic Agent Systems* in conjunction with *the 30th International Computer Software and Applications Conference (COMPSAC 2006)* in Chicago, Illinois, USA. Encouraged by five very successful annual workshops and two journal special issues on this theme, we felt confident that the research on semantic agents system has become mature enough to gather from experts and practitioners contributions for an edited volume in the series *Studies in Computational Intelligence* by Springer-Verlag.

The theme of this book, semantic agent systems, refers to the integration of semantic Web, artificial intelligence, and software agent technologies. Here semantic Web is described as a Web of semantically-linked data which aims to enable man, machine, and software to carry out more useful tasks. Central to this theme are software agents with the power to use linked data and its associated semantics through technologies such as RDF, SPARQL, OWL, and SKOS. When these software agents are elements of cooperating multi-agent systems, a whole host of new opportunities emerges. Research and development in this direction need more effort and dedication like the contributions in this book.

To appeal to a wide range of audiences, we organized this book along four main parts: The first part titled *Introduction to Agents and Semantics* aims to give the reader an idea of what a semantic agent system is in the first place. Here, Rule Responder is presented as a framework for semantic multi-agent systems which support collaborative teams. Then, handling complex obligations with semantic Web techniques in the context of multi-agent systems is presented. Finally, a noteworthy contribution of semantic Web technology to the field of distributed knowledge management is pointed out.

Part two titled *Engineering Semantic Agent Systems* deals with ontology development and management for agent systems in addition to agent-oriented software engineering. Here, we read about information governance, argumentation and reconciliation issues, complexity for MAS design, and composition of business processes, and human-robot interactions.

Part three titled *Applications of Semantic Agent Systems* deals with semantic Web applied to specific areas and pertinent lessons learned from these applications. Contributions in this section highlight domains as diverse as the cyber-physical world, manufacturing systems, context-aware mobile learning services, and user interests.

In the last part titled *Future Outlook*, we discover in great detail many of the intricacies of machine understanding and their potential in future research directions. The reader is led in this section to one of the foundations of the theme of this book: semantic agents with understanding abilities.

We are grateful to many authors who brought their contributions to this book with their valuable studies. We would like to express our appreciation to the reviewers who accepted to read and give their insightful comments on these contributions. Without their help and expert opinions, it would have been impossible to make decisions on each submitted paper and produce such a high-quality volume.

Our special thanks go to Thomas Ditzinger, Engineering Editor at Springer Verlag and responsible for the series *Studies in Computational Intelligence*, for his support, cooperation, patience and understanding. We kept in mind all along that this book was made possible through the foresight, timely initiative, and unfaltering support of Janusz Kacprzyk, the Editor-in-Chief of the *Studies in Computational Intelligence* Series.

November 2010   Atilla Elçi, Mersin, Turkey
Mamadou Tadiou Koné, Boston, MA, USA
Mehmet A. Orgun, Sydney, Australia

Reviewers

Contents

Part III: Applications of Semantic Agent Systems

Part IV: Future Outlook

Part I

Introduction to Agents and Semantics

Chapter 1
Rule Responder Agents
Framework and Instantiations*

Harold Boley[1] and Adrian Paschke[2]

[1] Institute for Information Technology, National Research Council Canada,
Fredericton, NB, Canada
harold.boley@nrc.gc.ca
[2] Freie Universitaet Berlin, Germany
paschke@mi.fu-berlin.de

Abstract. This chapter introduces Rule Responder and its applications. Rule Responder is a framework for specifying virtual organizations as semantic multi-agent systems that support collaborative teams. It provides the infrastructure for rule-based collaboration between the distributed members of such a virtual organization. Human members of an organization are assisted by (semi-) autonomous rule-based agents, which use Semantic Web rules to describe aspects of their owners' derivation and reaction logic. To implement different distributed system/agent toplogies with their negotiation/coordination mechanisms Rule Responder instantiations employ three core classes of agents - Organizational Agents (OA), Personal Agents (PAs), and External Agents (EAs). The OA represents goals and strategies shared by its virtual organization as a whole, using a rulebase that describes its policies, regulations, opportunities, etc. Each PA assists a group or person of the organization, semi-autonomously acting on their behalf by using a local knowledge base of rules defined by the entity. EAs can communicate with the virtual organization by sending messages to the public interfaces of the OA. EAs can be human users using, e.g., Web forms or can be automated services/tools sending messages via the multitude of transport protocols of the underlying enterprise service bus (ESB) middleware. The agents employ ontologies in their knowledge bases to represent semantic domain vocabularies, normative pragmatics and pragmatic context of conversations and actions, as well as the organizational semiotics.

1 Introduction

Rule Responder[1] extends the Semantic Web towards a Pragmatic Web infrastructure for collaborative rule-based agent networks realizing distributed inference services,

---

* Invited Chapter.
[1] http://responder.ruleml.org

A. Elçi, M.T. Koné, and M.A. Orgun (Eds.): Semantic Agent Systems, SCI 344, pp. 3–23.
springerlink.com   © Springer-Verlag Berlin Heidelberg 2011

where independent agents engage in conversations by exchanging messages and cooperate to achieve (collaborative) goals. Rule Responder can be characterized on three levels, from general to specific.

- It models a virtual organization of agents recursively as again being a single agent, forming what has been called [15] a hierarchy of *holons* (or, a *holarchy*).
- It supports different *interaction/coordination models*, where information is interchanged within a *pragmatic context* (e.g. language action speech acts, deontic norms, etc.).
- It provides a technical Web-based *multi-agent architecture* which supports different distribution models (distributed agent system topologies).

A virtual organization as a whole is represented by an Organizational Agent (OA), which uses ontologies and rules to assign and delegate incoming tasks (e.g., queries) to responsible Personal Agents (PAs). Rule Responder agents communicate in conversations that allow implementing different agent coordination and negotiation protocols. The interaction and interpretation is driven by the organizational semiotics which details how the information flow works within and between organizations. For instance, an OA can use a responsibility assignment matrix, represented as an ontology, to find an appropriate PA in its organization. The OA can then send a message (e.g., a query) to that PA and receive results (e.g., answers), typically using reaction rules. By means of pragmatic primitives, such as speech acts, deontic norms, etc., which are represented as ontologies, Rule Responder attaches the semantic and pragmatic context, e.g. organizational norms, purposes or goals and values, to the interchanged messages.

In its multi-agent architecture Rule Responder utilizes messaging reaction rules from Reaction RuleML[2] for communication between the distributed agent inference services. The Rule Responder middleware is based on modern enterprise service technologies and Semantic Web technologies for implementing intelligent agent services that access data and ontologies, receive and detect events (e.g., for complex event processing in event processing agent networks), and make rule-based inferences and (semi-)autonomous pro-active decisions for reactions based on these representations.

The core of a Rule Responder agent is a rule engine, such as Prova[3], OO jDREW, DR-Device (initially in Emerald), Euler, or Drools, which implements the decision and behavioral reaction logic of the agents' roles. An agent can employ vocabularies defined as Semantic Web ontologies (e.g., based on RDFS or OWL) to give its rules a domain-specific meaning. The vocabularies can be used within the conversation with other agents to enable a semantic and pragmatic interpretation of the messages. For the deployment of agents on the Web and for the communication in agent networks, Rule Responder uses the Mule-based enterprise service bus middleware, which supports a multitude of synchronous and asynchronous transport protocols (> 40) -- such as MS, SMTP, JDBC, TCP, HTTP, XMPP, Jade -- to transport rulebases, queries and answers between the

---

[2] http://reaction.ruleml.org
[3] http://prova.ws

agents. Reaction RuleML, the de facto standard for XML-serialized reaction rules, is used as a platform-independent rule interchange format for agent conversation.

In summary, Rule Responder can be seen to support a *digital ecosystem*, evolving from the Semantic Web [4] to the Pragmatic Web, which consists of all the semantic agents in one or more virtual organizations, as well as all the other components of this environment with which the agents interact, such as other services, tools, the ESB middleware, etc.

Several instantiations of Rule Responder have been developed, including the eScience infrastructure for Health Care and Life Sciences [11], the Rule-based IT Service Level Managment and Semantic BPM system [12, 13], multiple versions of the deployed SymposiumPlanner system [9], two versions of the WellnessRules prototype [5], and the PatientSupporter prototype.[4]

The rest of the chapter is organized as follows. Section 2 discusses the agent architecture and used technologies of the Rule Responder framework. Section 3 explains a typical distributed agent topology for virtual organizations and the types agents used to implement it. Section 4 focuses on interchange between the semantic agents which communicate by using (Reaction) RuleML as common rule interchange format. Section 5 demonstrates some application use cases of Rule Responder by means of selected Rule Responder instantiations. Section 6 concludes the paper.

2   The Rule Responder Framework

Three interconnected architectural layers consitute the Rule Responder framework, listed here from top to bottom:

- Computationally independent user interfaces such as template-based Web forms or controlled English rule interfaces.
- Reaction RuleML as the common platform-independent rule interchange format to interchange rules, events, queries, and data between Rule Responder agents and other agents (e.g., Semantic Web services or humans via Web forms).
- A highly scalable and efficient enterprise service bus (ESB) as agent/service-broker and communication middleware on which platform-specific rule engines are deployed as distributed agent nodes (resp. semantic inference services). These engines manage and execute the logic of Rule Responder's semantic agents in terms of declarative rules which have access to semantic ontologies.

In the following, the Rule Responder framework will be refined, and explained from bottom to top.

2.1   Mule Enterprise Service Bus

To seamlessly handle message-based interactions between the Rule Responder agents/services and other agents/services using disparate complex event processing (CEP) technologies, transports, and protocols, an enterprise service bus (ESB) --

---

[4] http://ruleml.org/PatientSupporter

the Mule open-source ESB [5] -- is used in Rule Responder as the communication middleware. This ESB allows deploying the rule-based agents as highly distributed rule inference services installed as Web-based endpoints on the Mule object broker and supports the communication in this rule-based agent processing network via a multitude of transport protocols (see Figure 1). That is, the ESB provides a highly scalable and flexible application messaging framework to communicate synchronously or asynchronously amongst the ESB-local agents and with agents/services on the Web.

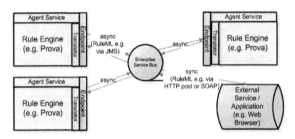

Fig. 1. Distributed Rule Responder Agent Services

Mule is a messaging platform-based on principles of ESB architectures, but goes beyond the typical definition of an ESB as a transit system for carrying data between applications by providing a distributable object broker to manage all sorts of service components such as the Rule Responder agent services. The three processing modes of Mule are:

- Asynchronous: many events (messages) can be processed by the same component at a time in various threads. When the Mule server is running asynchronously instances of a component run in various threads all accepting incoming events, though an event will only be processed by one instance of the component.
- Synchronous: when a component receives an event message, in this mode the whole request is executed in a single thread.
- Request-Response: this allows for a component to make a specific request for an event and wait for a specified time to get a response back.

The object broker follows the Staged Event Driven Architecture (SEDA) pattern [20]. The basic approach of SEDA is to decomposes a complex, event-driven application into a set of stages connected by queues. This design decouples event and thread scheduling from application logic and avoids the high overhead associated with thread-based concurrency models. That is, SEDA supports massive concurrency demands on Web-based services and provides a highly scalable approach for asynchronous communication.

Figure 2 shows a simplified breakdown of the integration of Mule into the Rule Responders framework.

---

[5] www.mulesoft.org

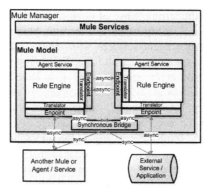

Fig. 2. Layering of Rule Responder on Mule ESB

Distributed agent services (see Figure 1), which at their core run a rule engine, are deployed as Mule components which listen at configured endpoints, e.g., JMS message endpoints, HTTP ports, SOAP server/client addresses or JDBC database interfaces, etc. Reaction RuleML is used as a common platform-independent rule interchange format between the agents (and possible other rule execution/inference services). Translator services are used to translate inbound and outbound messages from platform-independent Reaction RuleML into the platform-specific execution syntaxes of rule engines, and vice versa. XSLT and ANTLR based translator services are provided as Web forms, HTTP services and SOAP Web services on the Reaction RuleML Web page.

The large variety of transport protocols provided by Mule can be used to transport the messages to the registered endpoints or external applications/tools. Usually, JMS is used for the internal communication between distributed agent instances, while HTTP and SOAP is used to access external Web services. The usual processing style is asynchronous using SEDA event queues. However, sometimes synchronous communication is needed. For instance, to handle communication with external synchronous HTTP clients such as Web browsers where requests, e.g. by a Web from, are sent through a synchronous channel. In this case, a synchronous bridge component dispatches the requests into the asynchronous messaging framework and collects all answers from the internal service nodes, while keeping the synchronous channel with the external service open. After all asynchronous answers have been collected, they are sent back to the still connected external service via the HTTP-synchronous channel.

2.2 Selected Platform-Specific Rule Engines for Rule Responder Agents

The core of a Rule Responder agent, which is deployed as a service component on the Rule Responder ESB, is a platform-specific rule engine. These engines might differ, e.g., in their supported rule types, state representation, rule evaluation mechanism, conflict resolution and truth maintenance. Hence, depending on their expressiveness and functionalities, these rule engines might be capable of implementing agents in the strong sense of cognitive architectures for intelligent agents with goal/task-based, utility-based and learning-based functionalities, or in the weak sense of inference agent services with simple reflexive functionalities for, e.g., deductive query-answering capabilities. Following the general consensus defined by the strong notion of agency in [21], a Rule Responder agent, in addition to being (semi-)autonomous, should be capable of reactive, proactive, and communicative behavior. Additionally, it is often important that certain mentalistic notions[6] can be used in the rule language for describing the agent behavior in an abstract and intuitive way, e.g. in the interactions between agents to communicate the pragmatics of the interchanged information.

In the following, the interplay between our most often used rule engines Prova, OO jDREW, Euler will be discussed, although there are other engines such as DR-Device and Drools supported by Rule Responder.

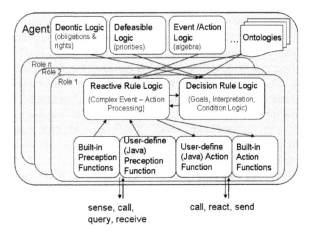

Fig. 3. Rule Responder Agent

Figure 3 shows the architecture of an intelligent cognitive Rule Responder agent which is implemented in Prova. Prova is an enterprise-strength, highly expressive distributed Semantic Web logic programming (LP) rule engine. The Prova rule engine supports different rule types:

- Derivation rules to describe the agent's decision logic
- Integrity rules to describe constraints and potential conflicts
- Normative rules to represent the agent's permissions, prohibitions and obligation policies
- Global ECA-style reaction rules to define global reaction logic which are triggered on the basis of detected (complex) events
- Messaging reaction rules to define conversation-based workflow reaction and behavioral logic based on complex event processing

---

[6] The term *mentalistic notions* aka *mental attitudes* refers to human-like properties such as beliefs, goals, etc. when transferred to describing machine agents.

Prova follows the spirit and design of the W3C Semantic Web initiative and combines declarative rules, ontologies and inference with dynamic object-oriented programming and access to external data sources via built-in query languages such as SQL, SPARQL, and XQuery.

File Input / Output

```
..., fopen(File,Reader), ...
```

XML (DOM)

```
document(DomTree,DocumentReader) :-   XML(DocumenReader),...
```

SQL

```
... ,sql_select(DB,cla,[pdb_id,"1alx"],[px,Domain]).
```

RDF

```
...,rdf(http://...,"rdfs",Subject,"rdf_type","gene1_Gene"),...
```

XQuery

```
..., XQuery = 'for $name in StatisticsURL//Author[0]/@name/text()
   return $name', xquery_select(XQuery,name(ExpertName)),...
```

SPARQL

```
...,sparql_select(SparqlQuery,...
```

One of the key advantages of Prova is its elegant separation of logic, data access, and computation as well as its tight integration of Java, Semantic Web technologies, with service-oriented computing and complex event processing. In particular, Prova supports external type systems such as, e.g., Java class hierarchies or Semantic Web ontologies (RDFS, OWL) via its typed order-sorted logic [18]. For instance, in the following example all agents from an external OWL ontology responsibility assignment matrix (RAM) are assigned to the typed variable *Agent* of type *Organizing_Committee* (with the namespace *ruleml*2010), where *Organizing_Committee* is a concept defined in the RAM ontology. The query then selects all agent individuals of type *ProgramChair* which is a subtype of *Organizing_Committee* , i.e. the query selects a subset with appropriate subtype from the bound variable.

```
% import external ontology representing responsibility assignment
matrix (RAM)
import("http://2010.ruleml.org/RuleML-2010.owl").
% bind all agent instances of type "Organizing_Committee" from the
RAM to the variable Agent
agent(Agent:ruleml2010_Organizing_Committee).
% query for all agents of type "ProgramChair"
:- solve(agent(Agent:ruleml2010_ProgramChair)
```

Prova can be run in a plain Java environment as stand alone application or rule inference service on the Rule Responder ESB, or as an OSGI component. Prova has a modular knowledge base to implement several different roles an agent might play in the same agent instance. Each role has its own set of reaction rules to autonomously react (potentially proactive) on detected situations (complex events) and its own set of decision rules to interpret goals and derive decisions according to conditional proofs. For instance, it is possible to consult (load) distributed rulebases from local files, a Web address, or from incoming messages transporting a rulebase.

```
%load from a local file
:- eval(consult("organization2009.prova")).
% import from a Web address
:- eval(consult("http://ruleml.org/organization2010.prova")).
```

The rulebases are managed as modules in the knowledge base. Their module label can be used for asserting or retracting complete modules from the knowledge base and for scoping queries/goals to a particular module, i.e. the query only applies to the particular scoped module. In the following example the subgoal *agent*(*Agent*) applies on the modules *organization*2009.*prova* and not on the module *http*://*ruleml.org*/*organization*2010.*prova*.

```
responsible(Agent, Task) :-
  @src("organization2009.prova") agent(Agent),
  ...
```

To sense the environment and trigger actions, query data from external sources such as databases, call external procedural code such as Enterprise Java Beans, and receive/send messages from/to other agents or external services, Prova provides a set of built-in functions and additionally can dynamically instantiate any Java object and call its API methods at runtime. For instance, the following simple rule creates a response sentence with the name using Java string computations and displays it via to the Java system out console.

```
hello(Name) :-
  S = java.lang.String("Hello, your name is "),
  S.append (Name),
  java.lang.System.out.println (S).
```

Additional libraries can be imported, e.g. to represent rights and obligations of agents, implement conflict handling rules, or describe complex events and actions. In its cognitive cycle a Prova agent follows the sense-reason-act pattern. However, Prova does not define one particular cognitive cycle, but allows configuring an agent with user-defined conversation-based negotiation and coordination protocols or workflow patterns. Via constructs for asynchronously sending and receiving event messages within rules, an agent interacts with the environment. The main language constructs of messaging reaction rules are: *sendMsg* predicates to send messages, reaction *rcvMsg* rules which react to inbound messages, and *rcvMsg* or *rcvMult* inline reactions in the body of messaging reaction rules to receive one or more context-dependent multiple inbound event messages:

```
sendMsg(XID,Protocol,Agent,Performative,Payload |Context)
rcvMsg(XID,Protocol,From,Performative,Paylod|Context)
rcvMult(XID,Protocol,From,Performative,Paylod|Context)
```

Here, *XID* is the conversation identifier (conversation-id) of the conversation to which the message will belong. *Protocol* defines the communication protocol. *Agent* denotes the target party of the message. *Performative* describes the pragmatic envelope for the message content. A standard nomenclature of performatives is, e.g., the FIPA Agents Communication Language (ACL). *Payload* represents the message content sent in the message envelope. It can be a specific query or answer or a complex interchanged rule base (set of rules and facts). For instance, the following rule snippet shows how a query is sent to an agent via the ESB and then an answer is received from this agent.

```
...
sendMsg(Sub_CID,esb,Agent,acl_query-ref, Query),
rcvMsg(Sub_CID,esb,Agent,acl_inform-ref, Answer),
...
```

Prova does not define a specific set of mentalistic notions as first-class programming constructs. Instead, interchanged messages besides the conversation's metadata and payload also carry the pragmatic context of the conversation such as communicative situations/acts, mentalistic notions, organizational and individual norms, purposes or individual goals and values. The payload of incoming event messages is interpreted with respect to the local conversation state, which is denoted by the conversation id, and the pragmatic context, which is given by a pragmatic performative. For instance, a standard nomenclature of pragmatic performatives, which can be integrated as external (semantic) vocabulary/ontology, is e.g., defined by the Knowledge Query Manipulation Language (KQML) (Finin et al. 1993), by the FIPA Agent Communication Language (ACL), which gives several speech act theory-based communicative acts, or by the Standard Deontic Logic (SDL) with its normative concepts for obligations, permissions, and prohibitions. Depending on the pragmatic context, the message payload is used, e.g. to update the internal knowledge of the agent (e.g., add new facts or rulebases), add new tasks (goals), or detect a complex event pattern (from event-instance sequences).

Several expressive logic formalisms are supported by Prova [17], e.g., for updating the knowledge base (transactional update logic), defining and detecting complex events (complex event algebra), handling situations/states (event calculus), as well as for reasoning (e.g., deontic logic for normative reasoning on permissions, prohibitions, obligations) and planning (abductive reasoning on plans and goals).

In summary, Prova agents can interchange event information, rules (tasks), and queries/answers in agent conversations, including information about the semantics and pragmatics of the interchanged information.

Besides Prova, Rule Responder supports rule engines such as OO jDREW, Euler, DR-Device, and Drools for implementing such query answering agents as inference services in Rule Responder.

3   Rule Responder Agents

With the support of Prova's agent conversations, various distributed coordination topologies can be implemented, from centralized orchestration, executed in star-like agent nodes, to decentralized ad-hoc choreography within the Rule Responder agent network. In the following, we describe a common hierarchical agent topology which represents a centralized star-like structure for virtual organizations (and many orchestrated distributed systems). Organizational Agents (OAs) act as central orchestration nodes which control and disseminate the information flow from and to their internal Personal Agents (PAs) and the External Agents/Services (EAs).

3.1   Organizational Agent

An Organizational Agent (OA) represents its virtual organization as a whole. An OA manages its local Personal Agents (PAs), providing control of their life cycle and ensuring overall goals and policies of the organization and its semiotic structures. OAs can act as a single point of entry to the managed sets of local PAs to which requests by EAs are disseminated. This allows for efficient implementation of various mechanisms of making sure the PAs functionalities are not abused (security mechanisms) and making sure privacy of entities, personal data, and computation resources is respected (privacy & information hiding mechanisms). For instance, an OA can disclose information about the organization to authorized external parties without revealing private information and local data of the PAs, although this data might have been used in the PAs to compute the resulting answers to the external requester.

OAs, which require high levels of expressiveness to represent the logic of cognitive agents, are implemented using the Prova Semantic Web rule engine. In the following we will discuss some of the expressive language constructs of Prova that are required to implement the Rule Responder framework.

For implementing the Rule Responder communication flows in the OAs, Prova messaging reaction rules are used. A typical coordination pattern implemented in a Rule Responder OA is the following messaging reaction rule (Prova variables start with an upper-case letter), which waits for an incoming query from an EA and delegates this query to an internal responsible PA.

```
% receive query and delegate it to another party
rcvMsg(CID,esb, Requester, acl_query-ref, Query) :-
  responsibleRole(Agent, Query),
  sendMsg(Sub-CID,esb,Agent,acl_query-ref, Query),
  rcvMsg(Sub-CID,esb,Agent,acl_inform-ref, Answer),
  ... (other goals)...
  sendMsg(CID,esb,Requester,acl_inform-ref,Answer).
```

When activated by an incoming request from an EA, e.g. an HTTP request coming from a Web form, this messaging reaction rule first selects the responsible role for the query. Then the rule sends the query in a new sub-conversation to the selected party and waits for the answer to the query. That is, the rule execution waits until an answer event message is received in the inlined sub-conversation, which activates the process flow again, e.g. to prove further `standard' goals, e.g. with information from the received answer, which is assigned to variables in the normal logic programming way, including also backtracking to other variable assignments. Finally, in this example, the rule sends back the answer to the original requesting EA.

The selection logic for the dissemination of queries to PAs is, e.g., implemented by a standard derivation rule which, e.g., accesses, via a Prova SPARQL query built-in, an external responsibility assignment matrix (RAM) (see section 3.4). The following rule selects responsible agents with a SPARQL query on a triple store Web interface, where the responsibility assignment matrix is stored.

```
% receive query and delegate it to another party
rcvMsg(CID,esb, Requester, acl_query-ref, Query) :-
  responsibleRole(Agent, Query),
  sendMsg(Sub-CID,esb,Agent,acl_query-ref, Query),
  rcvMsg(Sub-CID,esb,Agent,acl_inform-ref, Answer),
  ... (other goals)...
  sendMsg(CID,esb,Requester,acl_inform-ref,Answer).
```

RAMs (RACI matrices, Linear Responsibility Charts, etc.) are often in project management, when responsibilities are clearly defined for each role. It should be noted that Prova OAs can also implement other well-known agent coordination and negotiation mechanisms: for instance, a Contract Net coordination protocol, where PAs bid for the task offered by the OA and the OA selects the best PA according to the received bids, or a publish-subscribe protocol, where PAs are selected according to their subscriptions with the OA.

3.2 Personal Agents

Personal Agents (PAs) assist the local entities of a virtual organization. Often these are human roles in the orgnization. But, it might be also services or applications in, e.g. a service oriented architecture. A PA runs a rule engine which accesses different sources of local data and computes answers according to the local rule-based decision logic of the PA. Depending on the required expressivness to represent the PAs rule logic arbitrary rule engines can be used as long as they provide an interface to ask queries and receive answers which are translated into the common Reaction RuleML interchange format in order to communicate with other agents.

Importantly, the PAs might have local autonomy and might support privacy and security implementations. In particular, local information used in the PA rules becomes only accessible by authorized access of the OA via the public interfaces of the PA which act as an abstraction layer supporting security and information hiding. A typical coordination protocol is that all communication to EAs is via the OA, but the OA might also reveal the direct contact address of a PA to authorized external agents which can then start an ad-hoc conversation directly with the PA [6]. A PA itself might act as a nested suborganization, i.e. containing itself an OA providing access to a suborganization within the main virtual organization. This can be usefull to represent nested organizational structures such as departments, project teams, and service networks.

3.3 External Agents

External Agents (EAs) constitute the points-of-contact that allow an external user or service to query the Organizational Agent (OA) of a virtual organization. An EA is based, e.g., on a Web (HTTP) interface that allows such an enquiry user to pose queries, employing a menu-based Web form, which gets translated to an equivalent RuleML/XML message. An external agent -- from the point of view of a Rule Responder agent organization -- can be an external human agent, a service/tool, or another external Rule Responder organization, thus leading to cross-organizational Rule Responder communication.

3.4 Responsibility Assignment Matrix

As one possible way for coordination in a virtual organization the Rule Responder framework uses a `pluggable' Responsibility Assignment Matrix (RAM) to support the OA in its selection of a PA and its optional participating profiles underneath. A RAM describes the responsibility of agent roles in completing certain tasks or deliverables in a virtual organization. A standard RAM is a RAI matrix, with

- *R*esponsible -- agents who do the work to achieve the task. Typically, the PAs are the responsible roles.
- *A*ccountable (also Approver or final Approving authority) -- agent who is ultimately accountable for the correct and thorough completion of the deliverable or task, and the one to whom Responsible is accountable. Typically, this is the OA which receives the answer from the PA and further processes it before forwarding it to the EA.
- *I*nformed -- the agent who is kept up-to-date on progress, often only on completion of the task or deliverable; and with whom there is just one-way communication. Typically, this is the EA who is informed about the result by the OA.

In a simple star-like Rule Responder agent topology, a single RAI matrix can be used in the OA to map an incoming query to the PA whose local knowledge base is deemed to be best suited for answering it. The RAI matrix is represented as an OWL ontology (OWL Lite) and can be used by a Rule Responder agent via querying it with the Semantic Web built-ins of Prova, binding the respective roles and their responsibilities to typed variables in the agent's rule logic. Many variants of the RAM with different role distinctions are possible such as RACI (with Consulted agents), RASCI (with Supporting agents) etc. - see, e.g., table 1.

Table 1. Responsibility Assignment Matrix

	General Chair	Program Chair	Publicity Chair
Symposium	responsible	consulted	supportive
Website	accountable	responsible	
Sponsoring	informed, signs	verifies	responsible
Submission	informed	responsible	
...	...	...	...

For instance, the RAM has been split so that role responsibility assignment is done on the 'higher' level of a Group Responsibility Matrix (GRM) in the OA and on the 'lower' level of a Profile Responsibility Matrix (PRM) in the PAs.

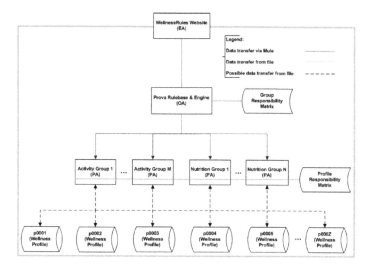

Fig. 4. Rule Responder architecture instantiated to WellnessRules

Figure 4 shows these two central matrices in the larger context of the Rule Responder architecture used in the WellnessRules instantiations (cf. Section 5.2), which has been further evolved for the PatientSupporter instantiation (cf. Section 5.3).

The GRM maps, many-to-one, relevant kinds of queries to a PA, who may represent a group. The GRM is usually specified as an OWL light ontology. The PRM lists a PA's profiles, participating in its group, along with the format (Prova, POSL, N3, etc.) each profile knowledge base is written in. The PRM is specified as an XML document.

4 Translation between Rule Responder Agents

Rule Responder permits agents to use local languages and engines, only requiring that all rulebases, queries, and answers will be translated to RuleML for transmitting them to other agents over the Mule ESB.

Reaction RuleML provides a translator service framework with Web form interfaces accepting controlled natural language input or predefined selection-based rule templates for the communication with external (human) agents on the computational independent level, as well as Servlet HTTP interfaces, and Web service SOAP interfaces, wich can be used for translation into and from platform-specific rule languages such as Prova.

On the computation-independent level, online user interfaces allow external human agents issuing queries to Rule Responder agents (typically the OA) in a controlled natural language or with template-driven Web forms and receive answers. The translation between the used controlled English rule language (Attempto Controlled English [14]) and Reaction RuleML is based on domain-specific language translation rules in combination with a controlled English translator service.

Queries to Rule Responder are formulated in Attempto Controlled English. The ACE2RML translator forwards the text to the Attempto Parsing Engine (APE), which translates the text into a discourse representation structure (DRS) and/or advices to correct malformed input. The DRS gives a logical/structural representation of the text. It is fed into an XML parser which translates it into a domain-specific Reaction RuleML representation of the query. Besides parsing and processing the elements of the DRS, the parser additionally employs domain-specific transformation rules to correctly translate the query into a public interface call of a Rule Responder OA.

On the platform-independent and platform-specific level, the translator services are using different translation technologies such as XSLT stylesheet, JAXB, etc. to translate from and to Reaction RuleML as a general rule interchange format. Reaction RuleML incorporates various kinds of production, action, reaction, and KR temporal/event/action logic rules as well as (complex) event/action messages into the native RuleML syntax. The general syntax of reaction rules is as follows:

```
<Rule style="active|messaging|reasoning" eval="strong|weak|defeasible|fuzzy">
     <oid>      <!-- object id -->                          </oid>
     <label>    <!-- meta data of the rule -->              </label>
     <scope><!-- scope of the rule e.g. a rule module --> </scope>
     <qualification> <!-- e.g. priorities, validity, fuzzy levels -->
                                                        </qualification>
     <quantification> <!- e.g. variable bindings-->      </quantification>
     <on>       <!-- event part -->                        </on>
     <if>       <!-- condition part -->                    </if>
     <then>     <!-- (logical) conclusion part -->         </then>
     <do>       <!-- action part -->                       </do>
     <after>    <!-- postcondition part after action, e.g.
                     to check effects -->                  </after>
</Rule>
```

Depending on which parts of this general rule syntax are used different types of reaction rules can be expressed, e.g. if-then (derivation rules), if-do (production rules), on-do (trigger rules), on-if-do (ECA rules). For communication between distributed rule-based (agent) systems Reaction RuleML provides a general message syntax:

```
<Message>
  <oid>        <!-- conversation ID-->                    </oid>
  <protocol>   <!-- used protocol -->                     </protocol>
  <agent>      <!-- sender/receiver agent/service -->     </agent>
  <directive><!-- pragmatic primitive, i.e. context --></directive>
  <content>    <!-- message payload -->                   </content>
</Message>
```

Using these messages agents can interchange events (e.g., queries and answers) as well as complete rule bases (rule set modules), e.g. for remote parallel task processing. Agents can be engaged in long running possibly asynchronous conversations and nested sub-conversations using the conversation id to manage the conversation state. The protocol is used to defines the message passing and coordination protocol. The directive attribute corresponds to the pragmatic instruction, i.e. the pragmatic characterization of the message context broadly characterizing the meaning of the message.

The Reaction RuleML translator services are configured in the transport channels of the inbound and outbound links of the deployed rule engines on the ESB. Incoming Reaction RuleML messages (receive) are translated into platform-specific rulebases which can be executed by the rule engine, e.g. Prova, and outgoing rulebases (send) are translated into Reaction RuleML in the outbound channels before they are transferred via a selected transport protocol.

The semantic agent architecture in Rule Responder supports privacy and security implementations. In particular, local information used in the PAs becomes only accessible by authorized access via the public interfaces of the OAs which act as an abstraction layer supporting security and information hiding. To achieve this, Prova supports an interface definition language (Reaction RuleML IDL) which allows descriptions of the signatures of publicly accessibly rule functions together with their mode and type declarations. *Modes* are states of instantiation of the predicate described by mode declarations, i.e. declarations of the intended input-output constellations of the predicate terms with the following semantics:

- " + " The term is intended to be input
- " − " The term is intended to be output
- " ? " The term is undefined/arbitrary (input or output)

For instance, the interface definition for the function $add(Arg1, Arg2, Result)$ is $interface(add(-,+,+))$, i.e. the function is a public interface which expect two input arguments and returns one output argument. $add(X,1,1)$ would be a valid query to this public function.

External agents can access the virtual organization only via these public interfaces, which often only reveal abstracted information to authorized users and hence hide local information of the organization and its PAs.

5 Rule Responder Instantiations

Early instantiations of Rule Responder include the Health Care and Life Sciences eScience infrastructure [11], the Rule-based IT Service Level Managment, and Semantic BPM system [12, 13]. Recent instantiations include multiple versions of the deployed SymposiumPlanner system [9], two versions of the WellnessRules prototype [5], PatientSupporter, a reputation management system, and a SCEP agent network. We will here highlight the principles of Rule Responder instantiations with an emphasis on the recent ones.

5.1 SymposiumPlanner

SymposiumPlanner is a series of deployed applications created with Rule Responder for the Q&A parts of the official websites of the RuleML Symposia.

Rule Responder started to support the organizing committee of the RuleML Symposium [8] and was further developed to assist the yearly RuleML Symposia since 2007. These applications embody responsibility assignment, automated first-level contacts for information regarding the symposium, helping the publicity chair with sponsoring correspondence, helping the panel chair with managing panel participants, and the liason chair with coordinating organization partners.

SymposiumPlanner utilizes a single organizational agent to handle the filtering and delegation of incoming queries. Each committee chair has a personal agent that acts in a rule-governed manner on behalf of the committee member. Each agent manages personal information, such as a FOAF-like profile containing a layer of facts about the committee member as well as FOAF-extending rules. These rules allow the PA to automatically respond to requests concerning the RuleML Symposium. Task responsibility for the organization is currently managed through a responsibility matrix, which defines the tasks committee members are responsible for. The matrix and the roles assigned within the virtual organization are defined by an OWL (Ontology Web Language) Lite Ontology.

Request users and personal agents can communicate by sending messages that transport queries, answers, or complete rulebases through the public EA interface of the OA (typically, an EA uses an HTTP port to which post and get requests are sent from a Web form). The Rule Responder instantiations to SymposiumPlanner are published and deployed online.[7]

5.2 WellnessRules

This is a Web 3.0 case study, where ontology-structured rules (including facts) about wellness opportunities are created by participants in rule languages such as

---

[7] http://ruleml.org/SymposiumPlanner

Prolog and N3, and translated for interchange within a wellness community using RuleML/XML. The wellness rules are centered around participants, as profiles, encoding knowledge about their activities, nutrition, etc. conditional on the season, the time-of-day, the weather, etc. This distributed knowledge base extends fact-only FOAF profiles with a vocabulary and rules about wellness group networking.

The communication between participants is organized through Rule Responder, permitting translator-based reuse of wellness profiles and their distributed querying across engines. WellnessRules interoperates between rules and queries in the relational (Datalog) paradigm of the pure-Prolog subset of POSL and in the frame (F-logic) paradigm of N3. These derivation rule languages are implemented in the engines OO jDREW and Euler, and connected via Rule Responder to support wellness communities.

WellnessRules is a system supporting the management of wellness practices within a community based on rules plus ontologies. The idea is the following. As in Friend of a Friend (FOAF)[8], people can choose a (community-unique) nickname and create semantic profiles about themselves, here about their wellness practices, for their own planning and to network with other people supported by a system that `understands' those profiles. As in FindXpRT [10], such FOAF-like fact-only profiles are extended with rules to capture conditional person-centered knowledge such as each person's wellness activity depending on the season, the time-of-day, the weather, etc. People can use rules of various refinement levels and rule languages ranging from pure Prolog to N3, which will be interoperated through RuleML/XML [3].

Interoperating with translators, WellnessRules thus frees participants from using any single rule language. In particular, it bridges between Prolog as the main Logic Programming rule paradigm and N3 as the main Semantic Web rule paradigm. The distributed nature of Rule Responder profiles, each queried by its own (copy of an) engine, permits scalable knowledge representation and processing.

WellnessRules has recently been developed to WellnessRules2, using a fourth kind of agent, the Computing Agent (CA), for accessing Google weather data. From the point of an OA, a CA can be queried similarly to a PA. However, while a PA is a personal assistant to a human owner, a CA is a pure machine agent, in WellnessRules2 acting as a wrapper for a Google service.

The Rule Responder instantiations to WellnessRules are further described and demoed online.[9]

5.3 PatientSupporter

Patients are increasingly seeking interaction in support groups, which provide shared information and experience about diagnoses, treatment, etc. PatientSupporter is an instantiation of Rule Responder that will permit a patient to query other patients' profiles for finding or initiating a matching group.

---

[8] http://www.foaf-project.org/
[9] http://ruleml.org/WellnessRules and http://ruleml.org/WellnessRules2

Rule Responder's External Agent (EA) is a Web-based patient-organization interface that passes queries to the Organizational Agent (OA). The OA represents the common knowledge of the virtual patient organization, delegates queries to relevant Personal Agents (PAs), and hands validated PA answers back to the EA. Each PA represents the medical subarea of primary interest to a corresponding patient group. The PA assists its patients by advertising their interest profiles employing rules about diagnoses and treatments as well as interaction constraints such as time, location, age range, gender, and number of participants.

PAs can be distributed across different rule engines using different rule languages (e.g., Prolog and N3), where rules, queries, and answers are interchanged via translation to and from RuleML/XML. The current implementation of PatientSupporter applies to a use case where the PA's medical subareas are defined through sports injuries structured by a partonomy of affected body parts.

PatientSupporter uses ontologies and rules for organizing geographically distributed patients -- here, suffering from sports injuries -- into virtual support groups around classes of an ontology of injuries -- here, a sports-injury partonomy. The prototype is designed to help patients with a similar sports injury to interact with a virtual support group having that common interest. Patients in an online PatientSupporter virtual organization create their semantic profile referring to classes in a disease ontology -- here a partonomy of body parts affected by sports injuries. Profiles contain rules about diagnoses and treatments as well as interaction constraints such as time, location, age range, gender, and number of participants. A patient can pose queries against the semantic profiles of other patients in his or her virtual organization to find or initiate a matching group.

PatientSupporter allows patients to have their profiles expressed in either Pure Prolog (Logic Programming rules) or N3 [2] (Semantic Web rules). Providing these quite different rule language paradigms permit patients to choose the language that best suits them. Rule Responder handles the interoperation between the rule languages of different patients using translators to and from RuleML/XML as the interchange format [7, 3].

Patients using the PatientSupporter Social Semantic Web portal are able to initiate the virtual support group about their sports injury on a global scale. They also benefit from PatientSupporter's interoperation facility in the background -- to transform patient profiles between Pure Prolog and N3 through RuleML/XML. The system employs a partonomy of sports-injury-affected body parts (a `body partonomy'), which makes it easy for patients to navigate hierarchically up or down to increase recall or precision, respectively. A patient's queries invoke other patients' interaction rules, allowing him or her to narrow down the search in a step-wise fashion. All of this saves a patient from browsing through a large set of irrelevant patient profiles and permits him or her to efficiently converge on a first Skype call.

The Rule Responder instantation to PatientSupporter is being described and demoed online.[10]

5.4 Reputation Management System

The Rule Responder reputation management system [1] is based on distributed Rule Responder rule agents, which use rules for implementing the reputation management functionalities as rule agents, and which use Semantic Web ontologies for representing simple or complex multi-dimensional reputation objects. This Semantic Web reputation ontology model enables reputation portability, eases the management of reputation data, mitigates risks in open environments, and enhances the decision making process in the reputation processing agents. The reputation management system computes, manages, and provides reputation about entities which act on the Web. It is implemented as a *Reputation Processing Network (RPN)* consisting of *Reputation Processing Agents (RPAs)* that have two different roles:

1. *Reputation Authority Agents (RAAs)*: Act as reputation scoring services for the reputee entities whose Reputation Objects (ROs) are being considered or calculated in the agents' rule-based Reputation Computation Services (RCSs). An RCS runs a rule engine which accesses different sources of reputation (input) data from the reputors about an entity and evaluates an RO based on its declarative rule-based computational algorithms and contextual information available at the time of computation.
2. *Reputation Management Agents (RMAs)*: Act as a reputation trust center offering reputation management functionalities. An RMA manages the local RAAs providing control of their life-cycle in particular, and also ensuring goals such as fairness. It might act as a Reputation Service Provider (RSP) which aggregates reputations from the reputation scores of local RAAs. Based on the final calculated reputation, it might also perform actions, e.g. compute trustworthiness, make automated decisions, or trigger reactions. It also manages the communication with the reputors, collecting data about entities from them, generates reputation data inputs for the reputation scoring, and distributes the data to the RAAs. It might also act as central point of communication for the real reputee entities (e.g., persons) giving them legitimate control over their reputation and allowing entities the governance of their reputations.

The agent-based approach to online reputation management ensures efficient automation, semantic interpretability and interaction, openness in ownership, fine-grained privacy and security protection, and easy management of semantic reputation data on the Web.

5.5 Semantic Complex Event Processing Agent Network

The Event Processing Network (EPN) [16] consists of Semantic Event Processing Agents (EPA) implemented as distributed Prova inference services which detect complex events using Prova's rule-based Semantic Complex Event Processing (SCEP) logic. [19]. The multi-agent approach allows for a highly-available distributed implementation with redundant Event-Calculus based state processing where events are processed concurrently in the EPN.

6 Conclusion

Rule Responder is a framework for specifying virtual organizations as semantic multi-agent systems. Characteristics of Rule Responder include

- the coverage of the distributed processing spectrum from Web Services to agents in one framework
- the recursive (holonic) modeling of a virtual organization of services and agents as a single agent,
- the use of ESBs, especially Mule, as a foundation for the Semantic and Pragmatic Web infrastructure,
- the use of Semantic-Pragmatic Web rules as the main knowledge representation, complemented by ontologies,
- the introduction of PAs as human-assisting agents into a virtual organization, besides the traditional computation-performing agents (CAs),
- the design of a `pluggable' agent-finding mechanism from role assignment to Semantic Service discovery.

The Rule Responder framework, with its increasing number of users and engines (Prova, OO jDREW, DR-Device, Euler, and Drools), is thus being proposed as a reference architecture for distributed knowledge representation and processing.

Acknowledgments. The international Rule Responder initiative has greatly helped us with work leading to this chapter. In particular, we want ot thank Alexander Kozlenkov, Benjamin Craig, Taylor Osmun, Derek Smith, Omair Shafiq, Mahsa Kiani, Kia Teymourian, Rehab Alnemr, Irfan ul Haq, Nick Bassiliades, Stratos Kontopoulos, and Kalliopi Kravari.

References

1. Alnemr, R., Paschke, A., Meinel, C.: Enabling Reputation Interoperability through Semantic Technologies. In: ACM International Conference on Semantic Systems. ACM, New York (2010)
2. Berners-Lee, T., Connolly, D., Kagal, L., Scharf, Y., Hendler, J.: N3Logic: A Logical Framework For the World Wide Web. Theory and Practice of Logic Programming (TPLP) 8(3) (2008)
3. Boley, H.: Are Your Rules Online? Four Web Rule Essentials. In: Paschke, A., Biletskiy, Y. (eds.) RuleML 2007. LNCS, vol. 4824, pp. 7–24. Springer, Heidelberg (2007)
4. Boley, H., Chang, E.: Digital Ecosystems: Principles and Semantics. In: Proc. IEEE Intl. Conf. Digital Ecosystems and Technologies, Cairns, Australia (2007)
5. Boley, H., Osmun, T.M., Craig, B.L.: Social Semantic Rule Sharing and Querying in Wellness Communities. In: Gómez-Pérez, A., Yu, Y., Ding, Y. (eds.) ASWC 2009. LNCS, vol. 5926, pp. 347–361. Springer, Heidelberg (2009)
6. Boley, H., Paschke, A.: Expert Querying and Redirection with Rule Responder. In: Zhdanova, A.V., Nixon, L.J.B., Mochol, M., Breslin, J.G. (eds.) Proceedings of the 2nd International ISWC+ASWC Workshop on Finding Experts on the Web with Semantics, CEUR Workshop Proceedings, Busan, Korea, November 12, pp. 9–22 (2007); CEUR-WS.org

7. Boley, H., Tabet, S., Wagner, G.: Design Rationale of RuleML: A Markup Language for Semantic Web Rules. In: Proc. Semantic Web Working Symposium (SWWS 2001), pp. 381–401. Stanford University, Stanford (2001)
8. Craig, B.L.: The OO jDREW Engine of Rule Responder: Naf Hornlog RuleML Query Answering. In: Paschke, A., Biletskiy, Y. (eds.) RuleML 2007. LNCS, vol. 4824, pp. 149–154. Springer, Heidelberg (2007)
9. Craig, B.L., Boley, H.: Personal Agents in the Rule Responder Architecture. In: Bassiliades, N., Governatori, G., Paschke, A. (eds.) RuleML 2008. LNCS, vol. 5321, pp. 150–165. Springer, Heidelberg (2008)
10. Li, J., Boley, H., Bhavsar, V.C., Mei, J.: Expert Finding for eCollaboration Using FOAF with RuleML Rules. In: Montreal Conference of eTechnologies 2006, pp. 53–65 (2006)
11. Paschke, A.: Rule responder HCLS eScience infrastructure. In: ICPW 2008: Proceedings of the 3rd International Conference on the Pragmatic Web, pp. 59–67. ACM, New York (2008)
12. Paschke, A., Bichler, M.: Knowledge representation concepts for automated SLA management. Decis. Support Syst. 46(1), 187–205 (2008)
13. Paschke, A., Kozlenkov, A.: A Rule-based Middleware for Business Process Execution. Multikonferenz Wirtschaftsinformatik (2008)
14. Sutcliffe, G., Goebel, R. (eds.): Proceedings of the Nineteenth International Florida Artificial Intelligence Research Society Conference, Melbourne Beach, Florida, USA, May 11-13. AAAI Press, Menlo Park (2006)
15. Koestler, A.: The Ghost in the Machine. Hutchinson & Co, London (1967)
16. Kozlenkov, A., Jeffery, D., Paschke, A.: State management and concurrency in event processing. In: DEBS (2009)
17. Paschke, A.: Rule-Based Service Level Agreements - Knowledge Representation for Automated e-Contract, SLA and Policy Management. Idea Verlag GmbH, Munich (2007)
18. Paschke, A.: A Typed Hybrid Description Logic Programming Language with Polymorphic Order-Sorted DL-Typed Unification for Semantic Web Type Systems. CoRR, abs/cs/0610006 (2006)
19. Teymourian, K., Paschke, A.: Towards semantic event processing. In: DEBS (2009)
20. Welsh, M., Culler, D., Brewer, E.: SEDA: An Architecture for Well Conditioned, Scalable Internet Services. In: Proceedings of Eighteeth Symposium on Operating Systems (SOSP-18), Chateau Lake Louise, Canada (2001)
21. Wooldridge, M.: An Introduction to MultiAgent Systems. John Wiley & Sons, Chichester (2001)

Chapter 2
Specifying and Monitoring Obligations in Open Multiagent Systems Using Semantic Web Technology

Nicoletta Fornara

University of Lugano, via G. Buffi 13, 6900 Lugano, Switzerland
nicoletta.fornara@usi.ch

Abstract. In nowadays open interaction systems where autonomous, heterogeneous and self-interested agents may interact, it is crucial to be able to declaratively specify the norms that regulate the actions of the interacting parties and to be able to monitor their behaviour in order to check whether it is compliant or not with the norms. In this chapter we propose and discuss the advantages of using semantic web languages, tools, and techniques for proposing an application independent model that should be used for the declarative specification and monitoring of obligations. Those obligations are characterized by a class of activation and deactivation events, a class of content actions that may satisfy the obligation and a deadline within which an action belonging to the content class has to be performed. The main contribution of this chapter is to show how it is possible to use semantic web technologies, and in particular OWL 2 DL as formal language for the specification and monitoring of complex obligations and to study how much it is feasible to use an OWL ontology to represent the state of a dynamic open interaction system.

1 Introduction

The specification of *open systems* for the interaction of autonomous agents is widely recognized to be a crucial issue in the development of innovative applications on the Internet, like e-commerce applications, or applications for the management of virtual enterprises. One possible approach to tackle this problem is to model open interaction systems as a set of artificial institutions [2, 1, 20, 11]. Those institutions are devised for the specification of the institutional context where the interaction among autonomous heterogeneous agents may take place. In particular the OCeAN meta-model [12, 9] is mainly composed by: a *communicative part* with the definition of an Agent Communication Language (ACL) whose semantics is defined in terms of social commitments and institutional power [8], a *normative part* for the specification of obligations, prohibitions and permissions [10], and an *organizational part* mainly devoted to the definition of roles.

A. Elçi, M.T. Koné, and M.A. Orgun (Eds.): Semantic Agent Systems, SCI 344, pp. 25–45.
springerlink.com   © Springer-Verlag Berlin Heidelberg 2011

In this chapter we will mainly focus on the *normative part* and we propose and discuss the advantages of using semantic web languages, tools, and techniques for defining an application independent model for the declarative formal specification and monitoring of *obligations*. In particular we want to be able to specify obligations with the following characteristics. They become active when an event belonging to a specified start event class or to its subclasses happens, this event can be viewed as a condition for obligations activation. A set of possible actions described by means of a more or less detailed class may fulfil those obligations if one of them happens before a given deadline. This is a crucial progress in the flexibility of the normative specification with respect to the solution proposed in [10] where (as better discussed in next section) the content of obligations was a specific action and the time interval for the performance of the action was delimited by fix instant of time. Finally those obligations become cancelled when an event belonging to an end event class happens.

The approach of specifying using a declarative formal language the normative part of a system has many crucial and interesting advantages. In particular it makes possible to represent the norms as *data*, instead of coding them in the software. This has the advantage of making possible to add, remove, or change the norms that regulate the interaction both when the system is off line, and at runtime, without the need to reprogram the interaction system or the interacting agents. Another interesting advantage is that it would be in principle possible to realize agents able to automatically reason on the consequences of their actions and able to interact within different systems without the need of being reprogrammed. Moreover it is possible to realize an application independent *monitoring component* able to keep trace of the state of obligations on the basis of the events that happens in the system and on the basis of agents' actions and capable of reacting to their fulfilment or violation. This is a fundamental component in the architecture of open interaction systems, and may be crucial also in the service oriented architecture [6] and for business process management systems [21]. Another important aspect is that designing a system by using the notion of norm may be very intuitive for human designers and those declarative norms may be more easily understood by human participants of socio-technical systems.

The choice of the formal language used for the declarative specification of normative systems is difficult, crucial, and many aspects have to be taken into account. The most important are: the expressivity of the language, its computational complexity, the fact that the underline logic is decidable, the diffusion of the language among software practitioners and research communities, its feasibility to be used for fast prototyping, and its adoption as an international standard. After many past experiments with other formal languages, in this chapter we decided to adopt OWL (in its OWL 2 DL version[1]), the description logic language recommended by W3C for Semantic Web applications, and more generally semantic web technologies. The main advantage of this choice is that Semantic Web technologies are increasingly becoming a standard for Internet applications and therefore, given

---

[1] http://www.w3.org/2007/OWL/wiki/OWL_Working_Group

that the OWL logic language is decidable, it is supported by many reasoners (like Fact++, Pellet, Racer Pro, HermiT), tools for ontology editing (like Protégé[2]) and library for automatic ontology management (like OWL-API). Given that it is a standard, it would be easier to achieve a high degree of interoperability of data and applications, which is indeed a crucial precondition for the development of open systems. Finally given that semantic web technologies are becoming very used in innovative applications it will become much easier to teach them to software engineers than convince them to learn and use a logic language adopted by a limited group of researchers.

There are some interesting and challenging problems that may arise from the fact that Semantic Web technologies are not devised for modelling dynamic systems (i.e. systems that changes in time). One is encountered when trying to perform full temporal reasoning; in fact OWL has no temporal operators. Another one is due to the fact that Semantic Web technologies have not been devised to check constrain for example on norm specification, but there are some interesting current studies on how to use the Pellet reasoner for "Simple Integrity Constraints"[3]. A third one is the open-world assumption of OWL logic, it may be a problem for successfully monitoring obligations, that is, when trying to deduce that when the deadline is elapsed an obligation has to be permanently fulfilled or violated.

The added value of this chapter is twofold: the first is to show how it is possible to use semantic web technologies, and in particular OWL 2 DL, as formal language for the specification and monitoring f obligations with activation and deactivation events and deadlines. This model may have many different kinds of applications like the specification of electronic commerce market places, or the monitoring of semantic web services execution, or the flexible specification and monitoring of business process where both software and human agents may interact. The second is to propose to use an OWL ontology not only for the *specification* of a normative systems but also for the *dynamic monitoring* of the state of the interaction among autonomous agents in an open and dynamic environment with respect to a specified set of norms. In particular with this work we are giving our contribution to the open problem of understanding how far the monitoring problem can be solved by using an OWL 2 DL ontology and when it is necessary to integrate it with Java programs.

This chapter is organized as follows. In Section 2 the proposed approach is compared with main alternative approaches. In Section 3 the formal language used in the paper is briefly described. In Section 4 the application independent ontology that can be used to represent and monitor obligations is introduced, discussed and exemplified. In Section 5 some obligations of a concrete case study are formalized using the proposed approach and finally in Section 6 some conclusions are drawn.

---

[2] See http://www.w3.org/2007/OWL/wiki/Implementations for a complete list of reasoners and tools

[3] http://clarkparsia.com/weblog/category/semweb/owl/pellet/integrity-constraints/

2 Other Approaches

The problem of modelling norms using formal languages is widely recognized as a crucial problem by the multiagent community [3, 19]. Moreover the problem of run-time monitoring those norms is becoming more and more an interesting open question for the multiagent community and for the web service community as demonstrated by various papers on this topic [7, 16, 23, 10]. In particular in [7] Faci et al. propose a framework for non-intrusive monitoring of the state of contract that, similarly to our proposal, is based on the observation of agents' message exchange. Their norms, having a structure quite close to the one proposed in this chapter, are specified using the XML language and their content is specified using ontologies. The main difference between the two approaches is on the monitoring component: in their work it is required to transform the XML representation of norms in another formalism: the augmented transition networks. This transformation presents all the drawbacks that may come from using two different formal languages to specify the same concept in term of consistency, performance, and required knowledge for the engineers who want to adopt this approach. In [16] Lomuscio et al. in order to monitor an agent "all its possible behaviours are represented as a timed automata with discrete data (TADD) and stored in the checker, the monitoring engine checks the snapshots against their TADD specification". One of the main advantages of this approach, as claimed by the authors, is its scalability, this is an important goal to be taken into account and that in our approach can be pursuit by splitting up the state of the interaction in sub-states holding only the information that in a certain moment is relevant for a given interaction. The reference architecture for contract monitoring in e-market scenarios presented in [23] is complementary to the model proposed in this chapter. Finally the main difference between the formalization proposed in this chapter and our previous work on the specification of norms using semantic web technology [10], is that in this chapter the content and the conditions of obligations are specified as classes of actions or events instead as specific action or event.

As discussed in the introduction the choice of using semantic web languages has many advantages and it is a crucial aspect when we compare our work with other ones on norms specifications and properties verification where other formal languages are adopted. Other formal languages are for example the Event Calculus [24, 9], the language for rule specification of the rule engine Jess [13, 4], a variant of Propositional Dynamic Logic (PDL) used to specify and verify liveness and safety properties of multi-agent system programs with norms [5], the Process Compliance Language (PCL) [14].

In literature there are few approaches that use semantic web languages for the specification of norms, even if their importance for the development of flexible security for dynamic and distributed environment is clearly recognized [15]. One interesting approach for policy specification and management is the KAoS framework [18]. In MAS community the word *norm* and *policy* have a similar meaning; a policy could be a positive or negative authorization to perform an action or an obligation. In KAoS, like in the model proposed in this chapter, policies are specified using a set of concepts defined in an OWL DL *core ontology* that could be extended with application dependent ontologies. A crucial difference between the two approaches is the fact that OWL 2 DL is more expressive that OWL DL. Another important difference is in the methods used for monitoring policies: in KAoS policies are usually regimented by means of "guards`` and are monitored by means of platform specific mechanisms.

3 OWL and SWRL

OWL is a practical realization of a Description Logic system known as $\mathcal{SROIQ}(\mathcal{D})$. It allows one to define *classes*, *properties*, and *individuals*. An OWL ontology consists of: a set of class axioms to describe classes, which constitute the *Terminological Box* (*TBox*); a set of property axioms to describe properties, which constitute a *Role Box* (*RBox*); and a collection of assertions to describe individuals, which constitute an *Assertion Box* (*ABox*). Properties can be either *object properties* or *data properties*. Classes can be viewed as formal descriptions of sets of objects (taken from a nonempty universe), and individuals can be viewed as names of objects of the universe. A class is either a *basic class* (i.e., an atomic class name) or a *complex class* build through a number of available *constructors* that express Boolean operations and different types of restrictions on the members of the class.

Through *class axioms* one may specify *subclass* or *equivalence* relationships between classes, that certain classes are *disjoint* (Discla), and that a class is defined by placing restrictions on properties (existential ($\exists$), universal ($\forall$), cardinality, "has-value" ($\ni$), and local reflexivity restrictions. *Property axioms* allow specifying that a given property is the inverse of another property ($^{-}$), or that a property is functional (Fun), or a transitive property (Tr), or that a property can be obtained by composing properties into *property chains* ($\circ$). Finally, *assertions* allows to specify that an individual belongs to a class, that an individual is related to another individual through an object property, that an individual is related to a data value through a data property, or that two individuals are equal or different.

OWL can be regarded as a decidable fragment of First Order Logic (FOL). The price to pay for decidability, which is considered as an essential preconditions for exploiting reasoning in practical applications, is limited expressiveness. Even in OWL 2 DL (the more expressive version currently under specification) certain useful first-order statements cannot be formalized. Given the limited expressivity of OWL the Semantic Web Rule Language (SWRL)[4] has been proposed to extend the set of OWL axioms to include Horn-like rules of the form of an implication between an antecedent (body) and consequent (head). Recently certain OWL reasoners, like Pellet, have been extended to deal with SWRL rules. To preserve decidability, however, rules have to be used in the *safe mode*, which means that before being exploited in a reasoning process all their variables must be instantiated by pre-existing individuals. An important aspect of SWRL is the possibility

---

[4] http://www.w3.org/Submission/SWRL/

of including *built-ins*, that is, Boolean functions that perform operations on data values and return a truth value. In what follows we use capital initials for classes and lower case initials for properties and individuals, we assume that all the individuals introduced are different.

4 An Application Independent Ontology for Modelling and Monitoring Agents' Interactions

In this section we introduce the classes, the properties, and the axioms of the *application independent* part of the ontology ("*upper ontology*") that one has to use to specify and monitor agents' obligations in those applications where the realization of an *open normative interaction system* is required. In order to completely formalize a real interaction system, as exemplified in Section 4.2, this ontology has to be extended with application dependent classes, properties, and axioms that are used to model the application dependent actions and events that appear in the content or in the condition of obligations.

In particular we first describe the OWL Time Ontology that we use in this chapter, the classes for representing *events* and *fluents* and their relationships with *obligations*. Subsequently we define one possible example of a domain dependent ontology that will be used in the examples contained in the paper. Then we introduce the part of the ontology that is necessary for representing *events* and the *elapsing of time*. Later on we present the part of the ontology used to represent the content, the condition, the deadline, and the expiration condition of *obligations*. Finally we introduce the part of the ontology and the mechanisms that have to be used to *monitor* the *time evolution* of obligations on the basis of the *actions* and *events* that happen in the system. At the end of this section the graphical representation of the proposed ontology is reported.

4.1 Modelling Time, Events, and Fluents

The first class that has to be introduced is the Agent class that is used to represent the agents involved in the interaction mediated by the open system. Secondly in order to be able to represent obligations with activation and deactivation events correlated to time and with temporal deadlines, we have to find a suitable and efficient way to represent instants and interval of time in the ontology. Given that OWL has not temporal operator, the simplest solution, which pursues also the goal of being interoperable with other ontologies, is to adopt the OWL Time Ontology[5]. Unfortunately the axiomatization of the OWL Time Ontology is very weak and therefore it will be impossible to perform certain type of interesting reasoning on the future evolution of the state of the system. Nevertheless, as we will see in the following subsections, we will try to partially overcome to this problem, in order to be able, at least, to represent and monitor the time evolution of the system. Here we report the list of classes and properties of the OWL Time Ontology that

---

[5] http://www.w3.org/TR/owl-time/

are relevant for the comprehension of this chapter (they are graphically represented in Figure 1 at the end of this section):

Instant ⊑ TemporalEntity, Interval ⊑ TemporalEntity,
ProperInterval ⊑Interval, TemporalEntity ≡ Instant ⊔ Interval,
hasBeginning: TemporalEntity → Instant,
hasEnd: TemporalEntity → Instant,
before: TemporalEntity → TemporalEntity, InvPro(after,before),
inDateTime: Instant → DateTimeDescription,
Discla(ProperInterval,Instant), Instant ⊑ = 1 inDateTime

In order to be able to represent *events* that happen at a certain *instant* of time, or
fluents, that is, state of affair that holds for a certain *interval* of time, we introduce
the class Eventuality and its two subclasses: Event, whose individuals are related
to an instant of time, and Fluent whose individuals are related to an interval:

Event ⊑ Eventuality, Fluent ⊑ Eventuality, Discla(Event,Fluent),
atTime: Eventuality → TemporalEntity,
Event ≡ ∃ atTime.Instant, Fluent ≡ ∃ atTime.Interval.

An event is before another event if the first one happens at an instant of time
that is before the instant of time of the second one:

evBefore: Eventuality → Eventuality,
atTime ∘ before ∘ atTime⁻ ⊑ evBefore, Tr(evBefore).

Two events that happens at the same instant of time are related by the evSame-
Time property:

evSameTime: Eventuality → Eventuality,
atTime ∘ atTime⁻ ⊑ evSameTime, Tr(evSameTime).

Actions are viewed as a particular type of events that have an actor, a recipient
and an object:

Action ⊑ Event, hasActor: Action → Agent,
hasRecipient: Action → Agent, hasObject: Action → Object,
Fun(hasActor), Fun(hasRecipient), Fun(hasObject).

Obligations are represented as particular type of *event*: Obligation ⊑ Event, and
they are characterized by the event that brings about their creation. Even if, in the
common sense perception, obligations are semantically different from events, this
choice gives us the flexibility to be able to specify class of actions as content of
the obligations and it makes the axiomatization of the notion of obligation fulfilment and violation simpler. An obligation has a *debtor* and a *creditor* as represented by the following properties:

hasDebtor: Obligation → Agent, hasCreditor: Obligation → Agent
Discla(Obligation,Action).

An obligation has also a content, an activation event, a deactivation event, and
a deadline, which are specified using classes, as discussed in Subsection 4.4.

4.2   An Example of a Domain Dependent Ontology

In order to be able to use in the content and in the condition of obligations concrete classes of actions and events, it is necessary to introduce in the ontology domain dependent classes and properties. Those classes have to be subclasses of the
class Action or of the class Event. For example we may need to introduce the class
of the actions of delivering a certain object to a certain recipient:

Deliver ⊑ Action ⊓ ∃ hasRecipient ⊓ ∃ hasObject,

the class of actions of paying a certain amount of money to a certain recipient:

Pay ⊑ Action ⊓ ∃ hasRecipient ⊓ ∃ hasObject,

and the class of actions of paying by means of a bank transfer, BankTransfer,
which is a subclass of the Pay class: BankTransfer ⊑ Pay. Those classes will be
used in Section 4.6 where different types of obligations for the electronic commerce domain will be presented.

For example the action of delivering a book book1 from agent Luca to agent
Marco performed at instant1 is described by the following assertions:

Agent(Luca), Agent(Marco), Object(book1), Instant(instant1),
Deliver(deliver1), hasActor(deliver1, Luca), hasRecipient(deliver1,Marco),
hasObject(deliver1,book1), atTime(deliver1,instant1).

4.3   Representing Events, Actions, and the Elapsing of Time

We want to use the specified OWL ontology to represent the evolution in time of
the state of the interaction between autonomous and heterogeneous agents in a
norm governed framework. This state has to be represented in every software that
is in charge of monitoring the behaviour of the interacting agents, a centralized,
mixed, or distributed one (the discussion of the advantages and problems due to
the choice of one or other architecture is crucial but due to its complexity it is beyond the scope of this specific paper), and may be represented inside the interacting agents in order to let them to reason and plan their future actions on the basis
of the rules of the system. It is moreover reasonable that the interacting agents
have a partial knowledge of the state of the interaction, which represents only the
interaction in which they are involved or that is relevant for the specific agent.

If the system evolution is simulated, the list of events that happen in the system,
the list of actions performed by the agents, and the instant of time when they happen, are known at design time and may be initially introduced in the ontology.
Differently, if an actual interaction between agents takes place at run-time, it is
necessary to tackle two problems. First of all it is required to support agents'
communication with an appropriate middle-ware, like for instance the widely used
JADE framework[6], or by using web services standard technologies[7]. Regarding
the agent communication language (ACL) we plan to adopt the commitment based

---

[6] http://jade.tilab.com/
[7] http//www.w3.org/standards/webofservices/description

one presented in [9] for the exchanged messages instead of the FIPA-ACL standard semantics[8] that presents a set of well known drawbacks [22]. Secondly it is necessary to write a program in charge of inserting in the ontology a representation of agents' actions and of the events observed, together with the corresponding instant of time when they happened, for example a typical type of action that needs to be recorded in the ontology is the exchange of messages between agents.

Either the interaction is simulated or it is actually happening at run-time, events or actions happen at certain instant of time and it is necessary to state what the temporal relation between those instants is. This can be simply done by asserting which instant comes after another using the after property. Then thanks to the transitivity of the after property, it is possible to deduce the temporal relation that subsists between all instants of time present in the ontology. Alternatively in order to be able to compare two instants of time and assert which one comes after the other the designer may decide to use an external Java program, or an SWRL rule with built-ins for comparisons, or simply inserting the instant of time in the ontology following their temporal order and asserting that the last instant inserted is after the last but one.

Certain subclasses of the class Event are used in the definition of specific obligations as explained in the following sections. In particular it will be certainly necessary to represent at least the following different *types of events*:

- *Time events* are used to represent the events related to the elapsing of time and belong to the TimeEvent ⊑ Event class. This class is disjoint from the Obligation and from the Action classes: Discla(Obligation,TimeEvent), Discla(Action,TimeEvent). A specific time event is related by means of its atTime property to the instant of time when it happens. Notice that a time event actually happens when its instant of time is asserted to belong to the class Elapsed that will be introduced later on.
- *Action events* are used to represent actions performed by the agents, they are represented as individuals of the class Action, for example the action of delivering a product. The action of exchanging a message is a common and very important type of action represented with the class ExchMsg ⊑ Action. It has an actor, the *sender* of the message, a recipient, the *receiver* of the message, an *illocutionary force* (see [9] for more details) connected to the message using the hasForce property whose range is the IllocutionaryForce class, and an object that is the *content* of the message.
- *Change events* are used to represent the events due to the change of the value of a property, they are represented as individuals of the ChangeEvent ⊑ Event class. For example the change in the state of an auction from close to open can be used as condition of the obligation for the auctioneer to declare the current price of the product to be sold. Usually a change event is characterized by the *entity* whose property is changed, the *previous value* and the *subsequent value* of the property, they are all represented as properties of change events. Obviously whenever the performance of an action, or the occurrence of an event,

---

[8] http://www.fipa.org/specs/fipa00037/SC00037J.pdf

has the effect to change the value of a property of an entity, and if the change event is relevant for one of the obligations represented in the ontology, it is necessary to introduce in the ontology an individual belonging to the ChangeEvent class with a suitable atTime value. This is a fundamental feature of the middle-ware, and it has to be strongly optimized because may be critical in terms of time consuming.

In general when a certain obligation has to be created it may happen that it is necessary to create new subclasses of those classes. Moreover if the new obligation is related to a specific time event (for example the obligation to deliver a book within a given deadline), a new individual, belonging to the TimeEvent class, has to be inserted in the ontology in order to represent such a time event.

In order to model the *elapsing of time* we need to have in the ontology a set of individuals used to represent all the relevant instants of time. An instant of time is *relevant* if an action, or an event, happens at that instant of time, or if such an instant of time is used to create a time event related to the specification of an obligation. The distance between an instant of time and the following one depends on the *time lag* chosen for the system: every type of interaction may have its own reasonable time lag that mainly depends on the frequency on which actions or events happen. During the evolution of the interaction, in order to model the elapsing of time, the individual corresponding to the actual instant of time (of the simulation or of the actual agents interaction) have to be asserted to belong to the Elapsed ⊑ Instant class, a special class introduced specifically for this purpose. Every instant of time that is before an elapsed instant of time is itself elapsed as expressed by the following axiom: ∃before.Elapsed ⊑ Elapsed.

In case the evolution of the system is simulated it is enough to repeatedly assert that the instant of time, subsequent to the current one, is elapsed, and then run the reasoner to deduce all the consequences of the events or actions happened at the current instant of time. Differently if the ontology is used to represent the state of an actual agents interaction, it is necessary to keep aligned the current instant of time represented in the ontology (the last that is asserted to be elapsed) with the external clock, that is, the clock of the world where the agents actually interact. Therefore an instant of time has to be asserted to be elapsed only when its inDateTime property is lower or equal to the time adopted by the interacting agents.

4.4 Representing Specific Obligations

In this chapter we specify how to formalize in the ontology used to represent the state of the interaction among agents their obligations and we describe how to monitor, using semantic web technologies, those obligations. An obligation exists between two specific agents that are the *debtor* and the *creditor* of the obligation. An obligation is characterized by the *instant of time* when the obligation is created, a class of events that may *activate* or *deactivate* it, a *content* described by means of another class, and a *deadline*. We assume (coherently with what is specified in the OCeAN meta-model [9]) that new obligations are created as the effect of the performance of certain communicative acts (like promises), or as consequence of the activation of a norm. A norm is activated whenever an agent, who is interacting with other agents within a certain institutional context, starts to play a role whose behaviour is regulated by the norm. Whenever a new obligation, obl-n, is created at a certain instant of time, instant-n, whose inDateTime property value is equal to the time when the obligation is created (in the following referred as now), the ABox of the ontology has to be automatically updated with the following assertions:

Obligation(obl-n), atTime(obl-n,instant-n), inDateTime(instant-n,now), hasDebtor(obl-n,agent1), hasCreditor(obl-n,agent2).

In addition it is necessary to update the TBox in the following way: the first change consists in defining the specific *activation*, *deactivation*, *content*, and *deadline* classes of the new obligation; secondly it will be necessary to write the axioms for deducing the state of a given obligation, with the goal of monitoring its fulfilment or its violation as described in the following subsection.

The StartEvent-n ⊑ Event class describes the type of events that may activate the obligation obl-n, that is, the conditional event that have to happen in order to make the obligation *activated*. For example in certain electronic commerce scenario an agent may start to be actively obliged to pay a certain amount of money after the reception of the ordered product. Certain obligation may be immediately activated without the need to specify any condition, in this case the StartEvent-n class coincides with the event that create the obligation:

StartEvent-n≡{obl-n}.

If it is possible to deduce that the StartEvent-n class is equal to the empty set ⊥, it means that the obligation obl-n will never be activated. This is a fundamental information for the agents when they are planning their future actions.

The EndEvent-n ⊑ Event class describes the type of events that may expire the obligation, that is, when an expiration event happens the obligation becomes *cancelled* and will not any more become active in the future. The specification of this class is crucial for those obligations that may be activated many times, for example an employer may have the obligation to pay the salary to his/her employees at the end of each month as long as they are employed in the company. Very often the EndEvent-n class is equivalent to the class of the actions that may be used to dismiss an agent from a specific role, the role indicated in the debtor or in the creditor field of the norm that generated the obligation. For example when an agent ceases to be an employer or an employee the obligation to pay the salary becomes cancelled. In some other cases the EndEvent-n class coincides with a fixed deadline, that is, with a certain time event, for example the instant of time when the contract of the employee terminates.

The Content-n ⊑ Action class describe the set of actions whose performance may fulfil or violate the obligation. An crucial aspect of the proposed model is the possibility that an action, belonging to a subclass of the Content-n class, satisfies the obligation. Moreover in the definition of the Content-n class it is also possible to use Boolean class constructors. The union of classes can be used for those cases when either an action belonging to one class or an action belonging to another class may fulfil an obligation.

When an agent has the obligation to perform an action it is necessary to define the deadline (i.e. the instant of time) within which the action has to be performed. For coherence with the other classes we introduce the class Deadline-n ⊑ Timevent even if it contains only one individual: the time event associated to the instant of time that represents the deadline of the obligation. Taking into consideration the existence for every obligation of a start and dead-line event it is natural to introduce a property hasInterval: Obligation→TemporalEntity that binds an obligation to the interval of time within which one action belonging to the Content class has to be performed. Such an interval has a beginning instant of time, an end instant of time, and a duration that can be obtained by means of the hasBeginning, hasEnd, hasDurationDescription properties. The instant of time when the interval of obl-n starts can be deduced on the basis of the instant of time when an individual belonging to the StartEvent-n class happens by introducing the following SWRL rule:

StartEvent-n(?e) ∧ atTime(?e,?inst) ∧ hasInterval(obl-n,?int) →
hasBeginning(?int, ?inst)

The Deadline-n class is equivalent to the class that contains only the time event that happens at the instant of time when the interval of the obligation finishes, as stated in the following axiom:

Deadline-n ≡ ∃ atTime.(∃ hasEnd⁻.(hasInterval⁻ ∋ obl-n))

It is important to remark that when the deadline of the obligation depends on the instant of time when the obligation is activated, the time event to be used as deadline is unknown when the obligation is created. In this case the Deadline-n class will become defined when the obligation becomes active. A example of this kind of obligations are those obligation where the deadline is equal to the instant of time when the obligation is activated plus a fixed amount of time, for instance the obligation to pay the product within 2 days from its reception. For these type of obligations it is necessary to insert in the ontology also the value of the duration of the interval associated with the obligation. Once the beginning instant and the duration of the interval are known, it is possible to use the following SWRL rule, which uses the swrlb:add built-in, to deduce the value of the end instant of time of the interval (we assume that the duration of the interval is expressed in days):

hasBeginning(?int,?inst1) ∧ inDateTime(?inst1,?dt1) ∧
dayOfYear(?dt1,?day1) ∧ Instant(?int2) ∧ inDateTime(?inst2,?dt2) ∧
dayOfYear(?dt2,?day2) ∧ hasDurationDescription(?int,?d) ∧ days(?d,?value)∧
swrlb:add(?day2,?day1,?value) → hasEnd(?int,?inst2)

For those obligations where the deadline event is a fixed time event that does not depend on the activation event (see for example in Section 4.6 the first type of obligations), it is important to check that the start event happens before the end event. This can be done with the following axiom that has to be written only for obligations whose StartEvent-n and Deadline-n classes are equivalent to a specific time event. In case the deadline time event is before or equal to the start time event the ontology becomes contradictory:

Deadline-n ⊓ (evBefore.StartEvent-n ⊔ evSameTime.StartEvent-n) ⊑ ⊥

In Section 4.6 specific examples will be used to illustrate the definition of the StartEvent, EndEvent, Content, and Deadline classes for different type of obligations.

4.5 Monitoring the State of Obligations

When a new obligation obl-n is created the second change to the TBox consists in introducing the four axioms that are necessary to deduce the state of a given obligation, that is, to deduce if it belongs to the Activated, Cancelled, Fulfilled, or Violated classes. Those classes are subclass of the class Obligation and the Fulfilled and Violated classes are disjoint:

Fulfilled ⊑ Obligation, Violated ⊑ Obligation, Activated ⊑ Obligation, Cancelled ⊑ Obligation, DisCla(Fulfilled, Violated).

The first axiom is the one to deduce that an obligation with a certain StartEvent class is activated. If an event e_s that belongs to the StartEvent-n class of an obligation obl-n happens after or at the same instant of time when the obligation is created, the time at which e_s happens is elapsed, and the obligation has not yet been cancelled, then the obligation becomes activated.

The main problem in writing this axiom is due to the negation that appears in the third condition. OWL reasoners operate under the open world assumption and therefore we cannot simply write in the axiom the condition "*not cancelled*". In fact the conclusion that an obligation is not cancelled can only be reached if the obligation can be definitely proved not to be member of the Cancelled class. To solve this problem we assume that our ABox contains complete information on the events happened or actions performed before the current time of the system. More specifically, we assume to use an external Java program that will always update the ABox whenever an event happens. Moreover we assume that such a program can only insert in the ABox the information that an event is happened at current time t, and that it is not possible to insert the information that an event is happened in the past. Starting from these assumptions we can adopt a closed-world perspective on the Cancelled class: an obligation "is not yet been cancelled" if it is not in the Cancelled class. Consequently in order to be able to perform some form of closed world reasoning on the Cancelled class (similarly to the solution proposed in [10]) we introduce in our ontology the explicit closure of such a class. More precisely, we introduce a new class, the KCancelled ⊑ Cancelled, which is meant to contain all obligations that, at a given time, are known to be in the Cancelled class. To maintain the KCancelled class as the closure of the Cancelled class, we define it periodically as equivalent to the enumeration of all individuals that can be proved to be members of the Cancelled class. This can be done by the external Java program that is also used to update the ABox to keep track of the elapsing of time and of the events that happen in the system. The axiom to deduce if an obligation obl-n is activated is therefore:

Axiom Activated Obl-n:
{obl-n} ⊓ ¬ KCancelled ⊓ (∃evBefore.(StartEvent-n ⊓ ∃atTime.Elapsed) ⊔
∃evSameTime.(StartEvent-n ⊓ ∃atTime.Elapsed)) ⊑ Activated

An obligation, when is not yet cancelled, may be activated more than once by different start events belonging the StartEvent class. It is important to be able to monitor the time evolution of the obligation for each one of its possible activation event. Therefore we assume that whenever an obligation is activated at instant i, an external program has to create a copy of that obligation and associate it to a creation time that is one instant of time later than the instant of time of the current activation i. This fact is crucial to avoid that the new copy of the obligation becomes active due to the current activation event.

If an event e_e that belongs to the EndEvent-n class of an obligation obl-n happens after the time when the obligation is created and the time at which e_e happens is elapsed, then the obligation becomes cancelled. It is important to underline that an obligation that is activated may be also cancelled (the Activated and Cancelled classes are not disjoint). This means that it can become fulfilled or violated but also that cannot be any more activated in the future by another start event. For example the obligation to pay the salary to an employee at the end of each month for an entire year becomes cancelled at the end of the year and is activated for twelve times, the last time that the obligation is activated it is also cancelled because the entire year is elapsed. If an end event happens before a start event the obligation is never activated. For example the obligation for a company to keep the streets of a city clear from the snow for a given winter will never be activated if the winter is particularly warm.

Axiom Cancelled Obl-n:
{obl-n} ⊓ ∃evBefore.(EndEvent-n ⊓ ∃atTime.Elapsed) ⊑ Cancelled

As mentioned before the Deadline-n class contains only one time event, the time event within which an action belonging to the Content-n class has to be performed.

If an event e_c that belongs to the Content-n class of an active obligation obl-n (created at i_n) happens at instant i_c, i_c is after or equal to i_n, i_c is before the deadline of obl-n, and i is elapsed, then the obligation becomes fulfilled as expressed by the following axiom.

Axiom Fulfilled Obl-n:
{obl-n} ⊓ Activated ⊓ (∃evBefore.(Content-n ⊓ ∃atTime.Elapsed) ⊔
∃evSameTime.(Content-n ⊓ ∃atTime.Elapsed)) ⊓
∃evBefore.(Content-n ⊓ ∃evBefore.Deadline-n) ⊑ Fulfilled

If the time event that represents the deadline of an active obligation obl-n elapses and the obligation is not yet fulfilled, the obligation has to become violated. Similarly to what we did for writing the axiom for the activation of obligations, in order to write the axiom to deduce that an obligation is cancelled we need to introduce the explicit closure of the Fulfilled class: the class KFulfilled ⊑ Fulfilled. The KFulfilled class is meant to contain all obligations that, at a given time, are known to be in the Fulfilled class. To maintain the KFulfilled class updated we define it periodically, by means of the external program, as equivalent to the enumeration of all individuals that can be proved to be members of the Fulfilled class. The axiom to deduce that an obligation obl-n is violated is:

Axiom Violated Obl-n:
{obl-n} ⊓ Activated ⊓ ¬ KFulfilled ⊓
∃evBefore.(Deadline-n ⊓ ∃ atTime.Elapsed) ⊑ Violated

Initially KCancelled ≡ KActivated ≡ KFulfilled ≡ KViolated ≡ Nothing then a Java external program has to update their extension on the basis of the deductions of the reasoner. In Figure 1 the graphical representation of the classes and properties introduced in the previous sections is depicted.

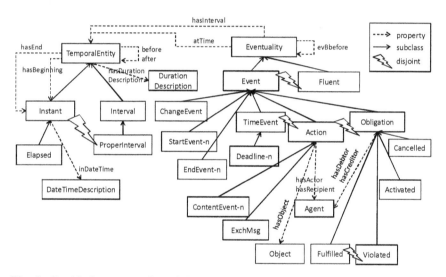

Fig. 1. Graphical representation of the ontology. Properties are represented with dotted lines, solid lines are used for subclasses.

4.6 Possible Type of Obligations

A first type of obligations are those obligations whose StartEvent and Deadline classes are equivalent to a specific time event. It means that the obliged action described with the Content class has to be performed between two specific instants of time. An example of an obligation of this type is the obligation obl-1 created at instant1 from agent Marco to agent Luca to pay 5 euro between instant of time instant2 and instant4 having certain specific dates as inDateTime properties.

To model the obligation obl-1 it is necessary to add to the ABox the following assertions:

Obligation(obl-1), Agent(Marco), Agent(Luca), Thing(5euro),Instant(instant1),
atTime(obl-1,instant1),hasDebtor(obl-1,Marco), hasCreditor(obl-1,Luca),
ProperInterval(interval1), hasInterval(obl-1,interval1),
hasEnd(interval1,instant4),TimeEvent(tevent4), Instant(instant4),
atTime(tevent4,instant4),

For this kind of obligations the StartEvent-1 classes consist of only one element: the time event that happens at instant2:

TimeEvent(tevent2), Instant(instant2), atTime(tevent2,instant2),
after(instant2,instant1), after(instant4,instant2),
StartEvent-1 ≡ {tevent2},
Content-1≡Pay ⊓ hasActor∃Marco ⊓ hasRecipient∃Luca ⊓ hasObject∃5euro.

The four axioms for deducing the state of obligations contextualized to this specific obligation have to be inserted in the ontology. Given that this obligation can become active only one time, it is not interesting to define the EndEvent class.

A crucial aspect of the proposed approach is that it is more flexible than other ones, in fact, given that the content of the obligations is expressed using a class of possible actions, the interacting agents have the flexibility to choose which one to perform. Moreover, if an event that belongs to one of the subclasses of the Content class happens, the obligation may equally become fulfilled. For example, if the bank transfer event (represented with the individual bankTr1∈BankTransfer where BankTransfer ⊑ Pay) from Marco to Luca of an amount of 5 euro happens after the activation event and before the deadline event, the obligation obl-1 becomes fulfilled.

The content of an obligation could also be the performance of either one class or another class of actions. This type of Content class can be represented using the union of two or more classes of actions. For example the obligation from Marco to Luca to either pay 5 euro to Luca or donate 6 euro to Unicef between instant2 and instant4 is identical to the previous obligation except for the Content-1 class that becomes:

Content-1 ≡
(Pay ⊓ hasActor∃Marco ⊓ hasRecipient∃Luca ⊓ hasObject∃5euro) ⊔
(Pay ⊓ hasActor∃Marco ⊓ hasRecipient∃Unicef ⊓ hasObject∃6euro)

A second type of obligations has the StartEvent class that can be interpreted as a condition for the activation of the obligation (a conditional obligation) and whose Deadline class depends on the time of its activation. An example of an obligation of this type is the obligation obl-2 created at instant1 from agent Marco to agent Luca to pay 5 euro within 2 days from the reception of the book (book1) on condition that the book was delivered from Luca to Marco. Besides the assertions previously introduce we have to add in the ABox those ones:

Obligation(obl-2), atTime(obl-2,instant1),
hasDebtor(obl-2,Marco), hasCreditor(obl-2,Luca), Object(book1),
ProperInterval(interval2), hasInterval(obl-2, interval2),
hasDurationDescription(interval2,duration2), days(duration2, 2),
StartEvent-2 ≡ Deliver ⊓ hasActor∋Luca ⊓ hasRecipient∋Marco ⊓
hasObject∋book1,
Content-2 ≡Pay ⊓ hasActor∋Marco ⊓ hasRecipient∋Luca ⊓ hasObject∋5euro,
EndEvent-2 ≡ {teventk}.

Reasonably the EndEvent-2 class is equivalent to the time event teventk whose instant property can be calculated as the time of creation of the obligation plus 3 months. This means that if the book is not delivered within 3 months Marco is not any more conditional obligated to pay for the book after its reception. As usual the four axioms presented in the previous section for deducing the state of an obligation, contextualized to this specific obligation, have to be inserted in the ontology.

A third type of obligations has not condition, that is, their StartEvent class is equivalent to the time of the creation of the obligation. Due to this fact the deadline of this type of obligations can be set when the obligation is created on the basis of the duration of the interval. An example of an obligation of this type is the obligation obl-3 created at instant1 from Marco to Luca to pay 5 euro before tomorrow, where tomorrow is computed at the creation of the obligation to be represented by the instant of time instant4. This obligation can be represented with the following assertions and axioms:

Obligation(obl-3), atTime(obl-3,instant1), hasDebtor(obl-3,Marco),
hasCreditor(obl-3,Luca) TimeEvent(tevent1), atTime(tevent1,instant1)
ProperInterval(interval3), hasInterval(obl-3,interval3),
hasEnd(interval3,instant4),
StartEvent-3 ≡ {tevent1}
Content-3 ≡ Pay ⊓ hasActor∋Marco ⊓ hasRecipient∋Luca ⊓ hasObject∋5euro

As already explained the four axioms for deducing the state of this obligation have to be inserted in the ontology.

5   A Case Study: Obligations in Vehicle Repair Contracts

In this Section we formalize and monitor the vehicle repair contract described in [16] using the model presented in this chapter. The scenario is as follows: a *repair contract* regulates the interactions between a client agent called cl and a vehicle repair company, called rc. A repair contract specifies details concerning a particular repair. The interaction between cl and rc is described as follows: when rc receives a request from cl to undertake a repair job, it has to send a repair contract within x days. In response, cl sends an acceptance or rejection message within y days. If accepted, cl has to send the vehicle within kl day from the acceptance. rc then waits for the vehicle to arrive, failing which it sends two reminders to cl. If the vehicle fails to arrive, it takes an offline action. As per the contract, if the vehicle arrives rc is obliged to assess the damage, repair the vehicle, and send a report to cl within $k2$ days from the reception of the vehicle. On receiving the report, cl is obliged to send payment to rc within $k3$ days from the reception of the report. If the payment is not sent, rc sends two reminders to cl and then takes an offline action. If the payment is sent cl has to pick-up the vehicle within $k4$ days from the reception of the report.

Every action has to be performed within a certain number of days, and the actual deadline is computed on the basis of the time when a certain event happens, the maximum duration of each activity is defined in the contract and may vary from one contract to another. Almost all these obligations are conditional obligations with deadline computed on the basis of the time of their activation; therefore they are similar to the second type of obligations presented in Section 4.6. Initially the interaction between rc and cl is devoted to the definition of the properties of a specific repair contract that is characterized by the type of the repair, the price, four duration of time used to compute the deadlines of the obligations for agent cl, and two duration of time used to compute the deadlines of the obligations for agent rc. We represent such a contract as an individual of the class VehicleRepair-Contract having the properties hasRepairType, hasPrice, hasDuration1,..., hasDuration6. This is another example of a domain dependent ontology. If the contract is accepted by both parties six conditional obligations start to hold, four for agent cl and two for agent rc. Subsequently the interaction is devoted to the execution of the contract. Given that the interacting agents belong to different owners having different interests, their behaviour has to be monitored to verify its compliance with the obligations.

In order to define the contract and reach an agreement on the value of the properties used to characterize the contract the two agents need to interact at least two times, but can interact also more times. A contract is *complete* if all its properties are set and therefore it belongs to the CompleteContract ⊑ VehicleRepairContract class as stated by the following axiom:

CompleteContract ≡ ∃ hasRepairType.TypeRepair ⊓ ∃ hasPrice ⊓
∃ hasDuration1 ⊓ ... ⊓ ∃ hasDuration6

The contract definition phase is regulated by two obligations: once is the obligation for rc to send a complete contract to agent cl within x days from the reception of the request from cl; the second is the obligation for agent cl to accept or reject a complete contract offer within y days. In case cl rejects the proposed contract the negotiation can continue with new requests and counter offers on the basis of the pro-activity of the two involved agents. If agent cl accepts the proposed contract then six new conditional obligations are created having as interval the duration specified in the contract. The first obligation for rc can be represented as:

Obligation(obl-4), atTime(obl-4,instant1), Instant(instant1),
ProperInterval(interval4), hasInterval(obl-4, interval4),
hasDurationDescription(interval4,duration4), days(duration4, x),
StartEvent-4 ≡ ExchMsg ⊓ hasActor∋cl ⊓ hasRecipient∋rc ⊓
hasForce∋request ⊓ ∃ hasObject.VehicleRepairContract.

The Content-4 class contains the actions of sending a request message from agent rc to agent cl with as content an individual belonging to the CompleteContract class:

Content-4 ≡ ExchMsg ⊓ hasActor∋rc ⊓ hasRecipient∋cl ⊓
hasForce∋request ⊓ hasObject.CompleteContract.

The obligation for agent cl to accept or reject a complete contract offer within y days can be represented as:

Obligation(obl-5), atTime(obl-5,instant1), Instant(instant1),
ProperInterval(interval5), hasInterval(obl-5, interval5),
hasDurationDescription(interval5,duration5), days(duration5, y),
StartEvent-5 ≡ Content-4.

The Content-5 class contains the actions of accepting or rejecting the contract whose proposal activated the obligation obl-5:

Content-5 ≡ (ExchMsg ⊓ hasActor∋cl ⊓ hasRecipient∋rc ⊓
hasForce∋accept ⊓ ∃ hasObject.(∃ hasObject⁻ StartEvent-5)) ⊔
(ExchMsg ⊓ hasActor∋cl ⊓ hasRecipient∋rc ⊓ hasForce∋reject ⊓
∃ hasObject.(∃ hasObject⁻ StartEvent-5))

Due to space limitation we will not describe in detail the formalization of all the other conditional obligations that will be created once the contract is accepted, they are similar to the second type obligations introduced in section 4.6. The application independent ontology described in this chapter with an ABox that contains the obligations described in the previous sections can be downloaded from the author's web page[9].

6   Conclusions and Future Works

In this chapter we presented a formal model for the specification and monitoring, using semantic web technology, of obligations whose content is a class of possible action, with activation and deactivation event and with deadline. The main goal of having this type of formal specification of obligations is to be able to have more flexible interactions among autonomous agents. This is possible because agents can decide at run-time which is the best action, among the ones belonging to the Content class, to perform in order to fulfil their obligations. This work is a first step in the broader project of formalizing, using semantic web technology, also prohibitions and permissions that present some crucial differences with respect to obligations. Another very important aspect of the formalization of normative concepts in open system is, besides their monitoring as explained in this chapter, their enforcement by the definition of sanctions and recovery actions.

Another interesting problem would be the definition of constrains for the validation of a normative specification and the introduction of mechanism for early

---

[9] http://www.people.lu.unisi.ch/fornaran/ontology/ObligationsOntology.html

detection of problematic situations. For example being able to point out that an agent is at the same time obliged to perform an action and obliged to perform another action that is inconsistent with the first one, like being in two different places at the same time. Another very interesting open problem is being able to demonstrate that a given set of obligations has some soundness properties [17].

Finally regarding the decision to adopt semantic web technology as formal language, there is still the open problem of better understanding what part of the model it is better and possible to represent in the ontology in order to be able to reason on it and what part of the model it is better to represent in the external application because current semantic web standards do not support its representation.

Acknowledgments. We would like to thank Marco Colombetti for the interesting discussions concerning this work.

References

1. Arcos, J.L., Esteva, M., Noriega, P., Rodríguez-Aguilar, J.A., Sierra, C.: Engineering open environments with electronic institutions. Engineering applications of artificial intelligence 18(2), 191–204 (2005)
2. Artikis, A., Sergot, M., Pitt, J.: Animated Specifications of Computational Societies. In: Castelfranchi, C., Johnson, W.L. (eds.) Proceedings of the 1st International Joint Conference on Autonomous Agents and Multi-Agent Systems (AAMAS 2002), pp. 1053–1061. ACM Press, New York (2002)
3. Boella, G., Noriega, P., Pigozzi, G., Verhagen, H. (eds.): Normative Multi-Agent Systems, Dagstuhl, Germany. Dagstuhl Seminar Proceedings, vol. 09121. Schloss Dagstuhl - Leibniz-Zentrum fuer Informatik, Germany (2009)
4. da Silva, V.T.: From the specification to the implementation of norms: an automatic approach to generate rules from norms to govern the behavior of agents. Autonomous Agents and Multi-Agent Systems 17(1), 113–155 (2008)
5. Dastani, M., Grossi, D., Meyer, J.-J., Tinnemeier, N.: Normative multi-agent programs and their logics. In: Boella, G., Noriega, P., Pigozzi, G., Verhagen, H. (eds.) Normative Multi-Agent Systems, Dagstuhl, Germany. Dagstuhl Seminar Proceedings, vol. 09121, Schloss Dagstuhl - Leibniz-Zentrum fuer Informatik, Germany (2009)
6. Erl, T.: Service-Oriented Architecture: Concepts, Technology, and Design. Prentice Hall PTR, Upper Saddle River (2005)
7. Faci, N., Modgil, S., Oren, N., Meneguzzi, F., Miles, S., Luck, M.: Towards a monitoring framework for agent-based contract systems. In: Klusch, M., Pĕchouček, M., Polleres, A. (eds.) CIA 2008. LNCS (LNAI), vol. 5180, pp. 292–305. Springer, Heidelberg (2008)
8. Fornara, N., Colombetti, M.: Operational specification of a commitment-based agent communication language. In: Castelfranchi, C., Johnson, W.L. (eds.) Proceedings of the 1st International Joint Conference on Autonomous Agents and Multi-Agent Systems (AAMAS 2002), pp. 535–542. ACM Press, New York (2002)
9. Fornara, N., Colombetti, M.: Specifying Artificial Institutions in the Event Calculus. In: Handbook of Research on Multi-Agent Systems: Semantics and Dynamics of Organizational Models, Information science reference, ch. XIV, pp. 335–366. IGI Global (2009)

10. Fornara, N., Colombetti, M.: Ontology and time evolution of obligations and prohibitions using semantic web technology. In: Baldoni, M., Bentahar, J., van Riemsdijk, M.B., Lloyd, J. (eds.) DALT 2009. LNCS, vol. 5948, pp. 101–118. Springer, Heidelberg (2010)
11. Fornara, N., Viganò, F., Colombetti, M.: Agent communication and artificial institutions. Autonomous Agents and Multi-Agent Systems 14(2), 121–142 (2007)
12. Fornara, N., Viganò, F., Verdicchio, M., Colombetti, M.: Artificial institutions: A model of institutional reality for open multiagent systems. Artificial Intelligence and Law 16(1), 89–105 (2008)
13. García-Camino, A., Rodríguez-Aguilar, J.A., Sierra, C., Vasconcelos, W.: Constraint rule-based programming of norms for electronic institutions. Autonomous Agents and Multi-Agent Systems 18(1), 186–217 (2009)
14. Governatori, G., Rotolo, A.: How do agents comply with norms? In: Boella, G., Noriega, P., Pigozzi, G., Verhagen, H. (eds.) Normative Multi-Agent Systems, Dagstuhl, Germany. Dagstuhl Seminar Proceedings, vol. 09121, Schloss Dagstuhl - Leibniz-Zentrum fuer Informatik, Germany (2009)
15. Kagal, L., Hendler, J., Berners-Lee, T.: Introduction. In: Web Semantics: Science, Services and Agents on the World Wide Web, vol. 7(1), pp. vii–ix (2009); The Semantic Web and Policy
16. Lomuscio, A., Penczek, W., Solanki, M., Szreter, M.: Runtime monitoring of contract regulated web services (extended abstract). In: Proceedings of the 9th International Conference on Autonomous Agents and Multi-Agent systems (AAMAS 2010), Toronto, Canada, pp. 1449–1450. ACM, New York (2010)
17. Singh, M.P., Chopra, A.K.: Correctness properties for multiagent systems. In: Baldoni, M., Bentahar, J., van Riemsdijk, M.B., Lloyd, J. (eds.) DALT 2009. LNCS, vol. 5948, pp. 192–207. Springer, Heidelberg (2010)
18. Uszok, A., Bradshaw, J.M., Lott, J., Breedy, M., Bunch, L., Feltovich, P., Johnson, M., Jung, H.: New developments in ontology-based policy management: Increasing the practicality and comprehensiveness of KAoS. In: IEEE International Workshop on Policies for Distributed Systems and Networks, vol. 0, pp. 145–152 (2008)
19. van der Torre, G.E.L., Boella, G., Verhagen, H. (eds.): Special Issue on Normative Multiagent Systems. Autonomous Agents and Multi-Agent Systems, vol. 17. Springer, Netherlands (August 2008)
20. Vázquez-Salceda, J., Dignum, V., Dignum, F.: Organizing multiagent systems. Autonomous Agents and Multi-Agent Systems 11(3), 307–360 (2005)
21. Weske, M.: Business. In: Process Management Concepts, Languages, Architectures. Springer, Heidelberg (2008)
22. Wooldridge, M.: Verifiable semantics for agent communication languages. In: Demazeau, Y. (ed.) Proceedings of the Third International Conference on Multi-Agent Systems (ICMAS 1998), Washington, DC, USA. IEEE Computer Society, Los Alamitos (1998)
23. Xu, L.: A Framework for E-markets: Monitoring Contract Fulfillment. In: Bussler, C.J., Fensel, D., Orlowska, M.E., Yang, J. (eds.) WES 2003. LNCS, vol. 3095, pp. 51–61. Springer, Heidelberg (2004)
24. Yolum, P., Singh, M.: Reasoning about commitment in the event calculus: An approach for specifying and executing protocols. Annals of Mathematics and Artificial Intelligence 42, 227–253 (2004)

Chapter 3
Programming Semantic Agent for Distributed Knowledge Management

Julien Subercaze and Pierre Maret

Laboratoire Hubert Curien, Saint-Etienne F-42000, France
julien.subercaze@univ-st-etienne.fr,
pierre.maret@univ-st-etienne.fr

Abstract. At the beginning of the decade, the Agent Mediated Knowledge Management workshops series as well as Bonifacio's theoretical approach layed the foundations of a new eld of distributed knowledge management based upon the agent paradigm. The agent based approach enables key features for knowledge management. The local management of knowledge by agents allows to go beyond the limitations of centralized knowledge management. Thus, knowledge can be maintained in each agent at a coarse-grained level, with different representations. In the mean time the rise of the semantic web technologies enables a new range of possibilities for agents dedicated to knowledge management. In this chapter we investigate the integration of semantic web technologies into an agent architecture that allows agents to represent their knowledge and their behavior in a semantic manner. We present the semantic agent model, its implementation and we discuss the perpectives open by semantic agents.

1 Introduction and Motivation

At the beginning of the decade, the Agent Mediated Knowledge Management workshops series [23, 1, 13] as well as Bonifacio's theoretical approach [3] layed the foundations of a new field of distributed knowledge management based upon the agent paradigm. The agent based approach enables key features for knowledge management. The local management of knowledge by agents allows going beyond the limitations of centralized knowledge management. Thus, knowledge can be maintained in each agent at a coarse-grained level, with different representations. Interactions between agents permit us to take into account the social aspect of knowledge and are well suited to represented organizational memories [9]. Van elst also outlined the importance of pro- activeness for knowledge management [22].

A. Elçi, M.T. Koné, and M.A. Orgun (Eds.): Semantic Agent Systems, SCI 344, pp. 47–65.
springerlink.com   © Springer-Verlag Berlin Heidelberg 2011

In the mean time, the publishing of the agent roadmap in 2003 [17] pointed out the lack of connection between multi-agent systems and semantic web technologies. Since then, many applications and frameworks have been developed to bridge this gap. Semantic web languages and tools are now widely used to represent agents' knowledge. TAGA [25] uses OWL and RDF as knowledge representation in the field of a trading agent competition, using a FIPA compliant framework. AgentOWL [15] extends JADE agents with OWL support for their knowledge Base (KB). It also introduces an OWL based semantic agent model. Knowledge agents, introduced by [2], are used for domain specific web search. In this application, agents' KBs are represented in RDF. RDF is also used in CORESE [8] which is a semantic web search engine for corporate knowledge developed within the COMMA (Corporate Memory Management through Agents) European IST project. The JADE framework, which is currently the most used in research and industry supports natively RDF representing agents' knowledge. These examples show us that semantic web technologies are widely used for representing agent knowledge, and that we can clearly state that the connection between agent-based knowledge management and semantic web has been made.

However, this example presents the use of semantic for the representation of agent's knowledge and not for the dynamic part of the agent: its behavior. Katasonov proposed a Semantic Agent Programming Language (S-APL) [14], based on BDI reasoning, in which agent behavior are semantically described. Behaviors remain programmed in JAVA but are described in RDF syntax. A closed-world reasoned (CWM[1]) is used for BDI support. S-APL has three main drawbacks. First the reasoning is restricted to the closed world assumption whereas semantic web languages such as OWL make the open world assumption. Secondly each new function in S-APL has to be programmed in JAVA, which reduces the interoperability between agents. An agent having some new functions will not be able to transfer its behavior to another agent since the latter does not own the JAVA code implementing the functions. Using this approach, agent behavior programming is not taking advantage of the semantic web technologies and limits interoperability.

Our motivation is to program agent behavior using semantic web standards using a finite number of actions that will be used to build complex behaviors. This language should not refer explicitly to a lower level language and should support open world reasoning. The latest advances in the field of semantic web have enabled rule languages supporting open world reasoning. We base our approach on the use of semantic rule language to program semantic agents. We aim at designing agents having a knowledge base and behavior base represented using the same syntax. The use of OWL for knowledge base and semantic rules for behavior allows this feature. In the next section we first discuss the choice of the semantic rule language that will be used to program agents and then present the design of an agent programming language and the resulting agent architecture. In section 3 we describe the resulting ontological agent model. Section 4 shows a practical example of a SAM behavior and details the different steps of its execution. Section 5

---

[1] http://www.w3.org/2000/10/swap/doc/cwm.html

concerns the implementation of the SAM prototype. We discuss in section 6 the perspectives of application of semantic agents in distributed knowledge management. Our conclusions are presented in section 7.

2   Building Agents with Semantic Rules

Semantic rules are an important part of the Semantic Web project. Figure 1 shows the current status of specification in the Semantic Web layer cake. The logic part, which is of primary interest for us, is still work in progress. This part describes the semantic rule languages. Currently, for this layer, two proposals are pending. The most well known one is the Semantic Web Rule Language (SWRL)[2] [12]; it is based on a combination of the OWL-DL with the RuleML language. The second proposal is the Web Rule Language (WRL[3]) initiative that was influenced by the Web Service Modeling Language WSML. Whereas WRL is at a draft stage, the semantic web community is focusing its research towards SWRL. Indeed, software such as Protégé[4], Pellet[5] and Jess[6] already provide support for SWRL. Due to these advances in implementation, it is now possible to develop agents based on semantic rules. Thus our choice naturally went to SWRL for the design of the Semantic Agent Model (SAM).

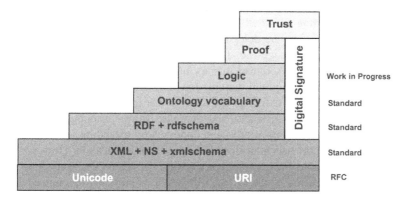

Fig. 1. The Semantic Web Layer Cake

SWRL presents two main advantages compared to other rule languages. First because it is an OWL based language, it allows writing the rules in terms of OWL concepts (i.e. classes, individuals, properties and data values). To these OWL concepts, the SWRL specification adds several built-ins functions for comparisons,

---

math, strings and time [12]. From a more agent programming point of view, concepts of the agent knowledge base can be directly treated in the rule language without loss of expressivity.

The second benefit of SWRL resides in its logical foundations. SWRL combines OWL-DL (decidable version of OWL) with Rule Markup Language (RuleML). Thus, it can be roughly considered as the union of Horn-Logic and OWL based on the description logic SHOIN. Consequently the expressivity of SWRL comes at the price of decidability [19]. SWRL is not decidable (SWRL full). However, a subset called DL Safe SWRL rules is decidable. For the agent development and for reasoning, DL Safe SWRL rules are more expressive compared to other rule languages. Indeed, most of the rule-based agents are based on Prolog supporting Horn Clauses. Researches are currently going on for implementing DL reasoning in Prolog but none is currently available for agents. Practical advantages for using Description Logic in the field of Multi-Agent Systems (MAS) has been shown in [18], especially in the field of information retrieval.

2.1 Architecture Design

Programming agent behavior using a rule language can be carried out in two ways. The first way consists in extending a logic programming language in order to support traditional agent features (i.e. message passing, threading, etc.). The second way consists in building a layered architecture using the rule language at an upper layer. Some agent features are delegated to a lower layer. In this type of architecture, the lower level language (i.e. Java, C++,etc.) is commonly used to handle communication, file access, thread management, etc. The main idea behind this approach is to reuse the required features for MAS that are already implemented in another language and to define an agent interpreter to support a particular architecture, such as BDI for instance. The literature shows examples of both approaches. Clark et al. [7] follows the first approach by extending Qu-Prolog with multi-threading support and inter-thread message communication. However, this approach is not scalable and does not comply with the Agen Communication Language (ACL) as specified by the FIPA[7]. FIPA-ACL is currently recognized as the standard for agent communication and ensures interoperability between MAS frameworks. S-APL that we discussed in the previous section follows the same approach but some direct calls to JAVA functions are directly inserted into the rules.

2.2 SAM Architecture

Standard MAS languages rely on the second approach. Agent0, the first agent dedicated language, which is an implementation of Shoham's Agent Oriented Programming was developed on top of LISP. Similarly, 3APL, 3APL-m, JA- SON and the BDI agent system Jadex are based on JAVA. Our architecture follows the

---

[7] http://www.fipa.org/repository/aclspecs.html

same approach. The specificity of our approach is to rely on a finite number of actions, in order to ensure interoperability of agents' behaviors.

In short, our approach results in the layered architecture (Fig.2):

1. Knowledge Base
2. Engine
3. MAS Framework and low level actions

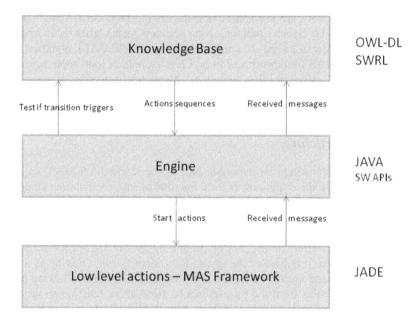

Fig. 2. SAM Agent architecture

2.2.1 Knowledge Base

The knowledge base is the upper layer of the SAM architecture. The knowledge base contains the knowledge of the agent which, in our approach, is composed of static knowledge and behavior. Agent behaviors are expressed using SWRL rules. As SWRL is based upon OWL, Terms (i.e. OWL concepts) of the knowledge base are directly manipulated in the rules. Terms of the knowledge base can appear in both antecedent and consequent of the rules. A formal specification of the rule syntax is given in section 2.5.

2.2.2 Engine

As SWRL built-ins do not cover all the requirements for agent programming, we have introduced additional low level actions (3rd layer) and a link between the rules and these actions. This link is given by a middle layer, which is the control structure that interfaces the rules contained in the knowledge base and the low level actions. Rules from the knowledge base are red by the engine, one at a time. If a rule requires a call to low level actions, the engine layer carries out this call.

2.2.3 Low Level Actions and MAS Framework

This layer contains the implementation of the low level actions that are complementary to SWRL built-ins. An extensive list of these actions is given in section 3. Notice that these actions are introduced as instances of OWL class Actions in the syntax of the rules (1st layer). Communication between agents relies on an existing MAS framework. Messages are structured following the FIPA-ACL standard, and consequently the MAS framework has to be FIPA compliant (our implementation is based upon JADE which is FIPA compliant). Messages from other agents are received through the MAS framework, then converted into an OWL representation and finally added to the knowledge base.

2.3 Control Structure

Rule-based agents constitute an important part of the research on MAS. In [11], Hindriks et al. define the requirements for a minimal agent programming language that includes rules and goals. They also defined formalization tools that were applied to three standard agent programming languages AGENT-0[21], AgentSpeak(L)[20] (that was later implemented and extended in JASON[4]) and 3APL[10]. Their definition of an agent program for goal directed agents includes a set of rules called the rule base of the agent. They identify rule ordering as a crucial issue in rule-based agents. However, this presents us with the following problem: when several rules from the rule set can be fired, there must be an order to determine the sequence of execution of those rules. So the order in which the rules will be sorted must be defined. Hindriks et al. [11] proposed that all rules fall into one of the following categories: reactive(R), means-end(M), failure(F) and optimization(O) with an order based on intuition:

$$R > F > M > O$$

As SWRL doesn't support rule ordering, we are also confronted with the same issue. However, instead of deciding an arbitrary order, we have decided to use another model of behavior, a slightly modified version of the Extended Finite State Machine (EFSM) model [6]. EFSM guarantees the execution of only one rule at a time. In EFSM, transitions between states are expressed using IF statements. A transition is fired when trigger conditions are valid. Once the transition has been fired, the machine is brought from the current state to the next state and the set of specified operations are performed. Our choice is to use a finite number of actions (called atomic actions) to fulfill basic MAS requirements. We differentiate two kinds of atomic actions, external and internal. Internal actions have an effect on the agent internal knowledge Base.

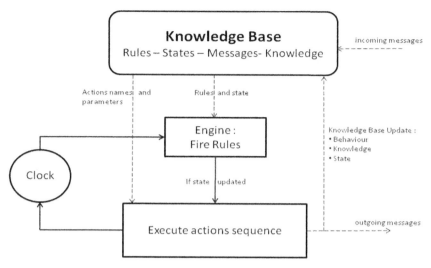

Fig. 3. SAM Agent Interpreter

External actions are the interactions of the agent within its environment. These actions include environment perception, action on the environment, message reception and emission. External actions are not included in SWRL built-ins whereas a subset of internal actions is. In section 3 we detail the list of atomic actions that are not SWRL built- ins. A deterministic EFSM is a restriction of EFSM in which there is at most one possible transition for each state and set of triggering conditions. We used this restriction to ensure that on   one rule can be triggered at a time. A pseudo code algorithm for the interpreter is defined in Algorithm 1.

Algorithm 1: SAM Interpreter

---

begin
    $CurrentState \longleftarrow sBegin$
    while $CurrentState \neq sEND$ do
        $temp \longleftarrow nextStateValue()$
        if $temp \neq currentState$ then
            $removeProperty(currentState, stateValue)$
            $actionList \longleftarrow getActionList()$
            if $executeAction(actionList)$ then
            $addProperty(currentState, temp)$
            else $addProperty(currentState, errorState)$
end

2.4 Execution Stack

The behavior of an agent can be seen as program executed by a computer. In the same manner as for computer programs, agent behavior should be able call sub behaviors. We designed an execution stack to maintain the history of behavior calls, and the state of the behavior that issued the call. For example let us consider an agent currently at the state A, and its current behavior is *GetInfo*, called with the parameter *Bob*. If the next action is to load the behavior *SearchPicture* with the same parameter *Bob*, this behavior will become the current behavior and will be placed on top of the stack over *GetInfo*. The figure 4 depicts the stack before and after the transition. In the OWL implementation, the current behavior is set with the property *hasBehavior* on the individual *currentBehavior*.

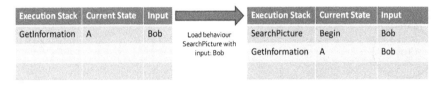

Execution Stack	Current State	Input		Execution Stack	Current State	Input
GetInformation	A	Bob	Load behaviour SearchPicture with input: Bob	SearchPicture	Begin	Bob
				GetInformation	A	Bob

Fig. 4. Stack Evolution after loading of a behavior

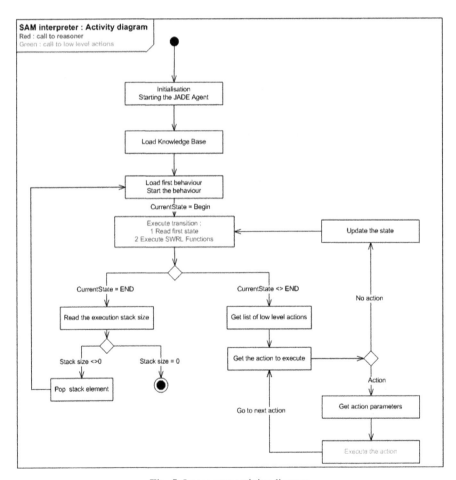

Fig. 5. Interpreter activity diagram

2.5 Language Syntax

The syntax of the rule language that we designed (given in figure 6) is expressed in Extended Backus-Naur Form (EBNF). This syntax is based on the existing SWRL EBNF syntax as specified in [12]. SAM grammar is included in the SWRL grammar. In the antecedent of a SAM rule (*SAMantecedent*) it is mandatory to specify to which state the rule applies. This is set up by the *hasStateValue* property. The previous property, *currentState*, ensures that the rule will be fired when the current state of the EFSM is the one to which the rule applies. The second part of the antecedent contains the triggering conditions. In this part, conditions under which the transition will be triggered are defined. The range of these conditions is the knowledge base of the agent. These conditions are represented by atom* which is not modified from the original SWRL specification. Conditions can test the validity of class belonging, property between classes or between individuals, including received messages.

```
SAMrule        ::= 'Implies(' { URIreference }
                   { annotation } SAMantecedent SAMconsequent ')'

SAMantecedent::= currentState('i-variable')'
                 hasStateValue' ('i-variable')' atom.

SAMconsequent::= hasNextState' ('i-variable')'
                 hasActionList' (a-list')' atom.

a-list         ::= hasValue(action) hasNext(a-list)
                 | endlist

action         ::= URIreference hasParameterName(a-name)

a-name         ::= hasParameterValue(i-object)

atom           ::= description '(' i-object ')'
                 | dataRange '(' d-object ')'
                 | individualvaluedPropertyID '(' i-object i-object ')'
                 | datavaluedPropertyID '(' i-object d-object ')'
                 | sameAs '(' i-object i-object ')'
                 | differentFrom '(' i-object i-object ')'
                 | builtIn '(' builtinID { d-object } ')'

builtinID      ::= URIreference

endlist        ::= URIreference

i-object       ::= i-variable | individualID

d-object       ::= d-variable | dataLiteral

i-variable     ::= 'I-variable(' URIreference ')'

d-variable     ::= 'D-variable(' URIreference ')'
```

Fig. 6. EBNF interpreted by SAM

The rule consequent term (*SAMconsequent*) specifies the destination state of the transition and the sequence of atomic actions to be executed. Each action has different parameters. Parameters are passed using two properties, *hasParameter-Name* and *hasParameterValue*. The first property applies to the action which is to be executed and it specifies the name of the parameter. Then property *hasParameterValue* is applied to the name of the parameter in order to specify its value.

3 Semantic Agent Model

The architecture, the control structure and the language syntax we have just presented before enable us to elaborate the semantic agent model. Using the previous given architecture, we built an OWL representation of the agent with different components (Figure 7). In accordance with the previous section, the model holds a finite number of states and of atomic actions, as well as the parameters for the actions. We defined two specials states, *sBegin* and *sEnd* that specify the beginning and end states of the EFSM. Every agent's behavior must start with *sBegin* and end with *sEnd*. Environment interactions are described within the received messages queue.

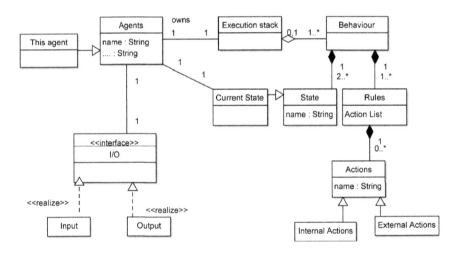

Fig. 7. The Semantic Agent Model

As mentioned, possible actions that are not SWRL built-ins are divided into two categories: internal and external actions. Here we detail the different atomic actions that are required in both categories (Figure 8 presents the different actions by layer).

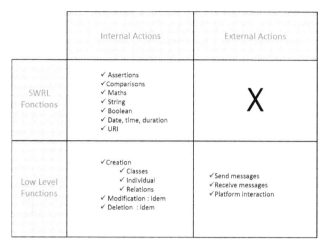

Fig. 8. Actions by layer

Internal Actions: agent knowledge is expressed using OWL concepts: classes, properties, individuals and data value. For each concept, three basic operations are needed: creation, modification, deletion. Unfortunately only the first one is supported by SWRL built in. SWRL supports assertion but does not support negation. In practical terms, it is possible to assert that properties apply to individuals or classes in the rule consequent. The following example is taken from the SWRL proposal document and shows the assertion of the uncle property by composing parent and brother properties:

$$parent(?x,?y) \land brother(?y,?z) \rightarrow uncle(?x,?z) \qquad (1)$$

However the following rules (2,3) are not possible since SWRL neither supports negation as a failure (2) nor non-monotonicity (3). Hence it is not possible to withdraw information using the rule consequent.

$$\neg Person(?x) \rightarrow NonHuman(?x) \qquad (2)$$

$$parent(?x,?y) \land brother(?y,?z) \rightarrow \neg aunt(?x,?z) \qquad (3)$$

As only creation is possible using SWRL (at a higher level), we define additional actions at lower level:

- modify/remove property
- modify/remove class belonging from a resource
- modify/delete individual
- modify/delete datarange property

Internal actions, belonging to SWRL built-ins are executed by the rule engine. Other internal actions, the low level actions are called by the agent interpreter.

External Actions refer to the agents' interactions with their environment. We restrict our scope to software agents that evolve in an electronic environment. Interactions are then limited to message exchanges between agents. We rely on the FIPA ACL specification for the message structures. Received messages are stored in the message list. In the agent's KB, messages are put in a list *ReceivedMessages* that is an instance of OWLList. Eventually there are two basic external actions, *sendMessage* and *receiveMessage*. Following the ACL specification, forging a message requires several parameters; among them we can cite sender, receiver, ontology used, performative and so on. From those simple actions, it is possible to build complex interactions between actions. For instance FIPA ACL specifies an extensive communicative act library including query-answer, contracting, proposal, subscribing. Different fields of the message are represented in the OWL knowledge Base using properties, i.e. *hasPerformative, hasContent, hasSender*.

3.1 Defining New Actions

The agent model contains a finite list of basic actions for communication and knowledge base management purpose. In SAM there are two approaches to define new actions. The first is to extend the set of available of low level actions. The second one is to define new actions by combining the existing ones.

Defining new atomic actions require implementing them in a low level language. This approach is then of low interoperability and is discouraged by the authors. It should be applied only in case of an extension of the model. The regular approach consists in defining new actions as a sequence of atomic ones. We denoted these actions as composed actions (Fig. 9).

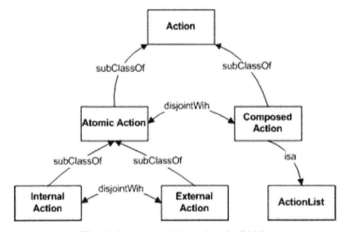

Fig. 9. Ontology of the actions in SAM

Actually, behavior of agents is a kind of composed action since it is composed by a sequence of actions, triggered by transition. To define new composed actions, we use the same representation as for agents' behaviors. Composed actions are a set of rules that represent an EFSM. These rules should only be active when the composed actions are called.

Therefore these rules are not stored as SWRL rules in the knowledge base of the agent but they are instances of the class Rule and their value is a string representation of the rule (In Manchester Syntax[8]). The process of execution of a composed action is

---

[8] http://www.co-ode.org/resources/reference/manchester_syntax/

the following. Let us assume that the agent is firing a transition between state A and B. During this transition a composed action called comp is to be executed. First the engine removes the rules of the current behavior from the knowledge base and stores them using a string representation. The engine also keeps tracks of the current state and transition sequence that was executed.

The engine sets the current state of the agent to an intermediate state *sBegin*. Then it extracts the string representation of the rules from comp and adds them to the knowledge base. The composed action is then executed following the same way as an agent behavior. Once the action finished, the engine removes the rules and sets back the agent's behavior context. Note that this process is recursive and a composed action can call another composed action.

4 Example

To illustrate the mechanism behind semantic agents, we take a simple example and process the several steps of the execution. The example has following content: start the agent Alice, register it with the directory facilitator of the framework, and send a query to the agent *Bob*. If a received message is from *Bob* and if this message has the performative answer then Alice adds the content of the answer into its knowledge base. The resulting EFSM is depicted in figure 10.

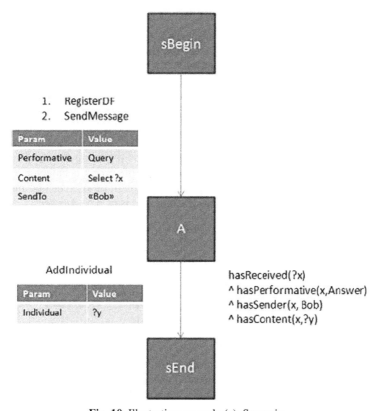

Fig. 10. Illustrative example (a), Scenario

The left column of the figure describes the low level atomic actions executed during a transition. Triggering conditions, contained in the antecedent of the rule, are on the right side of the picture. The first transition is conditions free. If *Alice* is in the *sBegin* state then the transition to the state *A* will occur. Actions related to the transition are executed as a sequence. The next actions are executed only if the previous succeeded. The action *registerDF* is executed first. If it is successful (returns true) a message with the query performative and containing a query is sent to agent Bob. The rule used to describe this transition is presented below in a human readable syntax:

CurrentStates(?x)
^hasStateValue(x,sBegin)
^NextState(?y)

→hasStateValue(y,A)
^hasContents(ActionSequence,registerDF)
^hasNext(ActionSequence,item)
^hasContents(item,SendMessage)
^hasParameterName(SendMessage,Sender)
^hasParameterValue(SendTo,Bob)
...same with other parameters
^hasNext(item,endList)

Within the architecture, the engine checks whether a transition occurs in requesting the *NextState* value to the knowledge base. If this value is different from the *CurrentState* then a transition is enabled. Then the engine retrieves the values of *ActionSequence*, with the respectives parameters. *ActionSequence* is a linked-list (Fig. 11) in which each item has the *hasParameterName* property. The value of the parameter is specified in the *hasParameterValue* property. The structure of the list of actions follows the OWL model depicted in figure 7. The *Sendmessage* instruction is linked to its parameters using properties as described in figure 11. The second transition contains triggering conditions regarding the received message. As *Alice* sent a query to *Bob*, the next step of *Alice's* behavior may be to handle the answer from *Bob*. Thus, we specify a condition on received messages to ensure that *Bob* is the sender and that the message is of type *Answer:*

CurrentStates(?x)
^hasStateValue(x,A)
^NextState(?y)
^hasReceived(?z)
^hasPerformative(z,Answer)
^hasSender(z,Bob)
^hasContent(z,?w)

→hasStateValue(y,sEnd)
^hasContents(ActionSequence,AddInvidual)
^hasNext(ActionSequence,EndList)
^hasParameterName(AddInvidual,name)
^hasParameterValue(name,w)

We will now detail the interactions between the different layers in the architecture during the execution of the first transition.

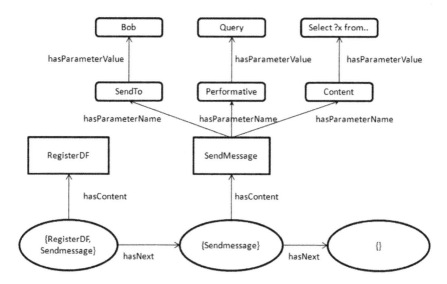

Fig. 11. Illustrative example (b), ActionList data structure

4.1 Execution Phase

Representing the action execution on a timeline following the architecture as in Section 2.1 is represented in Figure 12. It follows Algorithm 1.

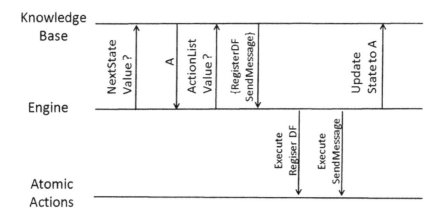

Fig. 12. Illustrative example (c), Flow chart of the transition from Begin to A

The SAM engine firstly enquires of a rule triggering, in this case, the knowledge Base query returns *NextState* = A. As A != *sBegin*, the engine retrieves the current list of actions containing *RegisterDF* and *SendMessage*. Actions are performed sequentially. First *RegisterDF* is executed and if it returns true then *SendMessage* is be executed. When both actions succeeded, the current state of the agent is updated to *NextState* value; in this case it is state A.

5   Implementation

We have developed a JAVA interpreter that communicates with the knowledge Base using the Protege-OWL API 10 Pellet is used in combination with Jena 11 as a SWRL reasoner. The JADE framework is used for the low level external actions. The framework handles agent registration, service discovery and message passing. It also provides an environment that is FIPA-ACL compliant. One implementation issue we encountered was that OWL does not support RDF lists. An OWL equivalent called OWLList has been developed and is used to represent action sequences and the queue of received messages. A first version of the open-source prototype is available online 12. Besides the validation of our model, the implementation prototype presents some limitations. Nowadays the status of SWRL reasoners is not satisfying because none of them fully support the SWRL specification. We have used Pellet as a SWRL reasoner, since it is currently the most advanced open- source implementation of SWRL. As developments stands at the moment, several important features are not supported by Pellet, for instance some SWRL built-ins are not yet available. The implementation results show the feasibility of the proposal and we intend to further develop the prototype to make it fully suitable for the development of applications. Semantic Web technologies is a field where advances take place. Current restrictions on SWRL support should no longer be an issue since advances in the field of the Semantic Web technologies occur very rapidly and regularly. Finally, this implementation of the prototype allowed us to validate our approach and to identify the limitations.

6   Perspectives

One of the primary advantages of agent based knowledge management over the classical centralized approach is the proactiveness of the agents [22]. Proactiveness is the ability of agents to initiate changes and to take initiatives. It is opposite to the reactive approach where agents react to stimulus or changes in their environment. Concretely, in the agent design, the proactiveness is implemented in different agent behaviors. The benefit of semantic agent programming is to enable the semantic description and exchange of the agents' behaviors. Thus, agents evolving in cooperative environments are able to learn behaviors from other agents. In common knowledge management frameworks the nature of knowledge exchanges between agents is limited to static knowledge. Semantic agent programming our proposal allows agents to share not only static knowledge but also dynamic knowledge insofar as agents are able to exchange their own behaviors.

This ability opens a broad scope of applications and questions. In the same way as for static knowledge exchanges, behaviors exchanges are subject to trust and security issues. Moreover, behaviors are executed by the agents and the execution of a malicious behavior could lead to serious security is- sues. We believe that existing cryptographic and trust mechanisms can easily be adapted to the exchanges of semantic behaviors.

From a larger point of view, this approach takes the opposite of the current software as a service trend. In the service approach, providers share (or sell) the use of their services, but not the implementation of the service. With the semantic behavior exchange approach, agents share the implementation of the service. Clearly this approach is valuable in cooperative environment, for example when agents belong to the same organizations. Several studies have shown that reactive and proactive agents lead to the same performance among organizations [16, 24, 5].

The consequence of semantic behavior sharing on proactive agents is important. Agents are now able to learn behaviors from other agents and to recombine, evaluate, modify these behaviors to enhance their proactive capabilities. These missing abilities were not taken into account in former studies and we believe it can greatly improve the performance of proactive agents.

7 Conclusion

In this chapter, we presented how the next generation of Semantic Web technologies can be applied in MAS programming. We discussed the limitations of current semantic approach and noticed the lack of semantic programming for agents' behaviors. State of the art frameworks are limited in terms of interoperability. To bridge this gap, we designed agent architecture to support behavior programming with semantic rules using a finite number of actions, identical for each agent. This approach allows the sharing of behaviors between agents, without relying on a specific lower level language. We detailed the three layer architecture, the language syntax and the interpreter. Afterwards we discussed the advantages of semantic agent programming in terms of knowledge management especially in the field of proactive cooperative agents.

References

1. van Elst, L., Dignum, V., Abecker, A. (eds.): AMKM 2003. LNCS (LNAI), vol. 2926. Springer, Heidelberg (2004)
2. Aridor, Y., Carmel, D., Lempel, R., Soffer, A., Maarek, Y.S.: Knowledge Agents on the Web. In: Klusch, M., Kerschberg, L. (eds.) CIA 2000. LNCS (LNAI), vol. 1860, pp. 15–26. Springer, Heidelberg (2000)
3. Bonifacio, M., Bouquet, P., Traverso, P.: Enabling distributed knowledge management: Managerial and technological implications. RMA TIQUE, p. 23
4. Bordini, R.H., Hübner, J.F.: BDI agent programming in agentSpeak using *jason* (Tutorial paper). In: Toni, F., Torroni, P. (eds.) CLIMA 2005. LNCS (LNAI), vol. 3900, pp. 143–164. Springer, Heidelberg (2006)

5. Carley, K.M., Prietula, M.J., Lin, Z.: Design versus cognition: The interaction of agent cognition and organizational design on organizational performance. Journal of Artificial Societies and Social Simulation 1(3), 1–19 (1998)
6. Cheng, K.T., Krishnakumar, A.S.: Automatic functional test generation using the extended finite state machine model. In: Proceedings of the 30th International Conference on Design Automation, pp. 86–91. ACM, New York (1993)
7. Clark, K., Robinson, P.J., Hagen, R.: Multi-threading and message communication in Qu-Prolog. Theory and Practice of Logic Programming 1(03), 283–301 (2001)
8. Corby, O., Dieng-Kuntz, R., Faron-Zucker, C.: Querying the semantic web with corese search engine. In: ECAI, vol. 16, p. 705 (2004)
9. Gandon, F., Dieng, R., Corby, O., Giboin, A.: A multi-agent system to support exploiting an XML-based corporate memory. In: Proceedings PAKM 2000, Basel (2000)
10. Hindriks, K.V., De Boer, F.S., Van der Hoek, W., Meyer, J.J.C.: Agent programming in 3APL. Autonomous Agents and Multi-Agent Systems 2(4), 357–401 (1999)
11. Hindriks, K.V., de Boer, F.S., van der Hoek, W., Meyer, J.-J.C.: Control structures of rule-based agent languages. In: Papadimitriou, C., Singh, M.P., Müller, J.P. (eds.) ATAL 1998. LNCS (LNAI), vol. 1555, pp. 381–396. Springer, Heidelberg (1999)
12. Horrocks, I., Patel-Schneider, P.F., Boley, H., Tabet, S., Grosof, B., Dean, M.: SWRL: A semantic web rule language combining OWL and RuleML. W3C Member Submission 21 (2004)
13. van Elst, L., van Diggelen, J., Dignum, V., Abeckerm, A. (eds.): Proceedings of AAMAS 2005 Workshop AMKM 2005 (July 2005)
14. Katasonov, A., Terziyan, V.: Semantic agent programming language (S-APL): A middleware platform for the Semantic web. In: Proc. 2nd IEEE International Conference on Semantic Computing, pp. 504–511 (2008)
15. Laclavik, M., Balogh, Z., Babik, M., Hluchy, L.: Agentowl: Semantic knowledge model and agent architecture. Computers and Artificial Intelligence 25(5) (2006)
16. Lin, Z., Carley, K.: Proactive or reactive: An analysis of the effect of agent style on organizational decision-making performance. Intelligent Systems in Accounting, Finance and Management 2, 271–289 (1993)
17. Luck, M., McBurney, P., Preist, C.: Agent technology: Enabling next generation computing. In: AgentLink II (2003)
18. Mller, R., Haarslev, V., Neumann, B.: Expressive description logics for agent-based information retrieval. In: Treur (ed.) Knowledge Engineering and Agent Technology. IOS Press, Amsterdam (2000)
19. Parsia, B., Sirin, E., Grau, B.C., Ruckhaus, E., Hewlett, D.: Cautiously Approaching SWRL. Technical report, Technical report, University of Maryland (2005)
20. Rao, A.S.: AgentSpeak (L): BDI agents speak out in a logical computable language. In: Perram, J., Van de Velde, W. (eds.) MAAMAW 1996. LNCS, vol. 1038, pp. 42–55. Springer, Heidelberg (1996)
21. Shoham, Y.: AGENT0: A simple agent language and its interpreter. In: Proceedings of the Ninth National Conference on Artificial Intelligence, vol. 2, pp. 704–709 (1991)
22. van Elst, L., Dignum, V., Abecker, A.: Towards Agent-Mediated Knowledge Management, pp. 1–30. Springer, Heidelberg (2003)
23. van Elst, L., Dignum, V., Abecker, A. (eds.): AMKM 2003. LNCS (LNAI), vol. 2926. Springer, Heidelberg (2004)

24. Xiao, J., Catrambone, R., Stasko, J.: Be quiet? Evaluating proactive and reactive user interface assistants. In: Human-Computer Interaction: INTERACT 2003; IFIP TC13 International Conference on Human-Computer Interaction, Zurich, Switzerland, September 1-5, p. 383. Ios Pr. Inc., Amsterdam (2003)
25. Zou, Y., Finin, T., Ding, L., Chen, H., Pan, R.: Using semantic web technology in multi-agent systems: a case study in the taga trading agent environment. In: ICEC 2003: Proceedings of the 5th International Conference on Electronic Commerce, pp. 95–101. ACM, New York (2003)

Part II
Engineering Semantic Agent Systems

Chapter 4
SBVR-Driven Information Governance: A Case Study in the Flemish Public Administration*,**

Pieter De Leenheer[1,2], Aldo de Moor[2,3,***], and Stijn Christiaens[2]

[1] Vrije Universiteit Amsterdam - De Boelelaan 1081a, 1081 HV Amsterdam
  pieter@collibra.com
[2] Collibra nv/sa - Ransbeekstraat 230, B-1120 Brussel
[3] CommunitySense - Cavaleriestraat 2, 5017 ET Tilburg

Databases persistently store data and provide standardised access to it for communities consisting of human as well as software agents. However, adequate information *governance* in such a bi-sortal setting requires more than that. Despite the rigorous formal structure that may have been imposed on a data set, if it cannot be disclosed to third parties, their value is practically zero. The ICT-outsourcing partnership between the Flemish Ministry of Education and Training (FMET) and EDS-Telindus, now HP, appreciated the subtle difference between data and information, and the need for more maturity regarding the governance of their vast (meta-)data landscape. This is shown by initiatives such as the development of a Data Warehouse and an Information Governance Organization.

Some claim that semantic technology will soon compete with, and eventually replace traditional business intelligence approaches as a means to achieve *just-in-time* information. However, critics complain that the fact that the killer Semantic Web application is still missing indicates a lack of convincing business drivers. Based on our own experience in large organisations, we observe that business drivers for semantically driven solutions are latent but are begging to be articulated. However, for a sustainable breakthrough, mere technology to reconcile and apply business semantics is not sufficient.

In this chapter, we describe the support of the ICT-outsourcing partnership in their next leap towards information maturity and governance. To this end, we adopt OMG's SBVR standard as a basis to agree on formal and detailed natural language *declarative* semantic description of complex business entities, including the governance structure of the organisation itself.

---

* Invited chapter.
** We would like to thank Frans Decuyper (FMET) and Guido Dedene (Katholieke Universiteit Leuven) for their support and input.
*** Aldo de Moor was working part-time for Collibra during the period this work was conducted.

A. Elçi, M.T. Koné, and M.A. Orgun (Eds.): Semantic Agent Systems, SCI 344, pp. 69–88.
springerlink.com   © Springer-Verlag Berlin Heidelberg 2011

1 Closed World Syndrome

The ability to unlock data is related to the ability to understand the hidden information in it. For this, human agents too often rely on their intuitive ability to understand the context, without guaranteed success and sometimes with disastrous consequences. To reduce cost and risk, for human as well as software agents, the semantic context of data has to be made explicit. The ICT-outsourcing partnership between FMET and HP (in this article referred to as *ICT-outsourcing partnership*) aims to give a meaningful answer to following questions about mission-critical data:

1. *semantics*: what is the meaning of my data?
2. *utilisation*: how is my data used?
3. *provenance*: where do my data come from?
4. *governance*: who is responsible for what data?
5. *quality*: what is the quality of my data?

Many information systems suffer from a *closed-world syndrome*. They were designed from a naive assumption they have already stored all possible facts about the domain. Facts not in the database are presumed to be false; hence there never will be a need for large-scale data exchange with other systems. Moreover, a database is usually designed from a strong *IT/IS (information system and technology)* perspective. Consequently, only the designer is familiar with its internal structure and rules, and changes are driven by technical fits rather than real business needs.

The technological nature of the syndrome lies in vendor *lock-in* of data caused by the fact that they are usually stored in proprietary (read: *closed*) formats. Obviously, this does not make any sense in today's networked value constellations (like the World Wide Web) where online information exchange across business processes becomes central.

2 Just-in-Time Information

To answer the above questions, the ICT-outsourcing partnership aims at *just-in-time* information (JITI) that we define as:

> *"JITI is the ability to interpret the latent information in exchanged data in the right context and in a timely manner, without the help of the original designer."*

The need for JIT information follows from the fact that information forms arguments during strategic decision making. Consider, for example, the Minister of Education ordering FMET to produce a business intelligence report on the possible influence of the mother's educational profile on her children's school performance. People are in constant need for relevant JITI in order to analyse this correlation and finally to allow the minister to make well-founded political decisions. Current business intelligence approaches deliver report updates in batches, hence intermediate JITI needs have to be addressed ad-hoc via the informal social networks leading to unnecessary overhead. JITI calls for a more *pragmatic* approach different from current approaches.

3 The Gap between Business and Technical Metadata

For JITI to be effected, the ICT-outsourcing partnership is convinced that they, as earlier for data itself, must pay attention to the structured recording and publishing of data *about* data, also called *meta-data*. The Flemish regulation[1] describes metadata as:

> *"Documentation that describes the content and frequency of updating an authentic data source, and technical manner in which that resource can be unlocked."*

This definition remains vague when it comes to defining metadata, but clearly hints that the underlying *ontology* allowing a meaningful interpretation of this metadata should, in addition to being shared and agreed, have a *dual utility* [1]:

1. in an *IT/IS context*, it serves as computer specification to build diverse semantic applications (such as data integration between software agents);
2. in a *business context*, it serves as a theoretical model referring to real-world objects aligning the strategic goals, values, and processes among (human) stakeholders.

This new requirement for duality in the specification of an ontology that does not merely describe facts from IT/IS owners, but also from business users, will push the introduction of ontology management in daily work practices. Ultimately, this will improve the co-evolution of business process changes and information system changes.

In business the somehow misleading term *metadata management* is preferred for ontology management. Enterprise applications, such as master data management and business intelligence claim to provide an integral solution for *metadata management*. In practice, however, they produce redundant or anomalous metadata because they do not take into account the interest of other applications (again the closed world syndrome). Furthermore, the metadata is of a mere *technical* nature, which at best provides only a partial answer to those questions mentioned earlier. Indeed, there are inherent limits on just how much technology can really help in this regard.

From an academic point of view, it turns out that in current ontology management practices, the underlying methodological principles are usually ignored [2]. Moreover, regarding the rather narrow focus on technical metadata, they systematically ignore the subtle gap that looms between (i) knowledge sharing between human agents on the business/social level (based on business metadata), and (ii) data exchange between computerized agents (such as information systems) at the operational/technical level (based on technical metadata).

---

[1] Decree for Electronic Adminstrative Data Exchange (in Dutch: Decreet voor Elektronische Bestuurlijke Gegevensverkeer) published by the Social Economic Council of Flanders (in Dutch: Sociaal-Economische Raad van Vlaanderen (SERV)): http://www.serv.be/uitgaven/1253.pdf

Large-scale ontologies, like Cyc[2], usually suffer from severe usage restrictions due to intellectual property rights. As only small parts of their content is publicly available, these ontologies have no chance being scrutinized under peer review, hence using them as standard would be problematic. The current proliferation of so-called ontologies (in fact merely small-scale vocabularies) on the Semantic Web may result in a Web rich in semantics but poor in ontological consistence [8]. This consistence must not merely be dependent on empirical validation but also on its pragmatic value (see our discussion on validation later).

In order for an ontology to become a sustainable, reliable and shared resource, De Leenheer et al. (2010), state that the basic principle of community-based ontology evolution lies in capturing the *co-evolution* of (a) social knowledge sharing and information needs that emerge from it, (b) the supporting computerized information systems, and (c) the semantics that enables (meta-)data exchange between these systems. Doing so, it imposes a conceptual bridge between business metadata and technology metadata. Before introducing our approach to this co-evolution, we elaborate on the business drivers that advocate the reconciliation of metadata, and the metadata landscape *as-is* within the ICT-outsourcing partnership.

4   Business Drivers to Bridge the Gap

Based on our interviews, within the ICT-outsourcing partnership metadata management is fuelled from seven business drivers.

4.1   Documentation

"Sometimes people do not know what data is out there. Knowing the whereabouts of data starts with many calls within and between business units over and over again." (Anton Derks)

Business (intelligence) metadata is key to the provision of relevant documentation. Employees must document their data in a systematic manner in order to minimize the loss of know-how in case they would leave the organization. This is especially important in the context of the ICT-outsourcing partnership in which a regime of high turnover dominates. Business analysts must be empowered to define relationships between files that document the reasoning steps of their strategic advices. The collective practice of documentation will progressively decrease the overhead brought by repetitive calls for often the same latent documentation. Ultimately, in combination with underlying social network data, it will provide a valuable resource of actionable knowledge for social analytics.

4.2   Communication

"No communication without metadata!" (Jan Dejonghe)

Metadata facilitates communication both internally as externally. E.g., vendors of administrative school software should be able to understand the semantics the

---

[2] http://www.cyc.com/

software should obey when reporting enrolment data back to FMET. This agreement can be formulated in the form of a technical data specification (e.g., in UML or XSD) that was generated from business metadata. Business metadata on its turn is compliant to certain administrative regulations. Doing so, metadata provides a "language" that can cope more effectively with communication problems between business and the in- and outflow of external ICT consultants.

4.3 Reuse

Metadata accelerates the retrieval of assets and promotes their reuse. Currently assets, including reports, queries, data, architecture, technology, and licenses are defined ad hoc. In the planned service-oriented architecture, metadata will facilitate the retrieval and reuse of software services.

Codification strategy where data is codified in a common format that makes it easy to exchange and reuse is not the silver bullet. Personalisation complements this via right tooling and culture, which allows next to codification, the emergence of personal networks, engendering reputation among its peers. Therefore, it is equally important to address the provenance of data. If one knows the owner of a data asset, one can use the personal network (with Web 2.0 tooling) to gain additional know-how through socialisation.

4.4 Impact Analysis

"You hear now and again, "they have changed something to the data table, and now it does not work anymore." (Frans Decuyper)

Metadata is crucial to capture complex dependencies between different systems, people and applications, and to calculate the impact in the event of a change transparently. A precise impact analysis allows a more precise cost-benefit analysis. Moreover, it reduces the likelihood that they are subsequently surprised by unexpected side effects. The ICT-outsourcing partnership as-is often cannot see the forest for the trees.

4.5 Disambiguation

"There are 180 000 teachers, more than one million students and thousands of educational institutions. FMET forms a large part of our society, hence it is certainly important that labels are attributed the same meaning." (Martin Maesen)

Metadata helps to get rid of inconsistencies or ambiguities. It is very valuable to know that a term has an unambiguous meaning. For example, the term "family" is fairly easy to understand for most people, but in the software application it has a strict sense, inferred from the legislation. Another problem is caused by (naturally occurring) homonyms. E.g., if one would check all decrees ranging from primary to higher education, the word "study area" is often attributed contradictory meanings, while one would expect this in FMET frequently used term should be intuitively obvious. There are also terms that are poorly defined, and require additional interpretation. An example is a term "part exemption" that is applied in regulation but apart from that has few leads to its definition. This brings about the question whether we should take the definition into account are not. These are issues that continually recur in discussions with institutions about changing data models ("yes, but… what about the part exemptions"). Several discussions are repeated over and over again, and would be unnecessary if the relevant terms would be disambiguated properly.

4.6 Uniformity

"In my early years I have worked on a glossary: the most difficult part was to obtain consensus." (Marleen Deputter)

Unambiguous metadata is not sufficient. It is crucial that metadata uniformly applies to the entire organization and its stakeholders. It is very valuable for everyone to be sure that what is said in one place, is also valid elsewhere. A major source for the lack of uniformity is the wave pattern that ICT-outsourcing partnership follows when it evolves: first, at the business level, there is a new decree (or set of related decrees). Next, this decree is "implemented" in a computer application. Inherent to its nature, legislation may change, giving rise to an organic growth in the implemented applications. An example is their Salary System and more specifically the lack of meaningful codes that are contained therein.

4.7 Compliance

Metadata plays an important role in regulatory compliance, a field that is not much explored so far:

- *Authentic sources*: where data are spread over different systems, it is difficult to determine which of the systems is the original data source.
- *Privacy*: some data are covered by privacy legislation. When it is impossible to learn whether a piece of data is covered by privacy rules, it is difficult for an organization to comply with such regulation.
- *Security*: FMET has currently a Security Officer to verify regulatory compliance. Using metadata provenance of data can be logged.

5 Metadata Landscape Dimensions

Just like with information management [12], the strategy, structure, and operationalization of metadata management is not trivial because it must be able to align the complex and rapidly evolving IT/IS needs to the ditto business needs. Technological support for metadata management in terms of software is necessary, but certainly not sufficient. To bridge the gap, we distinguish, alongside technology, three other dimensions in the metadata landscape: *methodology*, *organization*, and *culture*.

Crucial is the development of a teachable and repeatable methodology consisting of a number of coordinated methods and techniques that allows the organization to perform different metadata management activities effectively and efficiently. This methodology should be anchored in an organizational arrangement of social roles and responsibilities, appropriate to the establishment of the Information Governance[3]. Finally, in order to apply the methodology properly, it is of great importance that there is a right culture of joint understanding, managing and using metadata in the cross-process information chain. Our observations are supported by empirical investigations in knowledge management in general indicating that social *incentives* brought about by cultivation are essential to increase the usage of knowledge management tools and methods [11].

Designing a solution requires a metadata landscape analysis that deepens each of the four dimensions, and consequently focuses on the internal weaknesses and external threats. To this end, we have interviewed more than twenty people on both sides of the partnership.

6 Metadata Landscape SWOT Analysis

Based on our findings, we performed a strength-weakness (SWOT) analysis. The summarising SWOT diagram is shown in Figure 1.

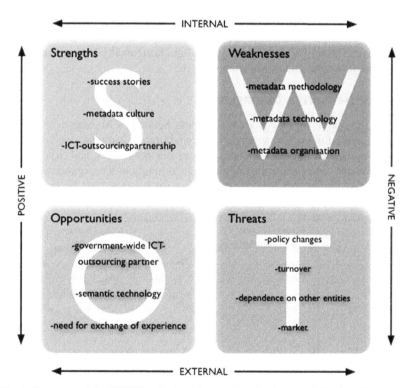

Fig. 1. Summary of the SWOT analysis of the metadata landscape within the ICT-outsourcing partnership.

---

[3] See internal report: Decuyper, F. (2009) *Information governance*. Technical report, O&V.

An important strength is that metadata culture within the partnership is already fairly mature. This is shown by the intention of its management and the curiosity of its employees who take grass root initiatives providing success stories. Moreover, there is already a targeted training on conceptual business modelling. Another important strength is the outsourcing partnership that already exists for many years.

On the contrary, the weakness lies in the fact that the current metadata management technology is not adequate. Moreover, a methodology and organisation is completely lacking. Threats for improvement are the constantly evolving policy changes that impede a smooth long-term rollout of information management. This is aggravated by the high turnover we already discussed. Moreover, if FMET does not act fast, there is a threat that other public administrations will enforce their own metadata standards, making FMET a mere dependent entity.

The opportunities include the growing availability now of robust commercial semantic technology that can be safely deployed to gain information maturity[4]. Moreover, there is an increasing demand for exchange of best practices. This demand is particularly visible in the context of e-government. Consider, for example, the large-scale, even competitive, initiatives to make government data public in the US[5] and the UK[6]. As a result, the "crowd" can build useful services and applications that - through semantics - reuse, link, and reason about these data resulting in interesting new business models. Finally, the fact that the ICT-outsourcing partner is present throughout all ministries is not only a strength, but also creates an opportunity to widely disseminate best practices horizontally but also vertically from regional, to federal and even European level.

7 Business Semantics Management

Inspired by the notion of co-evolution, Business Semantics Management (BSM) [7] provides methodology, technology, culture and organization that enable parties to (i) obtain consensus on (the semantics of) key business terms, and (ii) evaluate this consensus uniformly in various applications throughout the organization. Respectively, BSM consists of two complementary cycles: *semantic reconciliation* and *semantic application* (see Figure 2) that each groups a number of activities.

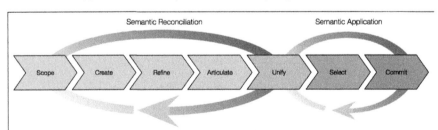

Fig. 2. Business Semantics Management consists of two complementary cycles: semantic reconciliation and semantic application. Both cycles communicate via the unify-activity.

---

[4] See, e.g., PwC Technology Forecast 2009:
   http://www.pwc.com/US/en/technology-forecast/spring2009/index.jhtml
[5] e.g., http://www.data.gov/
[6] e.g., http://data.gov.uk/

7.1 Fact-Orientation

BSM draws from best practices in ontology management [10] and ontology evolution [5]. The representation of business semantics was originally based on the DOGMA [13] ontology framework that follows a *fact-oriented* paradigm that was introduced in the conceptual database modelling approach NIAM[7] [15] (pre Object-Role Modelling)).

NIAM simplifies the design process by using natural language, as well as intuitive diagrams, which can be populated with examples, and by examining the information in terms of simple or elementary facts. By expressing the model in terms of natural concepts, like objects and roles, it provides a conceptual approach to modelling. Moreover, breaking down the domain into several elementary fact types reduces the problem complexity into smaller and thus more easily manageable sub-problems. This leverages the potential of domain experts to effectively externalise conceptions that were not revealed otherwise [9].

NIAM/ORM's attribute-free approach, as opposed to frame-based techniques such as UML or (E)ER, promotes semantic stability. *Semantic stability* is a measure of how well models or queries expressed in the language retain their original intent in the face of changes to the application [9]. Given the co-evolution principle, it is critical that the underlying ontology be crafted in a way that minimises the impact of these changes. Therefore regarding our objectives, fact-oriented models are more stable under business changes than e.g., OO or ER models.

7.2 Collaborative Business Semantics Modelling with SBVR

Recently, BSM adopted Semantics of Business Vocabulary and Business Rules (SBVR[8]), a recent OMG standard pushed by the business rule community and the fact-oriented modelling community. SBVR follows the meta-modelling principles of OMG's Meta-Object Facility (MOF[9]), which is essentially a set of concepts that can be used to define other modelling languages.

In the ICT-outsourcing partnership, UML is marginally adopted for business modelling. Some strongly advocate the suitability of UML for knowledge representation, especially in model-driven engineering [8]. However, UML has serious shortcomings like its lack of formal definition [3]. Constraints are expressed in a semi-formal language (i.e., OCL) and descriptions of the various elements are in plain (instead of structured English). Finally, UML is object-oriented, leaving out the possibility to refer to objects uniquely otherwise than with the auto-generated internal object identifier.

Driven by its success in conceptual data modelling, the fact-oriented approach of SBVR provides the basis for formal and detailed natural language *declarative* description of complex business entities. The derived formal vocabularies and rules can be interpreted and used by computer systems to develop Web, software and business intelligence applications. Additionally, the recent Ontology Definition

---

[7] Natural Information Analysis Method.
[8] OMG Semantics of Business Vocabulary and Business Rules, version 1.0:
   http://www.omg.org/spec/SBVR/1.0/
[9] OMG Meta Object Facility, version 2.0: http://www.omg.org/spec/MOF/2.0/

Metamodel[10] (ODM) provides (via MOF) a bridge to link SBVR to the Web Ontology Language for Services (OWL-S), Resource Description Framework Schema (RDFS), Unified Modeling Language (UML), Topic Map (TM), Entity Relationship Modeling (ER), Description Logic (DL), and Common Logic. Most of these extensions are outside the scope of this article. We refer to Gasevic et al., [8] for a detailed discussion in the context of model-driven engineering.

For the outsourcing partnership following domains are predominant.

1. *Software and Service engineering:* UML is a language to specify, visualise, and document software and service systems. Architects in the ICT-outsourcing partnership widely use UML.
2. *Business Intelligence*: Common Warehouse Model is an OMG standard for integrating tools for data warehousing and business analysis. In the ICT-outsourcing partnership, the Business Intelligence Competence Centre is considering the adoption of this standard.
3. *Web Engineering:* RDFS and OWL are W3C specifications for Web engineering. The ICT-outsourcing partnership has no short-term ambitions, but its uptake in the next long-term ICT plan is crucial.

Via MOF, business semantics (in SBVR) forms the basis for *forward* engineering of software (i.e. UML diagrams), business intelligence (i.e. OMG common warehouse model), and Web applications (W3C RDF(S) and OWL) and vice versa: existing models can be *reverse* engineered to feed the BSM process.

7.3 Business Semantics Structure

The structure of SBVR (illustrated in Fig. 3) allows implementing a business semantics system that follows the 6 principles of community-based ontology evolution earlier defined in De Leenheer et al. (2010).

1. **ICT Democracy:** An ontology should be defined by its community, and not by a single developer[11].
2. **Emergence:** Semantic interoperability requirements emerge from community evolution processes.
3. **Co-evolution:** Ontology evolution processes are driven by community evolution processes[12].
4. **Perspective Rendering:** Ontology evolution processes must reflect the various stakeholders perspectives[13].
5. **Perspective Unification:** In building the common ontology, relevant parts of the various stakeholder perspectives serve as input for the unified perspective.

---

[10] http://www.omg.org/spec/ODM/1.0/
[11] Because we assume the involvement of both business analysts and other non-technical ex- perts as well. This also presumes that the language use to represent the ontology is teachable to this type of contributers.
[12] In contrast to waterfall-like approaches that focus on a broad design upfront, agile methods perform short milestone-driven revision iterations in order to cope with dynamic environments.
[13] There is no generally applicable ontology, as each application will generate a contextualised model to match local needs and functionalities.

6. **Validation:** The explicit rendering of stakeholders perspectives allows us to cap- ture the ontology evolution process completely, and validate the ontology against these perspectives respectively.

In other words: it takes into account the existence of multiple perspectives on how to represent concepts (by means of vocabularies), and includes the modelling of a governance model to reconcile these perspectives *pragmatically* (read: insofar practically necessary) in order to come to an ontology that is agreed and shared (by means of communities and speech communities).

- A s*emantic community* is a group of stakeholders having a body of shared meanings. Stakeholders are people representing an organisation or a business unit. They already informally share knowledge via social network functionality.
- A *body of shared meanings* is a unifying and shared understanding (percep- tion) of the business concepts in a particular domain. Concepts are identi- fied by a URI. The scope of this body emerges from breakdowns during in- formal knowledge sharing.
- A *speech community* is a sub-community of a semantic community having a shared set of vocabularies to refer to the body of shared meanings. A speech community groups stakeholders and vocabularies from a particular natural language in multi-lingual community, or from a certain technical jargon.
- A *vocabulary* is a set of terms and fact types (called vocabulary entries) primarily drawn from a single language to express concepts within a body of shared meanings.

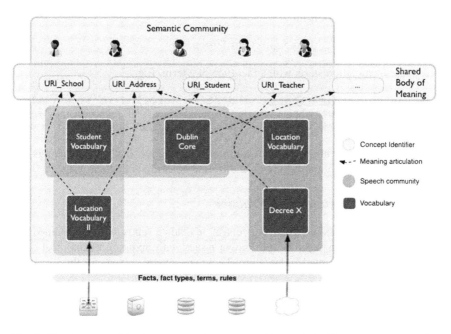

Fig. 3. The structure of business semantics: communities, stakeholders, concepts, vocabu- laries, facts, and rules.

As mentioned, the notion of vocabularies allows multi-linguality or within one language synonymous terms may refer to the same set of concepts, or a polysemous term may refer to different concept URIs depending on the vocabulary it is residing in. Following function maps a term in a vocabulary to a concept URI. For the full formalisation, we refer to De Leenheer et al. [4].

```
concept(vocabulary,term) = <URI>
```

E.g., consider a term "student" in a Dutch vocabulary and a term "étudiant" in a French vocabulary both meaning the same thing.

```
concept(Dutch, student)
= URI_STUDENT
= concept(French, étudiant)
```

A term is defined by at least a set of fact types and one or more rules. Terms form the building blocks for binary fact types, which can be read on two directions:

```
Student is enrolled in / enrolls School
School is located at / is location of Address
```

A fact type can also denote an ontological relationship such as a specialisation/generalisation relationship:

```
School is a / subsumes Institute
```

or an aggregation relationship:

```
Faculty part of / has part University
```

Every rule formulation starts with a fact type of a term. E.g.:

```
It is obligatory that a Student is enrolled in (/ enrolls)
exactly one School
Address is location of (/ locates) at most one School
```

Hence, vocabularies allow speech communities to give different meanings to terms and fact types, but also to impose different constraints on their usage.

7.4 Business Semantics in Practice

It is practically impossible to have a central "metadata repository" maintained by one person. E.g., consider the well-known issues with updating a canonical data model in a service-oriented architecture. There are several reasons for this:

- the historically grown inconsistent and difficult to unlock collection of metadata sources;
- the structural independence of the business units (called "entities") within FMET;
- the intended independence of FMET towards its ICT-outsourcing partner HP;

- the general economic trend towards dynamic value networks;
- the increasing presence of "Web 3.0" where data and services are decentralized and accessible to each other via URIs. A vision shared with Semantic Web and Web Science communities (cf. e.g., IEEE IS Jan. 2010).

However, note that a fully decentralized approach is not feasible within ICT-outsourcing partnership where business semantics are determined by regulations. Alternatively, BSM stands and falls with two initiatives: a *business semantics glossary* (BSG), and an *enterprise information model* (EIM).

8 Business Semantics Glossary

According to the English dictionary, a glossary defines a list of terms and their meanings in natural language. A BSG[14] is a glossary where the meaning of term is formally defined by fact types and rules. It provides a single point of reference for the ICT-outsourcing partnership different business vocabularies and rules, and a practical grouping according to their semantic and speech communities. The BSG supports the semantic reconciliation process.

Scope: Sets out the *scoped terms* that are actually needed to improve the information chain. Specific business drivers that want to resolve a weakness or threat in a certain application context fuel this activity. Regarding our considerations made above, we distinguish between IT/IS and business contexts.

1. In an IT/IS context a communication breakdown may be caused by an inadequate transformation of incoming personnel data from the more than 1,500 educational institutions to the data semantics of the central salary system. The breakdown here is caused by a lack of specification of terms as "personnel" and "salary". The derived need for manual translation (e.g., using XSLT) introduces a weakness, as defining the translation requires know-how about the resp. formats. Moreover, such a translation introduces even more legacy that is difficult to interpret.
2. In a business context, the lack of a uniform and unambiguous meaning of the term "study area" following externally imposed rules may form a legal threat. This observation initiates another semantic reconciliation cycle where metadata related to "study area" are to be reconciled. It is important to involve the relevant stakeholders in this process and assign them with appropriate roles and responsibilities.

Note that we have oversimplified the scoping process here. For supporting scoping techniques we refer to De Leenheer [6].

Create: During this activity, every scoped term is syntactically defined. E.g., Figure 4 illustrates the concept page (identified by a URI) in BSG for term "Home Address" (specialisation of the type "Address") in the BSG. The page consists of a

---

[14] Business Semantics Glossary is a product from Collibra. More information on http://www.collibra.com

gloss providing a natural language definition; a number of fact types (e.g., "home address has postal code") and a number of rules (e.g., home address has exactly one zip code). To each scoped term (within its context) there are also certain roles appointed such as a "concept steward" and a number of relevant stakeholders. The definition is fed by implicit know-how from the involved domain experts, or by to automatic extraction of facts from existing metadata (see [6] for a review of ontology extraction techniques).

Fig. 4. Screenshot of the term "Home Address" taken from the Business Semantics Glossary that currently deployed at FMET. Even though the concept definitions look like natural language. Thanks to the underlying MOF-compliant meta-model, one can automatically generate an enterprise information model from it that provides a formal specification.

According to our metadata landscape analysis, many FMET entities have isolated "grassroots" metadata initiatives. They manifest themselves in various forms such as taxonomies, keyword systems, glossaries, information about database fields, metadata in Web pages and content management systems (CMS). These use many proprietary formats as well as open formats such as XSD and UML. In SBVR, the business concepts can be defined in a natural way, while at anytime a formal enterprise information model can be automatically generated in any format. Note that each change is carefully logged in order to be able go back any time.

Refine: During this activity, fact types that were created during the Create activity are refined so they are understandable to both business and technology. E.g., The somewhat technical term Empl becomes Employee or EmplAddr is decomposed into a fact type Employee is located at / locates address. Coding conventions can be applied here to guide the process.

Articulate: Since multiple users concurrently render their perspective on a term, it may be that after the refine activity some fact types and rules impose contradicting statements. During this activity, conflicts and inconsistencies are removed. Specifically designed algorithms may help here. E.g., in the Netherlands, an address is uniquely identified by a combination of postcode and house number, while in Belgium a combination of postcode, street name and street number is required. Articulating these differences is crucial in order to be able to deal with different data integrity rules during information exchanges.

During **unification** is a new version of the enterprise information model is generated from the current version of the BSG.

8.1 Enterprise Information Model

An enterprise information model (EIM) is a "flattened" version of the BSG that is generated in a timely manner. The EIM is the product of semantic reconciliation and serves as a uniform technical specification to implement semantic applications. Through the underlying MOF framework, this EIM can be represented in many formats, such as UML, OWL, or XSD, serving a wide variety of applications. We distinguish two activities.

Select: Given an application context (such as a workflow or business artefact (Hull)), relevant concepts are selected from the EIM for a particular application. It may be required to add additional application-specific constraints that could not be agreed upon on the community level, or that are currently not supported by SBVR.

Commit: Information systems are improved using the selected concepts. Depending on the application context, this can be implemented in different ways. Concretely, this boils down to *data transformation, validation,* and *governance services.* For example, two or more XML structures can be virtually integrate by defining XSLT transformations to a shared XMLS-formatted EIM. The EIM may also be used to convert relational databases into RDF triple stores (cf. RDB2RDF initiative). Here we illustrate the application of an EIM to generate data transformation services. The Business Semantics Studio[15] (BSS) is a tool suite that supports these two activities. BSS provides mapping functionality to commit existing data sources and applications onto the EIM with Ω-RIDL [14]. Below we show two examples of such mappings: one committing a field in a database to a concept in the EIM and another path in an XML-document.

```
map TBLSchool.Street on Street of (/with) Address of (/ with)
School.
map /schools/school/street on Street of (/ with) Address of
(/ with) School.
```

These mappings can be used to automatically generate data transformations from one format into another by generating the appropriate queries (SQL, XPath, etc.). The example are intentionally kept simple for didactic reasons. For a more detailed description of Ω-RIDL we refer to Trog et al. (2007).

---

[15] http://www.collibra.com/products/business-semantics-studio

9 Full-Cycle BSM: Validation and Feedback

Once semantic applications are running, it must be possible to monitor and feed unexpected side effects or failures back, calling for a new iteration of BSM. We call this *full-cycle* BSM: the scope of the next version of the EIM is fed by the validation of the previous version in IT/IS contexts as well as business contexts. The BSG is the vehicle that serves the reconciliation of the newly scoped concepts. The overall picture of this full-cycle BSM is illustrated in Figure 5.

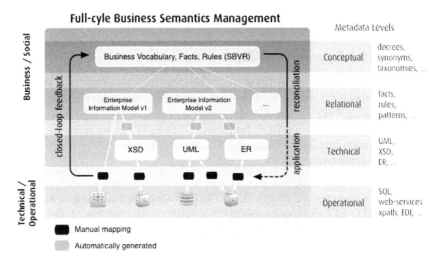

Fig. 5. Full-cycle business semantics management keeps the business semantics in line with changes in IT/IS and Business contexts.

The BSM cycle is repeated until an acceptable balance of differences and agreements is reached between the stakeholders that meets the requirements of the semantic community. Gradually, closed divergent metadata sources are replaced with metadata sources that follow an open standard, and are kept coherent via BSG. Referring to the external threats (discussed in SWOT analysis), once the ICT-outsourcing partnership (as semantic community) has internally standardized stable parts of its EIM, it stands stronger to push these up to federal[16] or even European level (e.g., Semic.eu[17]).

Regarding its dual utility, we distinguish between two complementary ways to validate business semantics: in the context of business applications or IT/IS applications.

9.1 IT/IS-Driven Validation

The formal specification of a semantic pattern can be empirically validated for its logico-computational "proof" and its computational performance within a predefined

---

[16] http://www.fedict.belgium.be
[17] http://www.semic.eu

IT/IS context. If a semantic pattern is sufficiently constrained then certain desired properties can only be derived strictly in a mathematical or logical fashion. E.g., the data structured according to the metadata of system A is transformed in a certain way into the format defined by metadata of system B. The performance of the computational implementation can be tested in simulations using very large data sets. De empirical evaluation feeds inconsistencies, inadequacies, and incoherencies back to the semantic reconciliation.

9.2 Business-Driven Validation

A more complex problem is the pragmatic evaluation of concept types within a business context. From case studies we learned that this requires a rather qualitative assessment of validity. We distinguish two categories of validity [1].

1. *Descriptive* validity: are the concept types indeed a substantial description of the business assets as they are perceived in the domain by a community?
2. *External* validity: can we declare the collected concept types sufficiently generally applicable en unambiguous so that they can be deployed in many situations or contexts?

Assessing the descriptive and external validity has to be done by a sufficiently large social arrangement of adequately skilled people, i.e., a community of practice. A social arrangement defines roles and responsibilities that collectively supervise the consistent implementation of semantic reconciliation and consequently produce qualitative concept types is often a wicked problem. This may be caused by the inability of domain experts to capture the domain properly and agreeing on a common representation of it effectively. Furthermore, domain experts may have difficulty using the tools that support these complex tasks. All together this advocates for a well-defined metadata governance model.

10 Metadata Architecture and Governance

Of particular importance is that BSM becomes structurally and gradually embedded in FMET's Enterprise Architecture by means of a *Metadata Architecture*. The basis for this is the EIM. Metadata Governance is concerned with establishment, modification, and implementation and monitoring of the Metadata Architecture by using BSM so that the ensuing business information systems will optimally contribute to the desired business results. While fueled by business drivers, the implementation of BSM is determined by a *metadata charter, principles* and *policies*.

- A metadata charter is a Memorandum of Understanding that provides motivation, goals, and key stakeholders. It provides a framework of roles and responsibilities and it identifies certain authorities.
- Metadata-principles establish start points for metadata management that must be respected. E.g., metadata must always be made publicly available and in an open standard format.

- Metadata policies contain clear guidelines for relevant actors within the organization to implement BSM in all its facets with sufficient quality and according to the principles.

Subject of these principles and policies are evolving concepts themselves, metadata applications, methodologies and culture, but also the relationship with the ICT outsourcing partner. E.g., a policy that implements the above principles is a clear choice for the RDF or XSD format for publishing metadata.

11 Conclusion

The embedding of BSM in FMET requires a planning to implement a coherent set of projects in line with the ICT Strategic Plan 2010-2014. Eventually this should bring the partnerships information management in 2014 to an acceptable level. BSM constitutes a powerful catalyst to align and fuel the information management processes from the business and supporting technical data management processes. Doing so, ICT can be used effectively and efficiently.

The yardstick that we use to measure information maturity is the Information Maturity Model (IMM)[18] (Figure 6). From implemented proof-of-concept we have reviewed some aspects of BSM. If we project our findings from these PoCs on the IMM, we conclude that FMET was at IMM level 2 at the beginning of our analysis, and that the organization is not far off from achieving level 3. The five years plan aims Level 4.

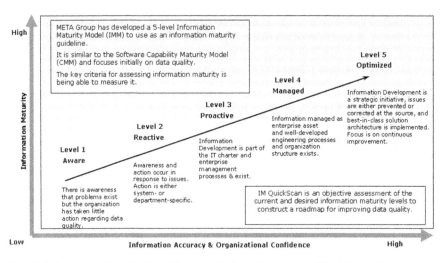

Fig. 6. Information Maturity Model: a practical yardstick to qualify information maturity and governance in an organisation (by courtesy of Sean McClowry).

---

[18] Defined by Meta Group (now Gartner):
http://mike2.openmethodology.org/wiki/Information_Maturity_Model

Achieving Level 4 IMM will provide a platform with many new capabilities such as the development of Semantic (Business) Intelligence and Semantics-driven SOA. This will be implemented in terms of data transformation, validation, and governance services. Moreover, the outreach of best-practice applications and associated metadata standards deliver a unique reputation to FMET. Pushing this to federal to even European levels will promote sharing and standardization of metadata for public administrations.

References

1. Akkermans, H., Gordijn, J.: Ontology Engineering, Scientific Method and the Research Agenda. In: Staab, S., Svátek, V. (eds.) EKAW 2006. LNCS (LNAI), vol. 4248, pp. 112–125. Springer, Heidelberg (2006)
2. Cardoso, J.: The semantic web vision: Where are we? IEEE Intelligent Systems, 22–26 (September/October 2007)
3. Cranefeld, S.: UML and the Semantic Web. In: Proc. of the Semantic Web Working Symposium, pp. 113–130. Stanford University, CA (2001)
4. De Leenheer, P., de Moor, A., Meersman, R.: Context Dependency Management in Ontology Engineering: A Formal Approach. In: Spaccapietra, S., Atzeni, P., Fages, F., Hacid, M.-S., Kifer, M., Mylopoulos, J., Pernici, B., Shvaiko, P., Trujillo, J., Zaihrayeu, I. (eds.) Journal on Data Semantics VIII. LNCS, vol. 4380, pp. 26–56. Springer, Heidelberg (2007)
5. De Leenheer, P., Mens, T.: Ontology evolution: State of the art and future directions. In: Hepp, M., De Leenheer, P., de Moor, A., Sure, Y. (eds.) Ontology Management for the Semantic Web, Semantic Web Services, and Business Applications. Springer, Heidelberg (2008)
6. De Leenheer, P.: Ontology Elicitation. In: Liu, L., Ôzsu, T. (eds.) Encyclopedia of Database Systems. Springer, Heidelberg (2009)
7. De Leenheer, P., Christiaens, S., Meersman, R.: Business semantics management: a case study for competency-centric HRM. Journal of Computers For Industry (2010) (forthcoming)
8. Gasevic, D., Djuric, D., Devedzic, V.: Model Driven Architecture and Ontology Development, 2nd edn. Springer, Heidelberg (2009)
9. Halpin, T.: Metaschemas for ER, ORM and UML data models: A comparison. J. Database Manag. 13(2), 20–30 (2002)
10. Hepp, M., De Leenheer, P., de Moor, A., Sure, Y. (eds.): Ontology Management for the Semantic Web, Semantic Web Services, and Business Applications. Springer, Heidelberg (2008)
11. Kankanhalli, A., Tan, B.C.Y., Wei, K.-K.: Contributing Knowledge to Electronic Knowledge Repositories: an Empirical Investigation. InMIS Quarterly 29(1), 113–143 (2005)
12. Maes, R.: Reconsidering information management through a generic framework. PrimaVera Working Paper 99-15 (1999)
13. Spyns, P., Meersman, R., Jarrar, M.: Data modelling versus ontology engineering. SIGMOD Record 31(4), 12–17 (2002)

14. Trog, D., Tang, Y., Meersman, R.: Towards ontological commitments with Ω-RIDL markup language. In: Paschke, A., Biletskiy, Y. (eds.) RuleML 2007. LNCS, vol. 4824, pp. 92–106. Springer, Heidelberg (2007)
15. Verheijen, G., Van Bekkum, J.: NIAM, an information analysis method. In: Proc. of the IFIP TC-8 Conference on Comparative Review of Information System Methodologies (CRIS 1982). North-Holland, Amsterdam (1982)

Chapter 5
Argumentation for Reconciling Agent Ontologies

Cássia Trojahn[1], Jérôme Euzenat[1], Valentina Tamma[2], and Terry R. Payne[2]

[1] INRIA & LIG
{cassia.trojahn,jerome.euzenat}@inrialpes.fr
[2] University of Liverpool
{v.tamma,t.r.payne}@liverpool.ac.uk

Abstract. Within open, distributed and dynamic environments, agents frequently encounter and communicate with new agents and services that were previously unknown. However, to overcome the ontological heterogeneity which may exist within such environments, agents first need to reach agreement over the vocabulary and underlying conceptualisation of the shared domain, that will be used to support their subsequent communication. Whilst there are many existing mechanisms for matching the agents' individual ontologies, some are better suited to certain ontologies or tasks than others, and many are unsuited for use in a real-time, autonomous environment. Agents have to agree on which correspondences between their ontologies are mutually acceptable by both agents. As the rationale behind the preferences of each agent may well be private, one cannot always expect agents to disclose their strategy or rationale for communicating. This prevents the use of a centralised mediator or facilitator which could reconcile the ontological differences. The use of argumentation allows two agents to iteratively explore candidate correspondences within a matching process, through a series of proposals and counter proposals, i.e., *arguments*. Thus, two agents can reason over the acceptability of these correspondences without explicitly disclosing the rationale for preferring one type of correspondences over another. In this chapter we present an overview of the approaches for alignment agreement based on argumentation.

1 Introduction

The problem of dynamic reconciliation of vocabularies, or *ontologies*, used by agents during interactions has recently received significant attention, motivated by the growing adoption of service-oriented and distributed computing. In such scenarios, agents are situated in open environments and may encounter unknown agents offering new services due to changes in a user's context or goal. These multi-agent systems are, by nature, distributed and heterogeneous, and as such, ontologies play a fundamental role in formalising the concepts that agents perceive, share, or encounter. However, as the heterogeneity that permeates these environments increases,

A. Elçi, M.T. Koné, and M.A. Orgun (Eds.): Semantic Agent Systems, SCI 344, pp. 89–111.
springerlink.com

fewer assumptions on the vocabulary and content of these ontologies can be made, hindering seamless interactions between the agents. Thus, mechanisms that can dynamically and autonomously reconcile the differences between ontologies are essential if agents are to communicate within such open and evolving environments.

Early systems avoided the problem of ontological heterogeneity by relying on the existence of a shared ontology, or simply assuming that a canonical set of ontology correspondences (possibly defined at design time) could be used to resolve ontological mismatches. However, such assumptions work only when the environment is (semi-) closed and carefully managed, and no longer hold in open environments where a plethora of ontologies exist. Moreover, the assumption of a common ontology forces an agent to comply with a fixed, but highly constrained view of the world, with respect to a set of predefined tasks and, as a consequence, abandon its own world view, which may have evolved due to interactions with other agents [8].

To facilitate the communication between two agents, they first need to establish a set of correspondences (or an *alignment*) between their respective ontologies. The reconciliation of heterogeneous ontologies has been investigated at length by research efforts in *ontology matching* [15], which tries to determine suitable correspondences between two ontologies. However the increased availability of mechanisms for ontology matching has facilitated the potential construction of a plethora of different correspondence sets between two ontologies, depending on the approach used. In addition, the majority of traditional matching approaches cannot be easily utilised as part of dynamic interaction protocols since they either require human intervention or they align the ontologies at design time. Even when alignments are pre-computed and stored within some alignment library, the selection of a possible correspondence that would be mutually acceptable to two transacting agents can be problematic, as the choice of correspondences can be highly dependent on the current task or available knowledge. For example, an agent may prefer correspondences which have been approved by its own institution and another one may prefer a correspondence designed for the same task. These may not be easy to reconcile. Hence, some correspondences may be preferable to some agents, but may be unsuitable or untrustworthy to others. In addition, it may not always be desirable for an agent to disclose a preference for a given type of correspondence as this may reveal its goal, and thus compromise it's ability to negotiate strategically with other agents within a competitive environment. Thus it is not always possible to utilise a collaborative approach, or exploit the use of a third party mediator to determine a mutually acceptable set of correspondences.

The agreement on a mutually acceptable alignment is an important problem that therefore arises when different parties need to reconcile private, yet potentially conflicting preferences over candidate correspondences. Such an agreement can be achieved through a negotiation process whereby agents iteratively exchange proposals and counter-proposals [28, 20] until some consensus is reached. Argumentation can be seen as a qualitative negotiation model based on the construction and comparison of arguments [12, 29, 3], either supporting or refuting a set of possible propositions. Thus, by considering these propositions as correspondences (with justifications that support their use), agents can strategically argue in favour of (or against) possible correspondences given their individual strategies or preferences.

This chapter presents an overview on the approaches for alignment agreement based on argumentation. The different approaches are presented following two scenarios. In the first one, agents with different preferences need to agree on the alignment of their ontologies in order to communicate with each other. For the second scenario, specialised matcher agents rely on different matching approaches and argue on their individual results in order to obtain a consensual alignment.

The remainder of this chapter is structured as follows. First, we provide the basic definitions of ontology matching and argumentation frameworks (§2). Next, two argumentation frameworks for alignment agreement are introduced (§3). The different proposals on argumentation for alignment agreement are then presented (§4). The limitations and challenges in this domain are discussed (§5). Finally, related work (§6) and final remarks (§7) are presented.

2   Foundations: Alignment and Argumentation Frameworks

2.1   Ontology Matching

An **ontology** typically provides a vocabulary describing a domain of interest and a specification of the meaning of terms in that vocabulary. As different agents within an open multi-agent system may be developed independently, they may commit to different ontologies to model the same domain. Whilst these different ontologies may be similar, they may differ in granularity or detail, use different representations, or model the concepts, properties and axioms in different ways.

In order to illustrate the matching problem, let us consider an e-Commerce marketplace, where two agents, a *buyer* and a *seller*, need to negotiate the price of a digital camera. Before starting the negotiation, they need to agree on the vocabulary to be used for exchanging the messages. They use the ontologies o and o', respectively (Figure 1). These ontologies contain subsumption statements (e.g., DigitalCamera $\sqsubseteq$ Product), property specifications (e.g., price domain Product) and instance descriptions (e.g., ThisCamera price 250$).

Ontology matching is the task of finding correspondences between ontologies. Correspondences express relationships supposed to hold between entities in ontologies, for instance, that an Electronic in one ontology is the same as a Product in another one or that DigitalCamera in an ontology is a subclass of CameraPhoto in another one. In the example above, one of the correspondences expresses an equivalence, while the other one is a subsumption correspondence. A set of correspondences between two ontologies is called an alignment. An alignment may be used, for instance, to generate query expressions that automatically translate instances of these ontologies under an integrated ontology or to translate queries with respect to one ontology in to query with respect to the other.

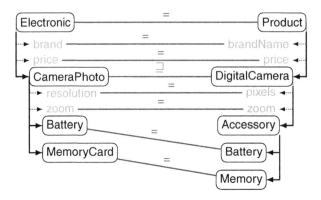

Fig. 1. Fragments of ontologies o and o' with alignment A.

Matching determines an alignment A' for a pair of ontologies o and o'. There are other parameters that can extend the definition of the matching process, namely: (i) the use of an input alignment A, which is to be completed by the process; (ii) the matching parameters, for instance, weights, thresholds, etc.; and (iii) external resources used by the matching process, for instance, common knowledge and domain specific thesauri.

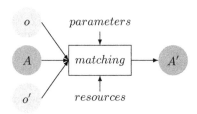

Fig. 2. The ontology matching process (from [15]).

Each of the elements featured in this definition can have specific characteristics which influence the difficulty of the matching task. As depicted in Figure 2, the matching process receives as input three main parameters: the two ontologies to be matched (o and o') and the input alignment (A). The input ontologies can be characterized by the input languages they are described (e.g., OWL-Lite, OWL-DL, OWL-Full), their size (number of concepts, properties and instances) and complexity, which indicates how deep is the hierarchy structured and how dense is the interconnection between the ontological entities. Other properties such as consistency, correctness and completeness are also used for characterizing the input ontologies. The input alignment (A) is mainly characterized by its multiplicity (or cardinality, e.g., how many entities of one ontology can correspond to one entity of the others) and coverage in relation to the ontologies to be matched. In a simple scenario, the input alignment is empty. Regarding the parameters, some systems take advantage of external resources, such as WordNet, sets of morphological rules or previous alignments of general purpose (Yahoo and Google catalogs, for instance). ·

Different approaches to the problem of ontology matching have emerged from the literature [15]. The main distinction between each is due to the type of knowledge encoded within each ontology, and the way it is utilized when identifying correspondences between features or structures within the ontologies. *Terminological* methods lexically compare strings (tokens or n-grams) used in naming entities (or in the labels and comments concerning entities), whereas *semantic* methods utilise model-theoretic semantics to determine whether or not a correspondence exists between two entities. Approaches may consider the *internal* ontological structure, such as the range of their properties (attributes and relations), their cardinality, and the transitivity and/or symmetry of their properties, or alternatively the *external* ontological structure, such as the position of the two entities within the ontological hierarchy. The instances (or extensions) of classes could also be compared using *extension*-based approaches. In addition, many ontology matching systems rely not on a single approach.

The output **alignment** A' is a set of correspondences between o and o'. Generally, correspondences express a relation r between ontology entities e and e' with a confidence measure n. These are abstractly defined in [15]. In this chapter, we will restrict the discussion to simple correspondences.

Definition 1 (Simple correspondence). *Given two ontologies, o and o', a simple correspondence is a quintuple:*

$$\langle id, e, e', r, n \rangle,$$

such that:

- *id is a URI identifying the given correspondence;*
- *e and e' are named ontology entities, i.e., named classes, properties, or instances;*
- *r is a relation among equivalence ($\equiv$), more general ($\sqsupseteq$), more specific ($\sqsubseteq$), and disjointness ($\perp$);*
- *n is a number in the $[0,1]$ range.*

The correspondence $\langle id, e, e', n, r \rangle$ asserts that the relation r holds between the ontology entities e and e' with confidence n. The higher the confidence value, the higher the likelihood that the relation holds. Alignments may have different cardinalities; 1:1 (one-to-one), 1:m (one-to-many), n:1 (many-to-one) or n:m (many-to-many). An alignment is a 1:1 alignment, if and only if no two different entities in one of the ontologies are matched to the same entity in the other ontology.

Mechanisms that facilitate the construction of alignments require access to both ontologies. Whilst it may be desirable to embed such mechanisms within agents that operate in transparent and collaborative environments, exposing one's ontology may not always be desirable in competitive or adversarial environments, as this may allow other agents to infer, and exploit this knowledge in subsequent negotiations.

In addition, creating alignments can be costly, and thus the ability to cache or save previously generated alignments (possibly generated by trusted third parties) may be desirable. Thus, agents may rely on an external **alignment service**.

For example, the *Alignment server*, built on the Alignment API [13], provides functionality to facilitate ontology matching, as well as storing and retrieving alignments. In addition, it can provide assistance to agents when attempting to determine relationships between their ontologies, so that they can understand and interpret each other's messages. An agent plug-in has been developed to allow agents based on the JADE/FIPA ACL (*Agent Communication Language*) to interact with the server in order to retrieve alignments.

Such a service can provide alignments over which the agents will argue in order to choose the more suitable correspondences. Alignments, and the correspondences within such alignments, can be better qualified, through the inclusion of *metadata*, which may refer to the provenance and origin of alignments, confidence ratings, and the original purposes for which they were created. Other metadata may also include any manual (human-based) checks or endorsements provided by some authority. This type of metadata is used, for instance, by *Bioportal* [26], which is an alternative alignment web-service, where users can select correspondences based on providence-based alignment metadata.

2.2 Argumentation Frameworks

Argumentation is a decentralised, peer-based negotiation model for reasoning based on the construction and comparison of arguments. The central notion in argumentation systems is the notion of *acceptability*. Different argumentation frameworks have been specified presenting different notions of acceptability. The classical argumentation framework (AF) was proposed by Dung [12], whose notion of acceptability defines that an argument should be accepted only if every attack on it is attacked by an accepted argument. Dung defines an argumentation framework as follows:

Definition 2 (Argumentation Framework [12]). *An Argumentation Framework (AF) is a pair* $\langle \mathcal{A}, \ltimes \rangle$, *such that* $\mathcal{A}$ *is a set of arguments and* $\ltimes$ *(attacks) is a binary relation on* $\mathcal{A}$. *a* $\ltimes$ *b means that the argument a attacks the argument b. A set of arguments S attacks an argument b iff b is attacked by an argument in S.*

The key question about the framework is whether a given argument $a \in \mathcal{A}$ should be accepted or not. Dung proposes that an argument should be accepted only if every attack on it is attacked by an accepted argument. This notion then leads to the definition of acceptability (for an argument), admissibility (for a set of arguments) and preferred extension:

Definition 3 (Acceptable argument [12]). *An argument* $a \in \mathcal{A}$ *is* acceptable *with respect to set arguments S, noted* $acceptable(a, S)$, *iff* $\forall x \in \mathcal{A}, (x \ltimes a \longrightarrow \exists y \in S, y \ltimes x)$.

Definition 4 (Conflict-free set [12]). *A set S of arguments is* conflict-free *iff* $\neg \exists x, y \in S, x \ltimes y$. *A conflict-free set of arguments S is* admissible *iff* $\forall x \in S, acceptable(x, S)$.

Definition 5 (Preferred-extension [12]). *A set of arguments S is a preferred exten-sion iff it is a maximal (with respect to inclusion set) admissible set of $\mathcal{A}$.*

Thus, a preferred extension represents a consistent position within an argumentation framework, which defends itself against all attacks and cannot be extended without raising conflicts.

In Dung's framework, all arguments have equal strength, and therefore attacks always succeed. This is reasonable when dealing with deductive arguments, but in many domains, arguments may lack some coercive force: they provide reasons which may be more or less persuasive. For that purpose, preference-based argu-mentation has been designed [2] which assigns preferences to arguments, so that preferred arguments would successfully attack less preferred ones (but not vice versa). Bench-Capon [6] went one step further with the *Value Based Argumentation framework* (VAF[1]), which assigns to arguments the values they promote. Agents are distributed among different *audiences* which ascribe different preferences to such values. Hence, different audiences will have different preferences among the argu-ments and similarly, successful attacks for an audience are those made by arguments of highest values to the audience.

Definition 6 (Value-based AF [6]). *A Value-based Argumentation Framework (VAF) is a quintuple $\langle \mathcal{A}, \ltimes, \mathcal{V}, v, \succeq \rangle$ such that $\langle \mathcal{A}, \ltimes \rangle$ is an argumentation frame-work, $\mathcal{V}$ is a nonempty set of values, $v : \mathcal{A} \to \mathcal{V}$, and $\succeq$ is the preference relation over $\mathcal{V}$ ($v_1 \succeq v_2$ means that, in this framework, v_1 is preferred over v_2).*

To each audience, α corresponds a value-based argumentation framework VAF_α such that $v_1 \succeq_\alpha v_2$ states that audience α prefers v_1 over v_2. Attacks are then deemed successful, based on the preference ordering on the arguments' values. This leads to re-defining the notions seen previously:

Definition 7 (Successful attack [6]). *In a value-based argumentation framework, $\langle \mathcal{A}, \ltimes, \mathcal{V}, v, \succeq \rangle$, an argument $a \in \mathcal{A}$ defeats (or successfully attacks) an argument $b \in \mathcal{A}$, noted $a \dagger b$, iff both $a \ltimes b$ and $v(b) \not\succeq v(a)$.*

Definition 8 (Conflict-free set [6]). *A set S of arguments is conflict-free for an audience α iff $\forall x, y \in S$, $\neg(x \ltimes y) \lor v(y) \succeq_\alpha v(x)$.*

Acceptable arguments and preferred extensions are defined as before. In order to de-termine preferred extensions with respect to a value ordering promoted by distinct audiences, *objective* and *subjective* acceptance are defined. An argument is *sub-jectively acceptable* if and only if it appears in some preferred extension for some specific audience. An argument is *objectively acceptable* if and only if it appears in all preferred extension for every specific audience.

---

[1] We describe here as VAF what [6] calls an audience-specific value-based argumentation framework, but the result is equivalent.

3   Argumentation Frameworks for Alignment Agreement

In alignment agreement, arguments can be seen as positions that support or reject correspondences. Such arguments interact following the notion of attack and are selected according to the notion of acceptability. Argumentation frameworks for alignment agreement redefine the notion of acceptability, taking into account the confidence of the correspondences and the number of agents agreeing on a correspondence. In this section we first introduce the general definition of argument, which will be extended according to the scenario where argumentation is used (§4), and then we present the argumentation frameworks.

3.1   Arguments on Correspondences

The different approaches presented below all share the same notion of correspondence argument originally defined in [22]. The general definition of correspondence argument is as follows:

Definition 9 (Argument [22]). *An argument $a \in AF$ is a tuple $a = \langle c, v, h \rangle$, such that c is a correspondence $\langle e, e', r, n \rangle$; $v \in \mathcal{V}$ is the value of the argument and h is one of $\{+, -\}$ depending on whether the argument is that c does or does not hold.*

In this definition, the set of considered values may be based on: the types of matching techniques that agents tend to prefer; the type of targeted applications; information about various level of endorsement of these correspondences, and whether or not they have been checked manually. Thus, any type of information which can be associated with correspondences (see §2.1) may be used. For example, an alignment may be generated for the purpose of information retrieval; however, this alignment may not be suitable for an agent performing a different task requiring more precision. This agent may therefor prefer the correspondences generated by a different agent for web service composition. Likewise, another agent may prefer human curated alignments rather than alignments generated on the fly.

   Arguments interact based on the notion of attack relation:

Definition 10 (Attack [22]). *An argument $\langle c, v, h \rangle \in \mathcal{A}$ attacks another argument $\langle c', v', h' \rangle \in \mathcal{A}$ iff $c = c'$ and $h \neq h'$.*

Therefore, if $a = \langle c, v_1, + \rangle$ and $b = \langle c, v_2, - \rangle$, $a \bowtie b$ and vice-versa (b is the counter-argument of a, and a is the counter-argument of b).

3.2   Strength-Based Argumentation Framework (SVAF)

Bench-Capon's framework acknowledges the importance of preferences when considering arguments. However, within the specific context of ontology matching, an objection can still be raised regarding the lack of complete mechanisms for handling persuasiveness. Indeed, many ontology matchers generate correspondences with a strength that reflects the confidence they have in the similarity between the two entities. These confidence levels are usually derived from similarity assessments made during the matching process, e.g., from the edit distance measure between labels, or overlap measure between instance sets, and thus are often based on objective grounds. In order to represent arguments with *strength*, reflecting this confidence in a correspondence, [34] proposed the *Strength-based Argumentation Framework (SVAF)*, extending Bench-Capon's *VAF* by redefining the notion of acceptability.

Definition 11 (SVAF [34]). *A strength-based argumentation framework (SVAF) is a sextuple* $\langle \mathcal{A}, \ltimes, \mathcal{V}, v, \succeq, s \rangle$ *such that* $\langle \mathcal{A}, \ltimes, \mathcal{V}, v, \succeq \rangle$ *is a value-based argumentation framework and* $s : \mathcal{A} \to [0, 1]$ *represents the strength of the argument.*

As in value-based argumentation frameworks, each audience α is associated with its own framework in which only the preference relation $\succeq_\alpha$ differs. In order to accommodate the notion of *strength*, the notion of *successful attack* is extended:

Definition 12 (Successful attack [34]). *In a strength-based argumentation framework* $\langle \mathcal{A}, \ltimes, \mathcal{V}, v, \succeq, s \rangle$, *an argument* $a \in \mathcal{A}$ *successfully attacks (or defeats, noted* $a\dagger b$*) an argument* $b \in \mathcal{A}$ *iff*

$$a \ltimes b \wedge (s(a) > s(b) \vee (s(a) = s(b) \wedge v(a) \succeq v(b)))$$

3.3   Voting-Based Argumentation Framework (VVAF)

The frameworks described so far assume that candidate correspondences between two entities may differ due to the approaches used to construct them, and thus these argumentation frameworks provide different mechanisms to identify correspondences generated using approaches acceptable to both agents. However, different alignment generators may often utilise the same approach for some correspondences, and thus the approach used for that correspondence may be significant. Some large-scale experiments involving several matching tools (e.g. the OAEI 2006 Food track campaign [14]) have demonstrated that the more often a given approach for generating a correspondence is used, the more likely it is to be valid. Thus, the SVAF was adapted and extended in [19], to take into account the level of consensus between the sources of the alignments, by introducing the notions of support and voting into the definition of successful attacks. Support enables arguments to be counted as defenders or co-attackers during an attack:

Definition 13 (VVAF [19]). *A voting-based argumentation framework (VVAF) is a septuple* $\langle \mathcal{A}, \ltimes, \mathcal{S}, \mathcal{V}, v, \succeq, s \rangle$ *such that* $\langle \mathcal{A}, \ltimes, \mathcal{V}, v, \succeq, s \rangle$ *is a SVAF, and* $\mathcal{S}$ *is a (reflexive) binary relation on* $\mathcal{A}$, *representing the support relation between arguments.* $\mathcal{S}(x, a)$ *means that the argument* x *supports the argument* a *(i.e., they have the same value of* h*).* $\mathcal{S}$ *and* $\ltimes$ *are disjoint relations.*

A simple voting mechanism (e.g. plurality voting) can be used to determine the success of a given attack, based upon the number of supporters for a given approach.

Definition 14 (Successful attack [19]). *In a VVAF* $\langle \mathcal{A}, \ltimes, \mathcal{S}, \mathcal{V}, v, \succeq, s \rangle$, *an argument* $a \in \mathcal{A}$ *successfully attacks (or defeats) an argument* $b \in \mathcal{A}$ *(noted a†b) iff*

$$a \ltimes b \wedge (|\{x | \mathcal{S}(x,a)\}| > |\{y | \mathcal{S}(y,b)\}| \vee |\{x | \mathcal{S}(x,a)\}| = |\{y | \mathcal{S}(y,b)\}| \wedge v(a) \succeq v(b)).$$

This voting mechanism is based on simple counting. As some ontology matchers include confidence values with correspondences, a voting mechanism can exploit this confidence value, for example by simply calculating the total confidence value of the supporting arguments. However, this relies on the questionable assumption that all values are equally scaled (as is the case with the SVAF). In [19], a voting framework that normalised these confidence values (i.e. strengths) was evaluated, but was inconclusive. Another possibility would be to rely on a deeper justification for correspondences and to have only one vote for each justification. Hence, if several matchers considered two concepts to be equivalent because WordNet considers their identifier as synonyms, this would be counted only once.

4 Argumentation over Alignments

The use of argumentation has been exploited in two different scenarios presented below. In the first, agents attempt to construct mutually acceptable alignments based on existing correspondences to facilitate communication, based on their alignment preferences (which may be task specific). They therefore argue directly over candidate correspondences provided by an alignment service, with each agent specifying an ordered preference of correspondence types and confidence thresholds. The second scenario focuses on the consensual construction of alignments involving several agents, each of which specialises in constructing correspondences using different approaches. These matching agents generate candidate correspondences and attempt to combine these to produce a new alignment through argumentation. Thus, whilst the first scenario utilises argumentation as a negotiating mechanism to find a mutually acceptable alignment between transacting agents, this latter scenario could be viewed as offering a service for negotiating alignments.

4.1 Argumentation over Alignments for Communication in Multi-agent Systems

4.1.1 Meaning-Based Argumentation

Laera et al. proposed the meaning-based argumentation approach [22, 23, 21], to allow agents to propose, attack, and counter-propose candidate correspondences according to the agents' preferences, in order to identify mutually acceptable alignments. Their approach utilises Bench-Capon's VAF [6] to support the specification of preferences of correspondent types (as discussed in §2.1) within each argument. Thus, when faced with different, candidate correspondences who's type differ, each agents' preference ordering can be considered when determining if an argument for one correspondence will successfully attack another. Different audiences therefore represent different sets of arguments for preferences between the categories of arguments (identified in the context of ontology matching).

Each agent is defined as follows:

Definition 15 (Agent). *An agent Ag_i is characterised by a tuple $\langle O_i, F, \epsilon_i \rangle$, such that O_i is the ontology used by the agent, F is its (valued-based) argumentation framework, and ϵ_i is the private threshold value.*

Candidate correspondences are retrieved from an alignment service (see §2.1) which also provides the justifications G (described below) for each correspondence, based on the approach used to construct the correspondence. The agents use this information to exchange arguments supplying the reasons for their choices. In addition, as these grounds include a confidence value associated with each correspondence, each agent utilises a private threshold value ϵ to filter out correspondences with low confidence values[2]. This threshold, together with the pre-ordering of preferences, are used to generate arguments for and against a correspondence. It extends the notion of argument presented in §3.1:

Definition 16 (Argument [22]). *An argument is a triple $\langle G, c, h \rangle$, where c is a correspondence $\langle e, e', r, n \rangle$, G is the grounds justifying a prima facie belief that the correspondence does, or does not hold; and h is one of $\{+, -\}$ depending on whether the argument is that c does or does not hold.*

The grounds G justifying a correspondence between two entities are based on the five categories of correspondence types (as discussed in §2.1) - namely *Semantic* (S), *Internal Structural* (IS), *External Structural* (ES), *Terminological* (T), and *Extensional* (E). These classes are used as types for the values $\mathcal{V}$, i.e., $\mathcal{V} = \{M, IS, ES, T, E\}$, that are then used to construct an agent's partially-ordered preferences, based on the agents ontology and task. Thus, an agent may specify a preference for terminological correspondences over semantic correspondences if the ontology it uses is mainly taxonomic, or vice versa if the ontology is semantically rich. Preferences may also be based on the type of task being performed; extensional correspondences may be preferred when queries are about instances that are frequently shared. The pre-ordering of preferences $\succeq$ for each agent Ag_i is over $\mathcal{V}$, corresponding to the specification of an audience. Specifically, for each candidate correspondence c, if there exists one or more justifications G for c that corresponds to the highest preferences $\succeq$ of Ag_i (with the respect of the pre-ordering), assuming n is greater than its private threshold ϵ, an agent Ag_i will generate arguments $x = (G, c, +)$. If not, the agent will generate arguments against: $x = (G, c, -)$. The arguments interact based on the notion of attack, as specified in §3.1.

The argumentation process takes four main steps: (i) each agent Ag_i constructs an argumentation framework VAF_i by specifying the set of arguments and the attacks between them; (ii) each agent Ag_i considers its individual frameworks VAF_i with

---

[2] The use of confidence profiles has since been explored to specify correspondence-type specific thresholds, resulting in the agreement over a greater diversity of agreed correspondences, and consequently more inclusive alignments [9].

all the argument sets of all the other agents and then extends the attack relations by computing the attacks between the arguments present in its framework with the other arguments; (iii) for each VAF_i, the arguments which are undefeated by attacks from other arguments are determined, given a value ordering – the global view is considered by taking the union of these preferred extensions for each audience; and (iv) the arguments in every preferred extension of every audience are considered – the correspondences that have only arguments for are included in the a set called *agreed alignments*, whereas the correspondences that have only arguments against them are rejected, and the correspondences which are in some preferred extension of every audience are part of the set called *agreeable alignments*.

The dialogue between agents consists of exchanging sets of arguments and the protocol used to evaluate the acceptability of a single correspondence is based on a set of speech acts ($Support$, $Contest$, $Withdraw$). For instance, when exchanging arguments, an agent sends $Support(c, x_1)$ for supporting a correspondence c through the argument $x_1 = (G, c, +)$ or $Contest(c, x_2)$ for rejecting c, by $x_2 = (G, c, -)$. If the agents do not have any arguments or counter-arguments to propose, then they send $Withdraw(c)$ and the dialogue terminates.

To illustrate this approach, consider the two agents buyer b and seller s, using the ontologies in Figure 1. First, the agents access the alignment service that returns the correspondences with the respective justifications:

- m_1: $\langle zoom_o, zoom_{o'}, \equiv, 1.0 \rangle$, with $G = \{T, ES\}$
- m_2: $\langle Battery_o, Battery_{o'}, \equiv, 1.0 \rangle$, with $G = \{T\}$
- m_3: $\langle MemoryCard_o Memory_{o'}, \equiv, 0.54 \rangle$, with $G = \{T\}$
- m_4: $\langle brand_o, brandName_{o'}, \equiv, 0.55 \rangle$, with $G = \{T, ES\}$
- m_5: $\langle price_o, price_{o'}, \equiv, 1.0 \rangle$, with $G = \{T, ES\}$
- m_6: $\langle CameraPhoto_o, DigitalCamera_{o'}, \equiv, 1.0 \rangle$, with $G = \{ES\}$
- m_7: $\langle resolution_o, pixels_{o'}, \equiv, 1.00 \rangle$, with $G = \{ES\}$

Agent b selects the audience R_1, which prefers terminology to external structure ($T \succ_{R_1} ES$), while s prefers external structure to terminology ($ES \succ_{R_2} T$). All correspondences have a degree of confidence n that is above the threshold of each agent and then all of them are taken into account. Both agents accept m_1, m_4 and m_5. b accepts m_2, m_3, while s accepts m_6 and m_7. Table 1 shows the arguments and corresponding attacks.

Table 1. Arguments and attacks.

id	argument	attack	agent
A	$\langle T, m_1, + \rangle$		b, s
B	$\langle ES, m_1, + \rangle$		b, s
C	$\langle T, m_2, + \rangle$	D	b
D	$\langle ES, m_2, - \rangle$	C	s
E	$\langle T, m_3, + \rangle$	F	b
F	$\langle ES, m_3, - \rangle$	E	s
G	$\langle T, m_4, + \rangle$		b, s

id	argument	attack	agent
H	$\langle ES, m_4, + \rangle$		b, s
I	$\langle T, m_5, + \rangle$		b, s
J	$\langle ES, m_5, + \rangle$		b, s
L	$\langle ES, m_6, + \rangle$	M	s
M	$\langle T, m_6, - \rangle$	L	b
N	$\langle ES, m_7, + \rangle$	O	s
O	$\langle T, m_7, - \rangle$	N	b

The arguments A, B, G, H, I, and J are not attacked and then are acceptable for both agents (they form the *agreed alignment*). The arguments C and D are mutually attacked and are acceptable only in the corresponding audience, i.e., C is acceptable for the audience b and D is acceptable for the audience s. The same occurs for the arguments E, F, L, M, M, and O. The correspondences in such arguments are seen as the *agreeable alignments*.

4.1.2   The Approach by Trojahn and Colleagues

In order to provide translations between messages in agent communication, [33] formally defines an alignment as a set of correspondences between *queries* over ontologies. The alignment is obtained by specialised matcher agents that argue in order to agree on a globally acceptable alignment. The set of acceptable arguments is then represented as conjunctive queries in OWL-DL [18].

A conjunctive query has the form $\bigwedge (P_i(s_i))$, where each $P_i(s_i)$ represents a correspondence. For instance, $\langle CameraPhoto_o, DigitalCamera_{o'}, \equiv, 1.0 \rangle$ is represented as $Q(x) : CameraPhoto(x) \equiv DigitalCamera(x)$.

Consider the example where the agents "buyer b" and "seller s" interact to agree on the price of a digital camera, using the ontologies o and o' of Figure 1, respectively. Before the agents can agree on the price, they need to agree on the terms used to communicate to each other. This task can be delegated to a matcher agent m, that receives the two ontologies and sends them to an argumentation module. This module, made up of different specialised agents $a_1, ..., a_n$ (which can be distributed on the web), receives the ontologies and returns a set of DL queries representing the acceptable correspondences. These interactions are loosely based on the Contract Net Interaction Protocol [16]. The argumentation process between the specialised matchers is detailed in Section 4.2. Table 2 describes the steps of the interaction between the agents.

Table 2. Interaction steps [33].

Step	Description
1	Matcher agent m requests the ontologies to be matched to agents b and s
2	Ontologies are sent from m to the argumentation module
3	Matchers $a_1, ..., a_n$ apply their algorithms
4	Each matcher a_i communicate with each others to exchange their arguments
5	Preferred extensions of each a_i are generated
6	Objectively acceptable arguments o are computed
7	Correspondences in o are represented as conjunctive queries
8	Queries are sent to m
9	Queries are sent from m to b and s
10	Agents b and s use the queries to communicate with each other

In fact, only one of the agents should receive the DL queries, which should be responsible for the translations. We consider that the set of objectively acceptable arguments has the correspondences shown in Figure 3, with the respective queries.

Query ID	Correspondences
$Q_{b1}(x)$	b:CameraPhoto(x)
$Q_{s1}(x)$	s:DigitalCamera(x)
m_1	$Q_{b1} \equiv Q_{s1}$
$Q_{b2}(y)$	b:zoom(y)
$Q_{s2}(y)$	s:zoom(y)
m_2	$Q_{b2} \equiv Q_{s2}$
$Q_{b3}(y)$	b:resolution(y)
$Q_{s3}(y)$	s:pixels(y)
m_3	$Q_{b3} \equiv Q_{s3}$

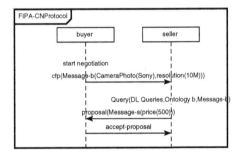

Fig. 3. Conjunctive queries.   **Fig. 4.** Interaction between buyer and seller agents.

Figure 4 shows an AUML[3] interaction diagram with the messages exchanged between the agents b and s during the negotiation of the price of the camera. The agents use the queries to search for correspondences between the messages sent from each other and the entities in the corresponding ontologies. In the example, the agent b sends a message to the agent s, using its vocabulary. Then, the agent s converts the message, using the DL queries.

4.1.3   Reducing the Argumentation Space through Modularization

Doran et al. [11] utilised modularization to identify the ontological descriptions relevant to the communication, and consequently reduce the number of correspondences necessary to form the alignment. The use of argumentation can be computationally costly, as the complexity can reach $\Pi_2^{(p)}$-complete in some cases. Thus, by reducing the number of arguments, the time required for generating the alignments can be significantly reduced; even when taking into account the time necessary for the modularization process itself. In an empirical study, the authors found that the use of modularization significantly reduced the average number of correspondences presented to the argumentation framework, and hence the size of the search space – in some cases by up to 97%, across a number of different ontology pairs. They also noted that three patterns emerged: i) where no reduction in size occurred (in 4.84% of cases within the study); ii) where the number of correspondences was reduced (55.14%); and iii) where modules of size zero were found (40.02%), corresponding to failure scenarios; i.e. where the subsequent transaction would fail due to insufficient alignment between the ontologies.

An ontology modularization technique extracts a consistent module M from an ontology A that covers a specified signature $Sig(M)$, such as $Sig(M) \subseteq Sig(O)$.

---

[3] AUML – Agent Unified Modelling Language [17].

M is the part of O that is said to cover the elements defined by $Sig(M)$. The first agent engaging in the communication specifies the $Sig(M)$ of its ontology O where M is an ontology concept relevant for a task. The resulting module contains the entities considered to be relevant for its task, including the subclasses and properties of the concepts in $Sig(M)$. The step-by-step interaction between two agents, following an argumentation based on modularization is presented in Table 3.

Table 3. Ontology modularization and argumentation for alignment agreement [10].

Step	Description
1	Ag_1 asks a query, $query(A \in Sig(O))$ to Ag_2.
2	Ag_2 does not understand the query, $A \notin Sig(O'))$ and informs Ag_1 they need to use a server.
3	Ag_1 produces, $om(O, Sig(A))$, an ontology module, M, to cover the concepts required for its task.
4	Ag_1 and Ag_2 invoke the server. Ag_1 sends its ontology, O and the signature of M, $Sig(M)$.
5	The alignment service aligns the two ontologies and filters the correspondences according to M. Only those features an entity from M are returned to both agents.
6	The agents begin the process of argumentation, with each agent generating arguments and counter-arguments.
7	The iteration terminates when the agents agree on a set of correspondences.
8	Ag_1 asks again Ag_2, using the agreed correspondences, $query(A \in Sig(O) \land B \in Sig(O'))$ where A and B are aligned.
9	Ag_2 answers the query using the agreed correspondences.

For communicating, only the initiating agent (Ag_1) is aware of its task and, consequently, which concepts are relevant to this task (Steps 1 and 2). These concepts will be included in $Sig(M)$, the signature of the resulting ontology module (Step 3). The set of candidate correspondences (Step 4) is filtered (Step 5) according to the filtering function $filter()$. $filter$ returns a subset Z of correspondences, where the entities e in these correspondence are in $Sig(M)$. The set Z is then used within the argumentation process. Modularization is therefore used to filter the correspondences that are passed to the argumentation process. The agents then argue (Steps 6-7) to reach an acceptable alignment.

The combination of argumentation and modularization reduces the cost of reaching an agreement over an alignment, by reducing the size of the set of correspondences argued over, and hence the number of arguments required. This greatly contributes to reduce the consumed time, at a minimal expense in accuracy.

Following the example of the buyer and seller agents, the buyer agent knows which concepts will be used for communicating and then a module of the ontology o is extracted containing such concepts (i.e., $CameraPhoto$, $resolution$, $zoom$, and $price$). The buyer agent then filters the correspondences in order to retrieve the subset containing only these concepts.

4.2   Solving Conflicts between Matcher Agents

In [34], alignments produced by different matchers are compared and agreed via
an argumentation process. The matchers interact in order to exchange arguments
and the SVAF model (§3.2) is used to support the choice of the most acceptable of
them. Each correspondence can be considered as an argument because the choice
of a correspondence may be a reason against the choice of another correspondence.
Correspondences are represented as arguments, extending the notion of argument
specified in §3.1:

Definition 17 (Argument). *An argument* $x \in AF$ *is a tuple* $x = \langle c, v, s, h \rangle$, *such
that* c *is a correspondence* $\langle e, e', r, n \rangle$; $v \in \mathcal{V}$ *is the value of the argument; s is the
strength of the argument, from n; and h is one of* $\{+, -\}$ *depending on whether the
argument is that c does or does not hold.*

The matchers generate arguments representing their alignments following a *negative arguments as failure* strategy. It relies on the assumption that matchers return
complete results. Each possible pair of ontology entities which is not returned by the
matcher is considered to be at risk, and a negative argument is generated ($h = -$).

The values v in $\mathcal{V}$ correspond to the different matching approaches and each
matcher m has a preference ordering $\succeq_m$ over $\mathcal{V}$ such that its preferred values
are those it associates to its arguments. For instance, consider $\mathcal{V} = \{l, s, w\}$, i.e.,
lexical, *structural* and *wordnet-based* approaches, respectively, and three matchers
m_l, m_s and m_w, using such approaches. The matcher m_l has as preference order
$l \succeq_{m_l} s \succeq_{m_l} w$. The basic idea is to obtain a consensus between different matchers,
represented by different preferences between values. Arguments interact based on
the notion of attack presented in §3.1.

The argumentation process can be described as follows. First, each matcher generates its set of correspondences, using some specific approach and then the set of
corresponding arguments is generated. Next, the matchers exchange with each others their set of arguments – the dialogue between them consists of the exchange of
individual arguments. When all matchers have received the set of arguments of each
others, they instantiate their SVAFs in order to generate their set of acceptable correspondences. The consensual alignment contains the correspondences represented
as arguments that appear in every set of acceptable arguments, for every specific
audience (objectively acceptable).

In order to illustrate this process, consider two matchers, m_l (lexical) and m_s
(structural), trying to reach a consensus on the alignment between the ontologies in
Figure 1. m_l uses an edit distance measure to compute the similarity between labels
of concepts and properties of the ontologies, while m_s is based on the comparison
of the direct super-classes of the classes or classes of properties. Table 4 shows the
correspondences and arguments generated by each matcher. The matchers generate
complete alignments, i.e., if a correspondence is not found, an argument with value
of $h = -$ is created. It includes correspondences that are not relevant to the task
at hand. For the sake of brevity, we show only the arguments with $h = +$ and
the corresponding counter-arguments (Table 5). We consider 0.5 as the confidence

level c for negative arguments ($h = -$). Considering $\mathcal{V} = \{l, v\}$, m_l associates to its arguments the value l, while m_s generates arguments with value s. m_l has as preference ordering: $l \succ_{m_l} s$, while m_s has the preference: $s \succ_{m_s} l$.

Table 4. Correspondences and arguments generated by m_l and m_s.

id	correspondence	argument	matcher
A	$c_{l,1} = \langle zoom_o, zoom_{o'}, \equiv, 1.0 \rangle$	$\langle c_{l,1}, l, 1.0, + \rangle$	m_l
B	$c_{l,2} = \langle Battery_o, Battery_{o'}, \equiv, 1.0 \rangle$	$\langle c_{l,2}, l, 1.0+ \rangle$	m_l
C	$c_{l,3} = \langle MemoryCard_o Memory_{o'}, \equiv, 0.33 \rangle$	$\langle c_{l,3}, l, 0.33, + \rangle$	m_l
D	$c_{l,4} = \langle brand_o, brandName_{o'}, \equiv, 0.22 \rangle$	$\langle c_{l,4}, l, 0.22, + \rangle$	m_l
E	$c_{l,5} = \langle price_o, price_{o'}, \equiv, 1.0 \rangle$	$\langle c_{l,5}, l, 1.0, + \rangle$	m_l
F	$c_{s,1} = \langle CameraPhoto_o, DigitalCamera_{o'}, \equiv, 1.0 \rangle$	$\langle c_{s,1}, s, 1.0, + \rangle$	m_s
G	$c_{s,2} = \langle zoom_o, zoom_{o'}, \equiv, 1.0 \rangle$	$\langle c_{s,2}, s, 1.0, + \rangle$	m_s
H	$c_{s,3} = \langle brand_o, brandName_{o'}, \equiv, 1.0 \rangle$	$\langle c_{s,3}, s, 1.0, + \rangle$	m_s
I	$c_{s,4} = \langle resolution_o, pixels_{o'}, \equiv, 1.0 \rangle$	$\langle c_{s,4}, s, 1.0, + \rangle$	m_s
J	$c_{s,5} = \langle price_o, price_{o'}, \equiv, 1.0 \rangle$	$\langle c_{s,5}, s, 1.0, + \rangle$	m_s

Table 5. Counter-arguments (attacks) for the arguments in Table 4.

id	correspondence	counter-argument	matcher
L	$c_{l,6} = \langle CameraPhoto_o, DigitalCamera_{o'}, \equiv, 0.5 \rangle$	$\langle c_{l,6}, l, 0.5, - \rangle$	m_l
M	$c_{l,7} = \langle resolution_o, pixels_{o'}, \equiv, 0.5 \rangle$	$\langle c_{l,7}, l, 0.5, - \rangle$	m_l
N	$c_{s,6} = \langle Battery_o, Battery_{o'}, \equiv, 0.5 \rangle$	$\langle c_{s,6}, s, 0.5, - \rangle$	m_s
O	$c_{s,7} = \langle MemoryCard_o, Memory_{o'}, \equiv, 0.5 \rangle$	$\langle c_{s,7}, s, 0.5, - \rangle$	m_s

Having their arguments $\mathcal{A}$, the matchers exchange them. m_l sends to m_s its set of arguments $\mathcal{A}_l$ and vice-versa. Next, based on the attack notion, each matcher m_i generates its attack relation $\ltimes_i$ and then instantiates its $SVAFs_i$. The arguments A, D, E, G, H and J are acceptable in both SVAFs (they are not attacked by counter-arguments with $h = -$). F, I, and B ($h = +$) successfully attack their counter-arguments ($h = -$) L, M and N, respectively, because they have highest confidence in their correspondences. C ($h = +$) is successfully attacked by its counter-argument O.

The arguments in the preferred extension of both matchers m_l and m_s are: A, D, E, F, G, H, J, F, I, B and O. While $\langle resolution_o, pixels_{o'}, \equiv, 1.0 \rangle$, $\langle Battery_o, Battery_{o'}, \equiv, 1.0 \rangle$ and $\langle CameraPhoto_o, DigitalCamera_{o'}, \equiv, 1.0 \rangle$ have been accepted, $\langle MemoryCard_o, Memory_{o'}, \equiv, 0.33 \rangle$ has been discarded.

5 Weaknesses and Challenges

As discussed above, argumentation for alignment agreement has been exploited in different ways, for different scenarios. However, there are still various challenges ahead for achieving a fully satisfying approach. We briefly consider some of them.

Confidence of arguments

In [34], the notion of attack between the arguments highly depends on the confidence associated to the correspondences. Such confidence levels are usually derived from similarity assessments made during the matching process, e.g., from edit distance measure between labels, or overlap measure between instance sets. However, there is no objective theory nor even informal guidelines for determining such confidence levels. Using them to compare results from different matchers is therefore questionable especially because of potential scale mismatches. For example, a same strength of 0.8 may not correspond to the same level of confidence for two different matchers.

Complete alignments

Generating complete alignments is at first sight quite unrealistic, but it can nevertheless be supported by the observation that most matchers try to provide as much correspondences as possible. However, dealing with a large number of arguments can become prohibitively costly. Following the approach from [10], the search space within the argumentation process can be reduced, by isolating only the correspondences that are relevant to the communication. Other authors isolate the subpart of the ontologies to be matcher relevant for the communication before matching only these pieces of ontologies instead of the whole ontologies [27]. These approaches have to be developed with guarantees that the isolated items are the relevant ones.

Inconsistent alignments

An important issue in such argumentation for alignment agreement is related to the potential inconsistency in the agreed alignment. Indeed, even if the initial alignments are consistent, selected sets of correspondences may generate concepts that are not satisfiable.

Solving the inconsistency problem in alignments has two possible alternatives:

- express the inconsistency within the argumentation framework [1];
- deal alternatively with the logical and argumentative part of the problem.

Integrating the logics within the argumentation framework seems a more elegant solution and it can be achieved straightforwardly when correspondences are arguments and incompatible correspondences can mutually attack each others. However, this works only when two correspondences are incompatible. When the set of incompatible correspondences is larger, the encoding is not so straightforward and may lead to the generation of an exponential amount of argument and attack relations. On the other side, alternating logical and argumentative treatments may also lead to prohibitive computational costs.

In this case, the solution seems to be a trade-off between the computational costs and the expected consistency.

Availability of justifications

The presented approaches argue for or against a correspondence based on justifications for the arguments. They are thus highly dependent on justifications for the arguments provided with the alignments. Although, alignment servers provide the necessary metadata for storing such justification with alignments (see §2.1), it is not common for people or for matchers to provide this information.

Ideally, matchers should provide such justifications, as a way to understand why a particular alignment is found or why a certain match is ranked higher than another. However, this is not common practice.

The development of such methods may therefore be slowed by the unavailability of justification metadata. It seems necessary to provide incentive to both automatic and manual matchers to generate this information. One such incentive could be, of course the ability to be involved in an argumentation process and then to provide better alignments. Another incentive would be to better help explain matcher results to users [30].

6 Other Related Work

This chapter has covered all the work carried out in the domain of alignment argumentation *per se*. However, in order to find alignments between ontologies used by agents, some work have proposed different techniques that we consider here.

[31] has proposed alignment negotiation to establish a consensus between different agents using the MAFRA alignment framework [24]. The approach is based on utility functions used to evaluate the confidence in a particular correspondence in the context of each agent. These confidence values are combined in order to decide if the correspondence is accepted, rejected or need to be negotiated. A meta-utility function is also applied to evaluate if the effort necessary to negotiate is beneficial or not; it may so automatically change the thresholds so that some correspondences are directly rejected or accepted. The approach is highly dependent on the MAFRA framework and cannot be directly applied to other environments.

Schemes for obtaining ontology alignments through the working cycles of agents have been developed. They either observe failure or success of the communication and statistically learn the alignments [7] or they use the interaction protocol of each agent for reducing the possible meaning of concepts used as performative [4].

[8] presents an approach for agents to agree on a common ontology in a decentralised way. The approach assumes that each agent adopts a private ontology and shares an intermediate ontology. The private ontology is used for storing and reasoning with operational knowledge, i.e., knowledge relevant to a particular problem or task at hand. The intermediate ontology is used for communication. Communication proceeds by translating from the speaker's private ontology to the intermediate ontology which the hearer translates back again into its own private ontology. The authors show how to establish such an intermediate ontology, which is the common goal for every agent in the system. In the approaches we have presented, on the other hand, the result of the negotiation is a set of correspondences between the terms of the different ontologies.

[5] presents an ontology negotiation protocol to provide semantic interoperability in multi-agent systems in an automated fashion at run-time. The ontology negotiation protocol enables agents to discover ontology conflicts or unknown terms. Then, it goes through (i) incremental interpretations of the unknown terms with the help of external resources, (ii) clarification, by proposing putative correspondences, (iii) evaluation, through the impact of such correspondences on some tasks, and (iv) update of the ontology with the correspondence. The final result of this process is that each agent will converge on a single, shared ontology. In contrast, in the approaches presented in this chapter, agents keep their own ontologies that they have been designed to reason with, whilst generating alignments with other agent's ontologies.

In [25], the authors propose an argumentation framework for inter-agent dialogue to reach an agreement on terminology, which formalizes a debate in which the divergent representations (expressed in description logic) are discussed. The proposed framework is stated as being able to manage conflicts between claims, with different relevancies for different audiences, in order to compute their acceptance. However, no detail is given about how agents will generate such claims.

[32] proposes a cooperative negotiation model, where agents apply individual matching algorithms and negotiate on a final alignment. Basically, the negotiation process involves the exchange of proposal and counter-proposals that represents correspondences. Each correspondence is negotiated individually. Three kinds of agents interact, lexical, structural, and semantic, and the communication is managed by a mediator agent.

7 Final Remarks

This chapter has presented an overview of the approaches for alignment agreement based on argumentation. Such approaches provide a way for agents, with different ontologies, to agree upon mutually acceptable ontology alignments to facilitate communication within a dynamic environment.

We have discussed how two agents commiting to different ontologies can align their ontologies in order to interoperate and how agents relying on different matching approaches can agree on a common alignment. The approaches for both scenarios are not fully satisfying and there are still various challenges ahead for achieving such maturity.

References

1. Amgoud, L., Besnard, P.: Bridging the gap between abstract argumentation systems and logic. In: Godo, L., Pugliese, A. (eds.) SUM 2009. LNCS, vol. 5785, pp. 12–27. Springer, Heidelberg (2009)
2. Amgoud, L., Cayrol, C.: On the acceptability of arguments in preference-based argumentation. In: Cooper, G., Moral, S. (eds.) Proceedings of the 4th Conference on Uncertainty in Artificial Intelligence (1998)
3. Amgoud, L., Cayrol, C.: A reasoning model based on the production of acceptable arguments. Annals of Mathematics and Artificial Intelligence 34(1-3), 197–215 (2002)

4. Atencia, M.: Semantic alignment in the context of agent interaction. Ph.D. thesis, Universita Autonoma de Catalunya, Barcelona (SP) (2010)
5. Bailin, S.C., Truszkowski, W.: Ontology negotiation between intelligent information agents. Knowledge Engineering Review 17(1), 7–19 (2002), DOI
   `http://dx.doi.org/10.1017/S0269888902000292`
6. Bench-Capon, T.: Persuasion in practical argument using value-based argumentation frameworks. Journal of Logic and Computation 13(3), 429–448 (2003)
7. Besana, P., Robertson, D.: How service choreography statistics reduce the ontology mapping problem. In: Aberer, K., Choi, K.-S., Noy, N., Allemang, D., Lee, K.-I., Nixon, L.J.B., Golbeck, J., Mika, P., Maynard, D., Mizoguchi, R., Schreiber, G., Cudré-Mauroux, P. (eds.) ASWC 2007 and ISWC 2007. LNCS, vol. 4825, pp. 44–57. Springer, Heidelberg (2007)
8. van Diggelen, J., Beun, R.J., Dignum, F., van Eijk, R.M., Meyer, J.J.: ANEMONE: An effective minimal ontology negotiation environment. In: Proceedings of the 5th International Joint Conference on Autonomous Agents and Multiagent Systems, pp. 899–906. ACM, New York (2006), DOI
   `http://doi.acm.org/10.1145/1160633.1160794`
9. Doran, P., Payne, T.R., Tamma, V., Palmisano, I.: Deciding agent orientation on ontology mappings. In: Patel-Schneider, P.F., Pan, Y., Hitzler, P., Mika, P., Zhang, L., Pan, J.Z., Horrocks, I., Glimm, B. (eds.) ISWC 2010, Part I. LNCS, vol. 6496, pp. 161–176. Springer, Heidelberg (2010)
10. Doran, P., Tamma, V., Palmisano, I., Payne, T.R.: Efficient argumentation over ontology correspondences. In: Proceedings of The 8th International Conference on Autonomous Agents and Multiagent Systems, pp. 1241–1242. International Foundation for Autonomous Agents and Multiagent Systems, Richland, SC (2009)
11. Doran, P., Tamma, V., Payne, T., Palmisano, I.: Dynamic selection of ontological alignments: A space reduction mechanism. In: International Joint Conference on Artificial Intelligence (2009),
    `http://www.aaai.org/ocs/index.php/IJCAI/IJCAI-09/paper/view/551`
12. Dung, P.: On the acceptability of arguments and its fundamental role in nonmonotonic reasoning, logic programming and n–person games. Artificial Intelligence 77(2), 321–357 (1995)
13. Euzenat, J.: An API for ontology alignment. In: McIlraith, S.A., Plexousakis, D., van Harmelen, F. (eds.) ISWC 2004. LNCS, vol. 3298, pp. 698–712. Springer, Heidelberg (2004)
14. Euzenat, J., Mochol, M., Shvaiko, P., Stuckenschmidt, H., Svab, O., Svatek, V., van Hage, W.R., Yatskevich, M.: Results of the ontology alignment evaluation initiative 2006. In: First International Workshop on Ontology Matching, Athens, GA, US (2006)
15. Euzenat, J., Shvaiko, P.: Ontology matching. Springer, Heidelberg (2007)
16. FIPA: Contract net interaction protocol specification. Tech. Rep. SC00029H, Foundation for Intelligent Physical Agents (2002)
17. FIPA: Modeling: Interaction diagrams. Tech. rep., Foundation for Intelligent Physical Agents (2003)
18. Haase, P., Motik, B.: A mapping system for the integration of OWL-DL ontologies. In: Proceedings of the 1st International Workshop on Interoperability of Heterogeneous Information Systems, pp. 9–16. ACM, New York (2005), DOI
    `http://doi.acm.org/10.1145/1096967.1096970`

19. Isaac, A., dos Santos, C.T., Wang, S., Quaresma, P.: Using quantitative aspects of alignment generation for argumentation on mappings. In: Shvaiko, P., Euzenat, J., Giunchiglia, F., Stuckenschmidt, H. (eds.) OM, CEUR Workshop Proceedings, vol. 431. CEUR-WS.org (2008)

20. Jennings, N., Faratin, P., Lomuscio, A., Parsons, S., Wooldridge, M., Sierra, C.: Automated negotiation: Prospects methods and challenges. Group Decision and Negotiation 10(2), 199–215 (2001)

21. Laera, L., Blacoe, I., Tamma, V., Payne, T., Euzenat, J., Bench-Capon, T.: Argumentation over ontology correspondences in MAS. In: Proceedings of the 6th International Joint Conference on Autonomous Agents and Multiagent Systems, pp. 1–8. ACM, New York (2007), DOI
http://doi.acm.org/10.1145/1329125.1329400

22. Laera, L., Tamma, V., Euzenat, J., Bench-Capon, T., Payne, T.R.: Reaching agreement over ontology alignments. In: Cruz, I., Decker, S., Allemang, D., Preist, C., Schwabe, D., Mika, P., Uschold, M., Aroyo, L.M. (eds.) ISWC 2006. LNCS, vol. 4273, pp. 371–384. Springer, Heidelberg (2006), doi:10.1007/11926078

23. Laera, L., Tamma, V.A.M., Euzenat, J., Bench-Capon, T.J.M., Payne, T.R.: Agents arguing over ontology alignments. In: Dunin-Keplicz, B., Omicini, A., Padget, J.A. (eds.) Proceedings of the 4th European Workshop on Multi-Agent Systems, CEUR Workshop Proceedings, vol. 223, CEUR-WS.org (2006)

24. Maedche, A., Motik, B., Silva, N., Volz, R.: MAFRA – A mApping fRAmework for distributed ontologies. In: Gómez-Pérez, A., Benjamins, V.R. (eds.) EKAW 2002. LNCS (LNAI), vol. 2473, pp. 235–250. Springer, Heidelberg (2002)

25. Morge, M., Routier, J.C., Secq, Y., Dujardin, T.: A formal framework for inter-agents dialogue to reach an agreement about a representation. In: Ferrario, R., Guarino, N., Prevot, L. (eds.) Proceedings of the Workshop on Formal Ontologies for Communicating Agents (2006)

26. Noy, N.F., Shah, N.H., Whetzel, P.L., Dai, B., Dorf, M., Griffith, N., Jonquet, C., Rubin, D.L., Storey, M.A.D., Chute, C.G., Musen, M.A.: Bioportal: ontologies and integrated data resources at the click of a mouse. Nucleic Acids Research 37(Web-Server-Issue), 170–173 (2009)

27. Packer, H., Payne, T., Gibbins, N., Jennings, N.: Evolving ontological knowledge bases through agent collaboration. In: Proceedings 6th European Workshop on Multi-Agent Systems, Bath, UK. Springer, Heidelberg (2008)

28. Parsons, S., Jennings, N.: Negotiation through argumentation-A preliminary report. In: Proceedings of the 2nd International Conference Multi-Agent Systems, Kyoto, Japan, pp. 267–274 (1996)

29. Prakken, H., Sartor, G.: Argument-based extended logic programming with defeasible priorities. Journal Applied Non-Classical Logics 7(1), 25–75 (1997)

30. Shvaiko, P., Giunchiglia, F., da Silva, P.P., McGuinness, D.L.: Web explanations for semantic heterogeneity discovery. In: Gómez-Pérez, A., Euzenat, J. (eds.) ESWC 2005. LNCS, vol. 3532, pp. 303–317. Springer, Heidelberg (2005)

31. Silva, N., Maio, P., Rocha, J.: An approach to ontology mapping negotiation. In: Proceedings of the Third International Conference on Knowledge Capture Workshop on Integrating Ontologies, Banff, Canada (2005)

32. Trojahn, C., Moraes, M., Quaresma, P., Vieira, R.: Using cooperative agent negotiation for ontology mapping. In: Proceedings of the 4th European Workshop on Multi-Agent Systems, CEUR Workshop Proceedings, vol. 223, pp. 1–10. CEUR-WS.org (2006)

33. Trojahn, C., Quaresma, P., Vieira, R.: Conjunctive queries for ontology based agent communication in MAS. In: Proceedings of the 7th International Joint Conference on Autonomous Agents and Multiagent Systems, pp. 829–836. International Foundation for Autonomous Agents and Multiagent Systems, Richland, SC (2008)
34. Trojahn, C., Quaresma, P., Vieira, R., Moraes, M.: A cooperative approach for composite ontology mapping. LNCS Journal on Data Semantic X (JoDS) 4900(1), 237–263 (2008), doi:10.1007/978-3-540-77688-8

Chapter 6
Measuring Complexity for MAS Design in the Presence of Ontology Heterogeneity

Maricela Bravo

CINVESTAV-IPN, Computing Science Department, México
`mbravo@computacion.cs.cinvestav.mx`

Abstract. Currently multiple agent-based solutions are being integrated and deployed to solve complex problems. This is possible because of the evolution of information technologies, such as Internet-based open standards, XML-based languages, protocols and middleware. However, one of the challenges that has to be tackled is to overcome inter-agent ontology communication heterogeneity. This chapter gives an overview of multi-agent systems (MAS) communication challenges, with special stress on the ontology heterogeneity problem. A set of measures are presented to give the MAS designer the key design guidelines and important considerations when selecting an architectural solution approach to overcome heterogeneity of the set of agents participating in the communication scenario. An overview of current solutions is discussed, analyzing the important aspects to consider for each solution approach. Finally, a comparison with various MAS examples is presented to observe the differences between architectures and their associated costs.

1 Introduction

The evolution of information technologies, such as Internet-based open standards, XML-based languages, protocols and middleware; have promoted the reutilization and integration of different autonomous agents (legacy or newly developed), to form intelligent systems which coordinate and/or cooperate to solve complex problems beyond their individual capabilities. Regardless of the existence of software platforms, tools, and integrated development environments for MAS; the interoperation of heterogeneous agents in open and dynamic environments still requires human intervention – from design and development to maintenance of integrated systems - warranting full interoperability. One of the most challenging tasks in MAS development and integration is to implement efficient communication interoperability between participating agents, especially when agents have been developed and deployed by different providers.

1.1 MAS Communication Overview

Communication in MAS is among the most researched topics in Distributed Artificial Intelligence (DAI). Many aspects of communication between agents

A. Elçi, M.T. Koné, and M.A. Orgun (Eds.): Semantic Agent Systems, SCI 344, pp. 113–132.
springerlink.com   © Springer-Verlag Berlin Heidelberg 2011

have been studied: social commitments [1, 2], dialogues [3, 4], conversations [5], protocols, agent communication language (ACL) semantics and pragmatics, among others. Communication in MAS comprises the following elements: agents, protocols, agent communication language, and messages.

Agents. One of the most referred definitions of agent was presented by Wooldgridge and Jennings [6], who state that "an agent is an encapsulated computer system that is situated in some environment, and that is capable of flexible, autonomous action in that environment in order to meet its design objectives". Of particular interest is the notion of an agent as an entity capable of showing flexible behavior for problem solving. The abilities of individual agents to solve problems and communicate are fundamental to integrate MAS.

Communication protocols. According to Endriss et.al [7] a protocol specifies the rules of interaction between agents by restricting the range of allowed utterances sequences for each agent at any stage during a communication interaction. Among the important aspects to study in agent communication protocols is their semantics and pragmatics. Correct semantic treatment of protocols allows a unified meaning across heterogeneous agents, while pragmatics studies the effect and contextual interpretation of an utterance, observing the intention of emitter, the effect on the receiver and the surrounding context of the conversation.

Agent Communication Language (ACL). Communication in MAS occurs in peer to peer connections, where agents exchange messages by means of an ACL. KQML [8] was the first standardized ACL from the Defense Advanced Research Projects Agency (DARPA) knowledge project. KQML consists of a set of communication primitives aiming to support interaction between agents. Another ACL [9] standard comes from the Foundation for Intelligent Physical Agents (FIPA) initiative. FIPA ACL is based on speech act theory, and the messages generated are considered as communicative acts. The objective of using a standard ACL is to achieve effective communication without misunderstandings.

Messages. According to FIPA specifications [10] an ACL Message Structure contains one or more of the parameters described in Table 1. The only mandatory parameter is performative.

Table 1. Elements of a message according to FIPA specification

Element type	Message parameters
Type of communicative act	Performative
Participant in communication	Sender, receiver, reply-to
Content of message	Content
Description of content	Language, encoding, ontology
Control of conversation	Conversation-id, reply-with, in-reply-to

In particular, this work is centered on the description of content, which requires the specification of the following parameters: language, to denote the language in which the content parameter is expressed; and ontology, to specify the ontology or ontologies used to give meaning to the symbols in the content expression. Both parameters are used in conjunction to support the interpretation of the message content by the receiving agent.

1.2   Ontologies for Inter-agent Communication

An important characteristic of intelligent agents is the use of ontologies to represent abstractions of their domain of knowledge. These ontologies are fundamental because they are part of the generated messages, and are helpful to communicate about the domain of knowledge of the particular agent. An ontology defines the basic terms and relations comprising the vocabulary of a topic area as well as the rules for combining terms and relations to define extensions to the vocabulary [11]. Each ontology represents the agent conceptualization of a particular domain, including hierarchical relations (subsuming, siblings, is-a); any specified semantic relations between concepts and individuals; axioms which restrict the population of the ontology; and sometimes a set of rules to execute inference over the concepts and individuals inside the ontology. Ontologies play an important role during inter-agent communication, because each agent uses its own ontology to generate messages and communicate its beliefs, desires and intentions to the rest of participating agents. One of the key design issues to achieve interoperation between agents is to create a good solution to overcome heterogeneity among their individual ontologies.

There are various reported works in literature addressing this problem. Weisman et.al [12] identified two types of heterogeneity: structural and semantic. Uschold [13] described two sources of problems for agent communication: language heterogeneity and terminological heterogeneity. Stuckenshmidt [14] analyzes three possible solutions: merging approach, mapping approach and translation with shared ontologies. The later is used as a preferred solution; the main disadvantage is that no attention is made to the learning capability of agents.

1.3   Problem Formulation

The main focus of this paper is to support MAS integrators in the arduous task of designing a solution to overcome the ontology heterogeneity problem, when a set of different agents are selected to participate in the solution of a complex problem. Given a MAS represented as the tuple $<A, P, CP, T>$, where

A – Represents the set of participating agents. $A = \{ a_1, a_2, a_3, ...,a_n \}$, where a_i is an independent agent, with i ranging from 1 to the number of agents n.

P – Represents the set of protocols

CP – Represents the set of all communication primitives

T – Represents the union of all domain concepts used for communication between all participating agents. The set of domain concepts is given by $T = \{ Ta_1, Ta_2, Ta_3, ...,Ta_n \}$, where each $Ta_n = \{ t_1, t_2, t_3,..., t_j \}$, represents the set of terms of agent n, and each term $t_j \in Ta_n$ is generated from the ontology of agent n.

Formalization of this problem is as follows:

Given a MAS with a set A of n autonomous agents with their respective sets of terms $Ta_1, Ta_2, ...,Ta_n$; there is a problem of inter-agent ontology heterogeneity if $|Ta_1 \cap Ta_2 \cap ... \cap Ta_n| < |Ta_1 \cup Ta_2 \cup ... \cup Ta_n|$.

The objective of this study is to provide design guidelines in the form of numerical measures, to support the MAS integrator with the decision of selecting a good solution.

The rest of this chapter is organized as follows: in Section 2, architectural design considerations are described; in Section 3, a set of basic measures are presented; Section 4 provides the set of measures for a centralized architecture; whereas Section 5 offers a set of measures for distributed architecture; in Section 6, an experimental case is described; in Section 7, results are analyzed and discussed; finally in Section 8, conclusions are presented.

2   MAS Architectural Design

MAS architectural design plays a crucial role towards efficient implementation of inter-agent communication support. In this section, the most important design considerations are presented: architectural considerations and associated costs (see Table 2). The goal is to provide the MAS designer with a set of measures that will guide the selection of an accurate solution approach to overcome heterogeneity.

Table 2. MAS Architectural design

Architectural Considerations	Architecture Type	a)	Fixed
		b)	Changing or adaptable
	Architecture Arrangement	a)	Centralized
		b)	Distributed
		c)	Hybrid
Associated Costs	Associated Costs	a)	Translation
		b)	Learning
		c)	Ontology maintenance

2.1   Architectural Considerations

Neches et.al [11] presented their vision of the challenges to enable knowledge sharing, they identified a range of potentially heterogeneous system models: centralized, distributed, hierarchical and mixed. Wermelinger and Fiadeiro [15] identified some issues related to the evolution of architectures over time. Specially, they presented a concept of dynamic reconfiguration, referring to architectures that may change before execution or at run-time. Based on these works and further analysis, the most important architectural considerations that should be analyzed are *architecture type* and *architecture arrangement*. *Architecture type* corresponds to the moment when the architecture is arranged, while the *architecture arrangement* represents the functional configuration of interrelationships between elements. Considering the architecture type, two possibilities exist:

a) *Fixed*. This architecture is pre-established at design time. The main purpose of the designer is to create a permanent architecture configuration that will support MAS interaction for a period of time, a pre-requisite is to analyze the heterogeneity between participating agents and all associated computational costs.

b) *Changing or adaptable*. This architectural configuration has the ability to adapt to changes in the MAS environment at run time. One of the most important considerations of adaptation is the cost associated to continuous monitoring of variables that affect the performance of interaction.

Considering the arrangement, the following configurations can be selected:

a) *Centralized*. A centralized configuration consists of a set of autonomous agents with their ontologies, a centralized ontology and a translator service. This configuration supports inter-agent communication through the invocation of a translator service whenever a misunderstanding occurs. The name of centralized is due to the central ontology, into which all terms from agents ontologies (classes and individuals) are allocated, enhanced, and aligned.

b) *Distributed*. A distributed configuration consists of a set of autonomous agents with their ontologies, in a distributed architecture agents are grouped into clusters in accordance to a similarity measure.

c) *Hybrid*. An hybrid configuration combines centralized and distributed organization of elements trying to maintain the MAS performance balanced. Different possibilities exist to arrange this architecture, for instance, the centralized ontology could allocate the most used and common terms across agents, while distributed ontologies would allocate particular inter-ontology alignments for specific peer to peer interactions.

2.2 Associated Costs

Associated computational costs are those important design derived costs which are intrinsic to the selected architectural configuration. The most important aspects are:

a) *Translation*. Translation is a key design issue, because the cost of translation strongly depends on the number of agents participating, the heterogeneity across ontologies and the arrangement of the architecture. For example in a centralized ontology, the translator should be designed and implemented to support heavy work loads, because it will serve all the agents participating in the MAS. Some of the functions that should provide are: ontology publication, inter-ontology alignment, multiple and concurrent translations at run time, among others. In the case of a distributed architecture, multiple translators are required, in this scenario the main task is the identification of groups of communication peers to construct translators, the objective is to distribute and balance work loads between translators.

b) *Learning.* Is the capability of an intelligent agent to acquire and associate new terms and concepts from another agent ontology into its own ontology. This characteristic is ideal to facilitate automated interaction, but there are still many obstacles to fully achieve this automation. For example, even when an agent is capable of learning, the accuracy and precision of the learning algorithm fails. Another shortcoming is the maintainability of the size of ontologies, because as more interactions occur with new agents, more terms will be acquired.

c) *Ontology Maintenance.* In the case of a centralized ontology, the cost of maintaining the ontology depends on the number of terms (classes and individuals) allocated and on the number of required semantic relationships between them. For a distributed scenario, the size of ontologies must be regulated, because agents may have limited resources. If agents are capable of learning new terms from their conversation counterparts, then for every different agent that they have interaction with, their ontologies will grow. Therefore, ontologies must be maintained according to recent interaction needs.

3 Basic Measures

The following represent a set of basic and important measures, which set the basis for further analysis and decision making about the best architectural solution.

The first measure that has to be calculated is the number of possible peer to peer links. Considering that every agent in the MAS environment may establish conversations links with the rest of agents. Given a set of n agents, the possible number of peer to peer communication links (nl) among them is n^2. However, the number of communication links where agents are equal needs to be extracted, which is n. It is also considered that a communication link between agents (a, b) has the same heterogeneity as a communication link of agents (b, a), thus the number of different communication links is reduced dividing by 2. Finally, the number of different communication links between n agents is given by

$$nl = (n^2\text{-}n)/2 \tag{1}$$

Another basic measure is the level of syntactical heterogeneity between agent ontologies. This measure provides a general ratio of the degree of syntactical differences between agent ontologies (considering all terms indistinctly). This measure can be defined as the number of different terms divided by the total number of terms.

Given the sets of domain concepts identified by Ta_1, Ta_2, Ta_3, ...,Ta_n from agents participating in the MAS, the total number of terms (**nt**) is obtained from the absolute value of the union operation of all sets of terms from each agent ontology, where **n** is the number of agents participating.

$$nt = |Ta_1 \cup Ta_2 \cup ... \cup Ta_n| \tag{2}$$

The number of different ontological terms (**ndt**) is obtained from the union of all sets of terms minus the intersection of all sets of terms (syntactically common terms).

$$ndt = |Ta_1 \cup Ta_2 \cup ... \cup Ta_n| - |Ta_1 \cap Ta_2 \cap ... \cap Ta_n| \tag{3}$$

Finally, the level of syntactical heterogeneity (**lsh**) results from dividing **ndt** by **nt**, which is the ratio that will serve as an indicator for evaluating heterogeneity.

$$lsh = ndt / nt \tag{4}$$

The **lsh** measure will return a value in the range from 0 to 1, where a 0 value indicates that all agents share identical terms, and returned value of 1 represents a fully syntactical heterogeneity.

4 Centralized Architecture

The architecture of this solution consists of a set of autonomous agents, a centralized ontology and a translator agent. To implement this solution, first the hierarchical class structure of the ontology must be designed and implemented. Once the basic taxonomy (class hierarchy) exists, all the terms from agent ontologies should be copied to the centralized ontology, together with all necessary semantic relationships between concepts and rules for inference. This process can be done manually, semi-automatically or fully automatically. This process is referred in literature as *ontology population*, and many solutions have been proposed.

The construction of a centralized ontology has to be accompanied by a translator agent. A translator is an intermediary agent which is invoked every time when a misunderstanding occurs due to differences in terms that are exchanged during a conversation between agents. When using this solution, an important associated cost is translation, which clearly depends on the heterogeneity between all agent ontologies. If the level of syntactic heterogeneity is high, then the number of required translations will be higher. Another important associated cost is the size of the centralized ontology. All terms of agents are copied to this centralized ontology, when more terms are copied, the ontology size is bigger. The main effect of the size of the ontology is on the performance of the translator agent, if more translations are required, with a bigger size of the ontology, the translator will work slower. These measures are designed to analyze these particular requirements.

4.1 Translation Costs

Considering a MAS with a set of n autonomous agents $A = \{ a_1, a_2, a_3, \ldots, a_n \}$, where every agent may establish conversations links with the rest of agents (see Formula 1), the necessary translations among them is the sum of terms that are unknown for each agent. For this formula, the number of required translations of each agent is considered independently, because translation is required independently of the translations of other agents.

This is, for each peer (a_i, a_j) of communication links the number of translations that agent a_i needs to have translated is equal to the set of terms from agent a_j, minus the set of terms that are common for both. Therefore, the number of required translations for all participating agents is the sum of required translations for each communication peer.

By using Formula 1, it is possible to obtain the set of heterogeneous communication links between agents, which will be identified as CL.

$$CL = \{ (a_i, a_j), (a_i, a_{j+1}), \ldots , (a_{n-1}, a_n) \},$$

$$\text{where } |CL| \leq nl,$$

$$\text{with } 0 < i < n, 1 < j \leq n, i \neq j.$$

It is important to note that a MAS is a dynamic environment where many communication interactions occur at a time, and the state of the MAS changes over time. Therefore, to model a MAS communication in a dynamic and changing scenario, and furthermore to calculate the number of required translations considering dynamic, random and concurrent communication links is too complex and is out of the scope of this study. Although, there is the need to measure translations, then a reference measure is given, assuming a short period of time, into which all communications links are enabled, and that all terms are to be exchanged causing the need for translation. The objective of these assumptions is to consider the worst scenario. Taking this into consideration, the number of required translations (rt), is calculated as follows:

$$\forall\, (a_i, a_j) \in CL,\, rt = \sum \left[\, |Ta_i \cup Ta_j| - |Ta_i \cap Ta_j|\, \right] \qquad (5)$$

$$\text{with } 0 < i < n, 1 < j \leq n, i \neq j.$$

The number of required translations has to be multiplied by a number of times (t), because every term exchanged can be translated more than once. A complex model based on conversation logs is required to know or predict the number of times that each term will be exchanged and therefore translated. However, it is important to include this number in the formula of translations to have a reference of its weigh in the cost.

Finally, the number of required translations (nrt) for a centralized architecture is calculated as follows:

$$nrt = rt * t \qquad (6)$$

For instance, consider an agent a_1 with a set $Ta_1 = \{meal\}$, an agent a_2 with a set $Ta_2 = \{food\}$, and a translator agent identified by *trans* using a general ontology. Assuming that the following conversation takes place: agent a_1 requests to a_2 the price of a *meal* for 1 person, agent a_2 requests to *trans* the translation of the term *meal*, then *trans* agent returns the term *food*, finally agent a_2 returns the answer to agent a_1. The sequence of messages occurs from top to bottom as shown in Table 3.

Using Formula 5 to calculate the number of required translations (*rt*), the result is 2. However, as the example shows, agents are not capable of learning and/or retaining translated concepts, therefore, the number of required translations has to consider the number of times that each term is translated during a conversation (Formula 6) in this case $t = 2$ and $nrt = 4$.

Table 3. A simple conversation example

Sender	Receiver	Performative	Content	Ontology	nrt
a_1	a_2	Request	Price(meal, 1, ?dollars)	Ta_1	
a_2	trans	Request	Translate(meal, a_1, ?translation)		1
trans	a_2	Inform	Translate(meal, a_1, food)		
a_2	a_1	Inform	Price(food, 1, $89.00)	Ta_2	
a_1	trans	Request	Translate(food, a_2, ?translation)		2
trans	a_1	Inform	Translate(food, a_2, meal)		
a_1	a_2	Request	Price(meal, 3, ?dollars)	Ta_1	
a_2	trans	Request	Translate(meal, a_1, ?translation)		3
trans	a_2	Inform	Translate(meal, a_1, food)		
a_2	a_1	Inform	Price(food, 3, $267.00)	Ta_2	
a_1	trans	Request	Translate(food, a_2, ?translation)		4
trans	a_1	Inform	Translate(food, a_2, meal)		

4.2 Ontology Costs

Costs associated to ontology are those related to the main ontological elements: concepts (classes and individuals), and semantic relationships between them.

In this work, each ontology is treated as a set of terms or concepts that belong to a common domain of interest. Hierarchical relations and in general any inter-concept relations (different to the synonym) are not considered. The main reason for this consideration is the fact that during an inter-agent communication session, exchanged messages generally include terms extracted from their particular ontologies, not a hierarchy of concepts. Additionally, the assumption that all participating agents will share and make public their ontologies is not realistic; some agents will share only part of their ontologies for security and privacy reasons.

For a centralized architecture, the number of concepts (*nc*) in the ontology is calculated as the union operation of all agents terms, where Ta_i represents the set of terms of agent i, with i ranging from 1 to n number of agents participating in the MAS.

$$nc = |Ta_1 \cup Ta_2 \cup ... \cup Ta_n| \tag{7}$$

In order to facilitate semantic mappings between ontological terms, the number of concepts has to be augmented with their synonyms. Therefore, the extended number of concepts (enc) in the ontology is calculated with the number of concepts (nc) multiplied by an average of s synonyms per concept.

$$enc = nc * s \tag{8}$$

Finally, another important measure is the number of semantic relationships (nsr) between concepts in the ontology. Considering that the purpose of the ontology is to support translation for multiple conversations between agents, then the most relevant semantic relationship is the synonymy. The easiest way to calculate this measure is to multiply the number of all concepts twice, taking into account that every term in the ontology may be related with the rest. However, only the number of different terms (ndt) needs to establish synonymy relationships, minus all possible relations with themselves (there is no need for synonymy relation between term a and term a), and finally dividing by 2, because, synonym relations are symmetric (if term a is synonym of term b, then the synonym relation in the opposite direction holds). Therefore, the number of possible semantic relationships (nsr) between terms is given by:

$$nsr = (ndt^2 - ndt)/2 \tag{9}$$

The size of the centralized ontology is the sum of the extended number of concepts plus the number of semantic relations, as follows:

$$sco = enc + nsr \tag{10}$$

5 Distributed Architecture

A distributed architecture consists of a set of n agents with their ontology each. Agents in a distributed scenario are grouped in a given number of clusters (c) according to a similarity measure. The objective of a distributed architecture is to share and balance translation or learning workloads between clusters. According to the capability of agents to learn new concepts, two options exist:

a) In the case that agents cannot learn new terms, a set of translators are required, and for each translator it is necessary to implement an auxiliary ontology to support the translator when misunderstandings occur. In this case, the costs associated are similar to the centralized architecture, with the difference that these costs should be calculated for each cluster.

b) If agents are capable of learning, then their ontologies will need to be maintained, because for each new conversation new terms will be allocated in the ontology. The main advantage of this approach is that there is no need to implement translators. However, there is an associated cost for learning.

5.1 Distributed Architecture with Translators

For a distributed architecture, where agents are not capable of learning, there is the need to implement translators to support communication between groups of agents. Considering that the number of translators is equal to the number of clusters, translation is calculated similarly to the centralized architecture.

Given a set C of r clusters each containing n agents, where each cluster is identified by c_m, the set of $CL(c_m)$ of different communication links into each cluster is obtained as follows: each agent inside the cluster may establish communication links with the rest of agents in the same cluster, but considering only heterogeneous pairs. The set of $CL(c_m)$ is defined as follows:

$$C = \{c_1, c_2, \ldots, c_r\}, \text{ where each } c_m = \{a_1, a_2, \ldots, a_n\}$$

$$CL(c_m) = \{(a_i, a_j), (a_i, a_{j+1}), \ldots, (a_{n-1}, a_n)\},$$

$$\text{with } 0 < m \leq r, 0 < i < n, 1 < j \leq n, i \neq j.$$

Then for each cluster the number of required translations (***nrtpc***) is calculated as follows:

$$\forall (a_i, a_j) \in CL(c_m), nrtpc(c_m) = \sum \left[|Ta_i \cup Ta_j| - |Ta_i \cap Ta_j| \right] \tag{11}$$

$$\text{with } 0 < m \leq r, 0 < i < n, 1 < j \leq n, i \neq j.$$

The reference number of terms that are required to be translated (***nrt***) in the MAS is the average of the number of required translations per cluster (***nrtpc***). It is important to take into account that when agents are not capable of learning, the number of required translations has to be multiplied by a given number of times (***t***) that the translation is executed per term. That is, the same term may be translated more than once.

$$\forall (c_i) \in C, nrt = (\sum nrtpc (c_m) * t) / r \tag{12}$$

Additional costs are related to auxiliary ontologies. First, the size for each supporting ontology is calculated similarly to the centralized architecture. Given a set C of r clusters, the number of ontological concepts per cluster (***ncpc***) is calculated as the union operation of all sets of terms, multiplied by an average s of synonyms.

$$ncpc(c_m) = |Ta_1 \cup Ta_2 \cup \ldots \cup Ta_n| * s \tag{13}$$

$$\text{with } 0 < m \leq r.$$

The number of different terms per cluster ***ndtpc*** is obtained from the union of all sets of terms minus the intersection of all sets of terms.

$$ndtpc(c_m) = |Ta_1 \cup Ta_2 \cup \ldots \cup Ta_n| - |Ta_1 \cap Ta_2 \cap \ldots \cap Ta_n| \tag{14}$$

The number of semantic relations per cluster (***nsrpc***) is given by:

$$nsrpc(c_m) = (ndtpc(c_m)^2 - ndtpc(c_m))/2 \tag{15}$$

The size of ontology per cluster is calculated as follows:

$$sdopc(c_m) = ncpc(c_m) + nsrpc(c_m) \tag{16}$$

Finally, the average size of distributed auxiliary ontologies (*sdao*) per cluster is calculated.

$$\forall \ (c_i) \in C, sdao = (\ \Sigma \ sdopc \ (c_m) \ ) \ / \ r \tag{17}$$

5.2 Distributed Architecture with Learning Capabilities

If agents are capable of learning, their ontologies will grow every time they communicate with a new agent. Therefore, the cost of learning a new term is calculated as the time required to select and classify a new term into the ontology. This measure depends also on the number of terms that are to be acquired. Given the set *C* of *r* clusters each containing *n* agents, and the set of $CL(c_m)$ different communication links into each cluster, the number of terms that need to be learnt per each cluster (*ntlpc*) is calculated as follows:

$$\forall \ (a_i, a_j) \in CL(c_m), ntlpc(c_m) = \ \Sigma \ \big[\ |Ta_i \cup Ta_j| \ - |Ta_i \cap Ta_j| \ \big] \tag{18}$$

with $0 < m <= r, 0 < i < n, 1 < j \leq n$, i $\neq$ j.

The total number of terms that are required to be learnt (*ntl*) in the MAS is the average of the number of terms per cluster (*ntlpc*).

$$\forall \ (c_i) \in C, ntl = (\ \Sigma \ ntlpc \ (c_m) * l \ ) \ / \ r \tag{19}$$

Finally, the associated cost of learning for the entire MAS is calculated multiplying the number of terms that are necessary to be learnt (*ntl*) by a given learning cost (*l*), which clearly depends on the learning algorithm used.

An important aspect to measure in a distributed architecture with learning capabilities is the size of ontologies. In the case that agents can learn new terms, then a measure to know ontology extension per agent (*oe*) is crucial.

$$\forall \ (a_i) \in c_m, oe \ (a_i) = |Ta_1 \cup Ta_2 \cup ... \cup Ta_n| \ - |Ta_i| \tag{20}$$

with $0 < m \leq r, 0 < i < n$.

The number of semantic relations per ontology extension (*nrpoe*) is given by:

$$nrpoe(a_i) = (oe(a_i)^2 - oe(a_i))/2 \tag{21}$$

The average size of ontology extensions per cluster is calculated as follows:

$$\forall \ (c_i) \in C, \forall \ (a_i) \in c_m, oea(c_i) = (\ \Sigma \ oe \ (a_i) + \Sigma \ nrpoe \ (a_i)) \ / \ n \tag{22}$$

The average size of ontology extensions for the entire MAS is calculated as follows:

$$\forall \ (c_i) \in C, oeaMAS = (\ \Sigma \ oea(c_i) \ ) \ / \ r \tag{23}$$

5.3   Coordination or Intermediation Costs

An additional cost for a centralized architecture is the cost associated with coordination and/or intermediation. As the general objective of a MAS is to solve a complex problem, then an inter-cluster communication has to be established in order to compose a general solution. The general idea is to divide the problem into sub-problems and assign them to specialized clusters, where agents grouped according to their specialty communicate inside the cluster to generate a partial solution. However, when individual cluster solutions have been generated, there is the need to integrate a general solution. In this case, there are two possibilities: one is the implementation of an intermediary agent which communicates directly with each cluster representing agent to assign sub-problems and collect partial solutions; and the other is that each cluster representing agent communicates with other representing agents, and collectively coordinate to assign sub-problems and construct an integral solution for the MAS.

In case that an intermediary agent is used, the number of communication links that needs to establish is equal to the number of clusters. However, as in the rest of this study, if a central intermediary agent needs to communicate with the rest of representing agents, then his local ontology has to be augmented with the terms of his communication counterparts. Given a set R of n representing agents, one per cluster, the number of communication links that an intermediary agent needs to establish is equal to n, and the size of its local ontology (sio) is calculated with the union operation of all agent sets of terms as follows:

$$R = \{ ra_1, ra_2, ..., ra_n \} \tag{24}$$

$$\forall (a_i) \in R, sio = | Tra_1 \cup Tra_2 \cup ... \cup Tra_n|$$

In case that all representing agents communicate each other to coordinate for sub-problems assignment and final solution integration, the number of communication links is calculated as in Formula 1:

$$nl = (n^2 - n)/2 \tag{1}$$

Sizes of their ontologies need to be augmented with the rest of representing agent individual ontologies. Given the set R of n representing agents, the set CL different communication links between representing agents, the average number of terms that need to be acquired per representing agent ($siopr$) is calculated as follows:

$$\forall (a_i, a_j) \in CL, siopr = \sum \left[| Ta_i \cup Ta_j| - | Ta_i \cap Ta_j| \right] / n \tag{25}$$

$$\text{with } 0 < i < n, \ 1 < j \leq n, \ i \neq j.$$

6   Experimental Case

Distributed Problem Solving (DPS) is a sub-field of Distributed Artificial Intelligence [16] which deals with complex problems. DPS researchers implement

MAS systems which coordinate, cooperate and distribute knowledge to achieve a common goal. In this context, various MAS examples are presented to analyze architectural considerations and associated costs.

In order to apply formulas and calculate design costs, a set of public ontologies related to the travel booking domain were searched and retrieved. The set of selected ontologies, their domain and sets of terms are shown in Table 4.

Table 4. Solver agents, domain and terms

Agent	Description	Domain	Terms
a_0	Travel message ontology [17]	Flight reservation	Airplane, Airport, Airtravel, Booking, Cabin, City, Company, Airline, Contact, Flight, Meal, Person, Seat.
a_1	Itinerary ontology [18]	Travel itinerary	Aircraft, Class, Flight, HotelReservation, Itinerary, Meal, RentalCar, RecordLocatorNumber.
a_2	Travel ontology [19]	Travel	Accommodation, BedAndBreakfast, BudgetAccommodation, Campground, Hotel, LuxuryHotel, AccommodationRating, Activity, Adventure, Relaxation, Sightseeing, Sports, Contact, Destination, BackpackersDestination, Beach, BudgetHotelDestination, FamilyDestination, QuietDestination, RetireeDestination, RuralArea, UrbanArea.
a_3	QALL-ME ontology [20]	Tourism	Contact, Country, CreditCard, Currency, Destination, Event, EventContent, Facility, Genre, Language, Location, Period, PersonOrganization, Price, Room, Site, Transportation.
a_4	e-Tourism ontology [21]	Tourism	Accommodation, Activity, ContactData, DateTime, OpeningHours, Period, DatePeriod, TimePeriod, Season, Event, Infraestructure, Location, GPSCoordinates, PostalAddress, Room, ConferenceRoom, Guestroom, Ticket.
a_5	TAGA ontology [22]	Travel	Itinerary, Customer, Reservation, HotelReservation, AirlineReservation, EntertainmentReservation, ServiceProvider, TravelService, Cinema, Restaurant, Opera, Accommodation, Transportation.

Given a MAS integrated with agents from Table 4, using Formula 1, the number of different communication links between them is given by

$$n = 6, \; nl = (6^2\text{-}6)/2 = 15$$

The total number of terms *nt* is calculated using Formula 2.

$$nt = 76$$

The number of different ontological terms *ndt* is obtained from the union of all sets of terms minus the intersection of all sets of terms.

$$ndt = 76 - 0 = 76$$

The level of syntactical heterogeneity is calculated using Formula 4.

$$lsh = ndt / nt, \qquad lsh = 76 / 76 = 1$$

The resulting level of heterogeneity is high for this set of agents.

6.1 Cost of a Centralized Architecture

To calculate the cost of a centralized architecture, the following measures are considered: number of participating agents, the level of syntactic heterogeneity among them, translation related cost, and ontology related costs. The resulting set of communication links *CL*, with *6* agents is:

$$CL = \{ (a_0, a_1), (a_0, a_2), (a_0, a_3), (a_0, a_4), (a_0, a_5),$$
$$(a_1, a_2), (a_1, a_3), (a_1, a_4), (a_1, a_5), (a_2, a_3),$$
$$(a_2, a_4), (a_2, a_5), (a_3, a_4), (a_3, a_5), (a_4, a_5) \}$$

The number of required translations (*nrt*) is the sum of all communication links required translations. Recalling this is a reference measure of the worst case in a given period of time.

$$nrt = 19 + 34 + 29 + 31 + 26 + 30 + 25 + 26 +$$
$$19 + 37 + 38 + 34 + 31 + 29 + 30 = 438$$

The measures related to ontology costs are calculated with formulas 6, 7 and 8. First the number of concepts in the centralized ontology considering all participating agents is:

$$nc = 76$$

The extended number of concepts (*enc*) in the ontology is calculated with the number of concepts (*nc*) multiplied by an average of *5* synonyms per individual.

$$enc = 76 * 5 = 380$$

The number of possible semantic relationships (*nsr*) between terms is given by:

$$nsr = (ndt^2 - ndt)/2$$
$$nsr = (76^2 - 76)/2 = 2850$$

Further experiments were carried out with different combinations of agent terminologies. See table 5 for results.

Table 5. Results of centralized architecture cost with a set of different MAS

Num. agents	Agents	Required translations	Ontology size	Total
4	{a0, a2, a1, a3}	1740	1650	3390
5	{a0, a1, a2, a3, a4}	3000	2412	5412
6	{a2, a5, a0, a4, a1, a3}	4380	3078	7458
7	{a0, a1, a3, a2, a4, a3, a5}	6060	3078	9138
8	{a0, a2, a4, a1, a3, a5, a3, a2}	8380	3078	11458

6.2 Cost of a Distributed Architecture

For a distributed architecture, a set of clusters must be defined. There are some clustering algorithms reported in literature to automate this process. However, as the purpose of the experiment is to analyze different configurations and calculate their costs, then the MAS variations shown in Tables 6 and 7 are defined.

For each MAS, the set of formulas for a distributed architecture using translators and learning capabilities were calculated. Results are shown in Table 6 and 7.

Table 6. Results obtained with MAS examples for distributed architecture with translators

Num. agents	Clusters	Required translations per cluster	Ontology size	Intermediation	Total
4	{a0, a2}, {a1, a3}	290.00	561.50	13	864.50
5	{a0, a1}, {a2, a3, a4}	895.00	1185.00	34	2114.00
6	{a2, a5}, {a0, a4}, {a1, a3}	296.67	581.00	40	917.67
7	{a0, a1, a3}, {a2, a4}, {a3, a5}	466.67	671.33	49	1187.00
8	{a0, a2, a4}, {a1, a3, a5}, {a3, a2}	710.00	950.67	35	1695.67

Table 7. Results obtained with MAS examples for distributed architecture with learning

Num. agents	Clusters	Terms to learn per cluster	Ontology size	Intermediation	Total
4	{a0, a2}, {a1, a3}	58.00	124.50	13	195.50
5	{a0, a1}, {a2, a3, a4}	179.00	237.40	34	450.40
6	{a2, a5}, {a0, a4}, {a1, a3}	59.33	190.00	40	289.33
7	{a0, a1, a3}, {a2, a4}, {a3, a5}	93.33	229.10	49	371.43
8	{a0, a2, a4}, {a1, a3, a5}, {a3, a2}	142.00	333.70	35	510.70

7   Results Discussion

In order to evaluate the three architectural options, the same MAS variations were tested with the measures for a centralized architecture and distributed in both options: with distributed translators and with learning capabilities. Results are shown in Figure 1.

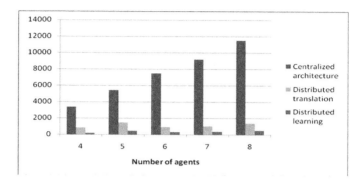

Fig. 1. General costs results for centralized and distributed architectures

Figure 1, shows an expected behavior for the architectural MAS variations. The centralized architecture resulted with the highest cost and a normal correlation tendency: the more agents participate, the higher is the cost. On the contrary, distributed architecture with translators reduced the cost of centralized in an average of 80%. The distributed architecture with learning capabilities reduced the cost in an average of 68% in relation with the distributed architecture with translators. This is the main reason to select distributed architectures over centralized for complex problems that are divisible into sub-problems.

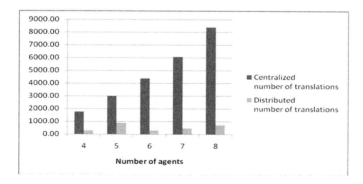

Fig. 2. Number of required translations for a centralized and distributed architecture

An expected result is observed from the number of required translations between centralized and distributed architectures shown in Figure 2. For the same number of agents with the same sets of terms, distributed translations reduce in an average of 83%. However, the number of translations of a distributed architecture does not represent the division of the centralized number of translations into the number of clusters nor into the number of agents, as it could be guessed.

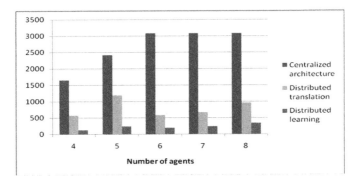

Fig. 3. Size of ontologies for centralized and distributed architectures

The size of ontologies is another important aspect, because depending on this size there are derived costs for retrieving, searching, browsing, reviewing logical consistency and in general maintainability of ontologies. In Figure 3, it is observable that for all cases the implementation of a central ontology has the highest cost due to the number of terms and inter-relationships that need to be allocated. In this graph, there are two important results to observe. First, for MAS variations with 6, 7 and 8 agents each, the size of the central ontology remains equal, even when the number of agents increases. The reason for this result is that some agents and their ontologies are repeated into the same MAS. Term representation into ontologies is required only once, there is no need for term redundancy when multiple homogeneous agents are participating. The second result to note is for the distributed architecture with translators, where the MAS integrated with 5 agents resulted in the highest size of ontology. The reason for this result is because one of the two clusters of this MAS is integrated with the three largest ontologies, therefore the size of the ontology resulted higher than the MAS with 6, 7 and 8 agents respectively.

8   Conclusions

In this chapter a set of measures are presented to give the MAS designer the key design guidelines and important considerations when selecting and implementing an architectural solution approach to overcome the ontology heterogeneity problem. MAS architectural design is based on a series of measures and formulas which support the analysis of current solutions and the important aspects which affect the cost.

A set of calculations were executed with various MAS examples, three different architectural solutions were analyzed: a centralized architecture, a distributed architecture based on distributed translators and a distributed architecture with agents that are capable of learning. As logically guessed, the highest cost results with the centralized architecture, followed by the distributed architecture with translators, and the less costly was the distributed architecture with agents that learn. It is clear that the distribution of workloads between clusters of agents offers the best option to implement a solution. However, the complexity associated with the implementation of a distributed architecture is high and multi-factorial, for instance: the definition of an optimal number of clusters is not as easy as it seems, it does not depend on a simple division. It depends on the domain of knowledge and agent individual capabilities, agents should form clusters in accordance to their functionality and convenience.

The set of MAS examples were integrated with ontologies with the highest level of heterogeneity among them. This is, that for any MAS there were no common terms between all ontologies, causing a high cost for a centralized architecture. However, with a low level of heterogeneity (few different terms between ontologies), even with a larger number of agents, the implementation of a centralized architecture is feasible because of its simplicity and the low cost derived.

More examples and exhaustive experimentation is desirable to measure scalability and performance with more agents. It is also of current interest to model and develop simulation environments to study dynamic changes during periods of time to establish prediction models.

Acknowledgements. Author acknowledges CONACyT, Mexico for the postdoctoral research support.

References

1. Castelfranchi, C.: Modeling Social Action for AI Agents. Artificial Intelligence 103(1-2), 157–182 (1998), doi:10.1016/S0004-3702(98)00056-3
2. Singh, M.P.: A Social Semantics for Agent Communication Languages. In: Dignum, F.P.M., Greaves, M. (eds.) Issues in Agent Communication. LNCS, vol. 1916. Springer, Heidelberg (2000)
3. Walton, C.D.: Model Checking Agent Dialogues. In: Leite, J., Zhang, S.-W., Torroni, P., Yolum, p. (eds.) DALT 2004. LNCS (LNAI), vol. 3476, pp. 132–147. Springer, Heidelberg (2005)
4. Orgun, B., Dras, M., Cassidy, S., Nayak, A.: DASMAS – Dialogue based Automation of Semantic Interoperability in Multi Agent Systems. In: Proceedings of Australian Ontology Workshop, AOW 2005, Sydney, Australia (2005)
5. Flores, R.A., Pasquier, P., Chai-draa, B.: Conversational Semantics Sustained by Commitments. Autonomous Agents and Multi-Agent Systems 14(2), 165–186 (2007), doi:10.1007/s10458-006-0011-1

6. Jennings, N.R., Wooldridge, M.J.: Agent-Oriented Software Engineering. Journal of Artificial Intelligence 117, 277–296 (2000)
7. Endriss, U., Maudet, N., Sadri, F., Toni, F.: Logic-based agent communication protocols. In: Dignum, F.P.M. (ed.) ACL 2003. LNCS (LNAI), vol. 2922, pp. 91–107. Springer, Heidelberg (2004)
8. Finning, T., Fritzon, R., McEntire, R.: KQML as an agent communication language. In: Proceedings of the 3rd International Conference on Information and Knowledge Management (1994)
9. FIPA – Foundation for Intelligent Physical Agents. FIPA Specifications (2003), http://www.fipa.org/specifications/index.html (accessed September 12, 2010)
10. FIPA ACL Message Structure Specification, (2002) http://www.fipa.org/specs/fipa00061/SC00061G.pdf (accessed September 12, 2010)
11. Neches, R., Fikes, R.E., Finin, T., Gruber, T.R., Patil, R., Senator, T., Swartout, W.R.: Enabling technology for knowledge sharing. AI Magazine 12(3), 16–36 (1991)
12. Wiesman, F., Roos, N., Vogt, P.: Automatic Ontology Mapping for Agent Communication. In: Proceedings of the First International Joint Conference on Autonomous Agents, pp. 563–564 (2002)
13. Uschold, M.: Barriers to effective agent communication. In: Proceedings of the CEUR Workshop on OAS 2001 Ontologies and Agent Systems (2001)
14. Stuckenschmidt, H.: Exploiting Partially Shared Ontologies for Multi-agent Communication. In: Klusch, M., Ossowski, S., Shehory, O. (eds.) CIA 2002. LNCS (LNAI), vol. 2446, pp. 249–263. Springer, Heidelberg (2002)
15. Wermelinger, M., Fiadeiro, J.L.: A Graph Transformation Approach to Software Architecture Reconfiguration. In: Proceedings of the Workshop on Graph Transformation Systems (2000)
16. Durfee, E.H.: Distributed Problem Solving and Planning. In: Multi-Agents Systems and Applications, pp. 118–149 (2001)
17. http://www.srdc.metu.edu.tr/webpage/projects/satine/ontologies/TravelMessageOntology.owl (accessed September 12, 2010)
18. http://www.daml.org/2001/06/itinerary/itinerary-ont.daml (accessed September 12, 2010)
19. Knublauch, H., http://protege.cim3.net/file/pub/ontologies/travel/travel.owl (accessed September 12, 2010)
20. Ou, S., Pekar, V., Orasan, C., Spurk, C., Negri, M.: Development and Alignment of a Domain-Specific Ontology for Question Answering. In: Proceedings of the 6th Edition of the Language Resources and Evaluation Conference (2008)
21. DERI, E-Tourism Ontology, http://www.sti-innsbruck.at/results/ontologies (accessed September 12, 2010)
22. Zou, Y., Finin, T., Ding, L., Chen, H., Pan, R.: TAGA: Trading Agent Competition in Agencities (2003), doi:10.1.1.12.7153

Chapter 7
Ontology-Based Matchmaking and Composition of Business Processes

Duygu Çelik[1] and Atilla Elçi[2]

[1] Computer Engineering Department, Istanbul Aydin University, Istanbul, Turkey
duygucelik@msn.com
[2] Software Engineering Program, Toros University, Mersin, Turkey
atilla.elci@toros.edu.tr

Abstract. This chapter shows how it is possible to use agents and Semantic Web technologies to deal with dynamic composition of business processes via an agent-based workflow system. The aim of the system is to discover composable processes at first among heterogeneous business processes that are running possibly under different Web servers and then execute them in the order specified by a planner to reach a complex requested goal. We proposed a framework of an Inference-based Semantic Composition Agent (SCA) of atomic business processes that employs process similarity matching and inference techniques. SCA synthesizes new services from existing ones in an automatic fashion. A powerful matching mechanism is needed to find fitting tasks in order to attain the required composition. An innovative Semantic Matching Step (SMS) of SCA helps to find the fitting tasks while constituting workflow to achieve required composition. Additionally, SCA composes available OWL-S atomic processes utilizing Revised Armstrong's Axioms (RAAs) in inferring functional dependencies. Experiments show that SCA System produces atomic process sequences as a workflow in achieving the required composition plan that satisfies user's requirements as a complex task. The novelty of the SCA System is that for the first time Armstrong's Axioms are revised and used for semantic-based planning and inferencing of services.

Keywords: Web Services matchmaking, Web Services composition, Semantic Agents, Armstrong's Axioms, and Semantic Web Services.

1 Introduction

Web Services have gained importance due to their interoperability and ease of use. Finding suitable Web services or composing appropriate new services from available set of services that fits the request of a user the best is still a major problem. At present, there is no such mechanism by which numerous interrelated Web services can be composed and recommended with semi-automated or fully automated

A. Elçi, M.T. Koné, and M.A. Orgun (Eds.): Semantic Agent Systems, SCI 344, pp. 133–157.
springerlink.com

fashion in order to fulfill an overall objective. Moreover, next generation of Web is expected to combine pre-existing web services with semantics to provide an advanced non-existing service to meet user demands. Such will be required in all domains, e-commerce for one, and service providers, search engines, web agents or spiders will have to deal with it. Therefore, the problem of discovery and composition of Web services has received much attention to support e-commerce or enterprise applications [1]. In this chapter, emphasis is on Web services composition utilizing atomic business processes of Web services. Web services composition addresses a situation where a client's request cannot be met by a single available service; it may however be satisfied by suitably combining multiple interconnected or composable services. Various AI planning techniques proposed solutions to *the composition problem* by using a planner. A planner uses a list of participating candidate Web services (specified as atomic or composite processes) and a complex goal (stated by the user in the form of a task description) in generating a composition plan. However, non-semantic AI-based planning methods can only compose services upon user's necessity description and this lacks the flexibility in meeting later change. In addition, semantic based methods for planning mechanisms may be helpful to a purposeful agent; it makes the agent more intelligent in lining up the individual services in more ways than otherwise possible.

Some of the most popular non-semantic and semantic planners in the literature are those based on the Hierarchical Task Network (HTN) planner using Situation Calculus (SC) as discussed by Sirin *et. al.* [2 to 4]; Event Calculus (EC) based composition approach as discussed by Aydin *et. al.* [5 and 6]; Planning Domain Definition Language (PDDL) [7] based composition approach as introduced by Yang *et. al.* [8]; Compositional/Process Algebra (PA) based composition approach as given by Hashemian *et. al.* [9]; and, SWRL Planner of Domenico *et. al.* [10]. Next section gives a brief explanation about the contributions of the proposed composition system and major differences between the above mentioned approaches and the proposed system.

Rest of this chapter is organized as follows: Section 2 mentions the fundamentals of SCA approach. Section 3 gives brief information for theoretical background to provide better understanding of system details. Section 4 presents architecture of the proposed SCA system. Section 5 describes the need of input and output matching during the composition of web services though workflows. Section 6 deals with the detail of Semantic Matching Step. Additionally, the section 7 takes up the Revised Armstrong's Axioms in more detail. Section 8 investigates the inferencing mechanism through RAAs in the SCA which concludes the composition task of SCA. Finally, section 9 concludes the chapter.

2 Contributions

In this chapter, we present a composition model involving an *inference-based semantic business process composition agent (SCA)* for verifying compatibility and composition of atomic business processes. The SCA System performs composition tasks by utilizing Semantic Web technologies in order to sequence execution of business processes in such an order that accomplishes a client's complex process requirement.

SCA System composes suitable atomic business processes of *Semantic Web Services (SWSs)* utilizing *Revised Armstrong's Axioms (RAAs)* in inferring functional dependencies.

Unlike other approaches mentioned above, SCA has extremely different planning and inferencing mechanism employing the *RAAs*. SCA uses the RAAs and OWL tags to describe task dependencies among semantically predetermined task instances in its own knowledge base. Those descriptions are then used while producing control/data flow of the task instances for achieving the client's composite plan. Briefly, we can say that during inferencing task dependencies in order to constitute a composition plan, the other approaches mentioned above employ SC, EC, PDDL, SWRL rules, and PA methods whereas we used RAAs. Although these approaches are alternatives, their solutions do not keep the produced compositions for re-use later. Additionally, the performed semantic matching among candidate processes can't be reasoned properly since the number of semantic descriptors used in their knowledge bases is limited. Thus the knowledge base of SCA is more powerful than others.

The *Armstrong's Axioms (AAs)* are a set of axioms (*or more precisely, Armstrong's inference rules*) that are used to infer functional dependencies on a relational database. They were developed by William W. Armstrong in his paper titled *Dependency Structures of Data Base Relationships* (Armstrong, 1974) [11]. The novelty of the SCA System is that for the first time Armstrong's Axioms are revised and used for semantic-based inferencing and planning of Web services. We revised AAs in order to shorten inference chains while preserving the integrity to derive functional dependencies of the processes, and used in this work during the inferencing and planning phase.

Furthermore, SCA System has five main parts which are *translator, planner, inference engine, execution engine,* and *monitoring agent* that deal with composition stage. Additionally, a well-organized matchmaking algorithm is considered to discover functional dependencies and similarities between a pair of processes in the candidate set of SWSs: *Semantic Matching Step (SMS)* scores similarity degree based on the assessment of similarity distance among concepts / parameters of these processes on focus.

The system has three ontology knowledge bases (*Tasks-TKBO, Concepts (or Domains-CKBO) and Processes-PKBO*). TKBO contains domain-specific task instances and semantic annotations of the RAAs. CKBO contains domain-specific concepts and their relations through an OWL property. PKBO is like a repository that keeps the entire atomic business processes of domain-related candidate SWSs. It is very possible that a semantic-based discovery agent can find suitable SWSs related to the client's request. Discovery of SWSs is required also for composing a sequence of processes in order to meet a client's request. This aspect was covered in our previous work [12 to 17]. Here it is assumed that the discovery of potential SWSs for required composition would have been performed before initiating the composition task of SCA.

SMS and other functional parts that are given above use these knowledge bases to produce a required composition plan. Framework of the SCA System was designed; a prototype was implemented and tested on a corpus of Web services in demonstrating it. Experiments show that the proposed SCA System produces process sequences as a composition plan that satisfies user's requirement for a complex task.

3 Theoretical Background

Web services interact by passing XML data, with data types specified using XML Schema. Simple Object Access Protocol (SOAP) can be used as the communication protocol [18], and the Input / Output (I/O) signatures for web services are given by Web Services Description Language (WSDL) [19]. UDDI stands for Universal Description, Discovery and Integration [20] and provides the means to publish and discover web services through a UDDI registry. However, these technologies do not support semantics and contain descriptions of the functionality of a web service. Accordingly, this situation creates difficulties in discovering and composing required web services by a client or software agent due to syntactical characterization and continuously rising number of web services.

Semantic Web Services (SWSs) [1] are developed through applying the semantic web technologies to web services. More particularly, through the use of semantic description frameworks, SWSs will prop up the provision of intelligent methods for the discovery, composition, monitoring, and execution of web services.

A fundamental problem of the Web services composition is to *Discover* existing functional dependencies of the different structured business processes and then coordinate them. By coordination, we mean all the work needed to sequence all these processes in order to fulfill the main goal in an efficient manner.

Therefore, we think that *Web Ontology Language (OWL)* [21] will be very useful to describe data dependencies between ontology-driven knowledge bases of service suppliers in B2B/B2C areas. This utility derives from the fact that the use of e-commerce requires transaction-mediated parameter information flows between the two end points. The transaction of parameter information may belong to a product, service, customer, price, availability, quality, tax, etc. These parameters can be used as Input / Output / Precondition / Effect (I/O/P/E) parameters that this type of structural information could be integrated into a *Web Ontology Language of Services (OWL-S)* [22]. Additionally, parameter matching is required in order to use knowledge bases that store and tie parameter information between agent/supplier ontologies. Each piece of tie information provides the relationship between the metadata of the agent and the supplier's parameters. Assume that a product's unit information is stored as (unit-u) in a supplier's ontology, but the agent needs the product amount as (dozen-d) due to the client request. Conversion is required to allow recognition of the product by the service search agent.

The following example displayed in Figure 1 below describes design aspects of process matching of e-business services through a case study that contains *Online Car Parts Product Selling*, *Price Quota* and *Currency Converter* web services. For the most part of this example, focus is on *Online Car Parts Product Selling* in order to facilitate a better understanding of semantic matching necessity during discovery and composition.

Many companies may sell car parts and offer car part sales packages that need to associate with an agent for this. A typical scenario would be a customer looking for a service to search/buy car parts that have *a specific model*, *year*, *id*, *length* and *price* (Figure 1).

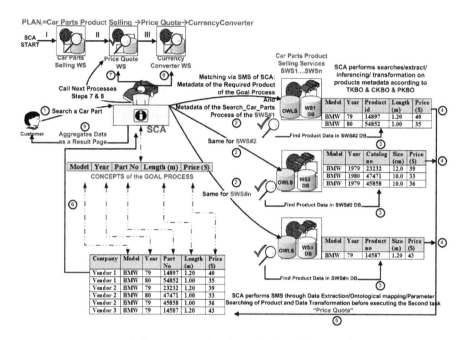

Fig. 1. A car parts broker (CPB) web service scenario

The customer starts by invoking the SCA system (Step 1 in Figure 1) to obtain a *Plan$_i$*. The service interprets the customer's query using upper specific domain ontology *(for instance Concept.owl or Vehicle.owl)* and then collects all the metadata of the required product information. Additionally, each company has its own OWL-S ontology that contains all the metadata of the processes for selling products in the form of I/O/P/E.

The concepts of service parameters that are under its OWL-S ontology might refer to a *concepts ontology knowledgebase (CKBO)* through OWL *parameter-Type*. For instance, a company might represent the *Product No* field as *PNo* or *Catalog No* in its own ontology. SMS of the SCA is responsible for making the appropriate conversions of transactional data or collecting all relational data *(Synonym, Is_a etc...)* between the metadata kept by the SCA *(such as an upper concepts ontology)* and the existing metadata in the OWL-S of the product service company (Step 2 to Step 5 in Figure 1).

The SCA collects all data from SMS then applies same steps for the next process and so on (Step 6 to Step 8). At the end of the execution of all processes in the plan (see **PLAN$_i$= Online Car Parts Product Selling→Price Quota→Currency Converter** in the Figure 1), SCA displays retrieved results to customer (Step 9). Next section presents the details of the proposed system architecture.

4 System Architecture

The SCA System has two main stages that are **Matching** and **Composition**. **Matching** stage, executing a **Semantic Matching Step (SMS)**, works independent of **Composition** stage. **Composition** stage contains five functional parts which are **translator, planner, inference engine, execution engine**, and **monitoring agent**. The system also has three ontology knowledge bases **(Tasks-TKBO[1], Concepts-CKBO[2] (or Domains-DKBO) and Processes-PKBO[3])**, and one **Rule knowledge base (RKB)** that contains the **Revised Armstrong's Axioms (RAAs)**. Parts of SCA are introduced below and depicted in Figure 2. The numbers in circles in Figure 2 signify the number of execution order and associations between parts of the SCA.

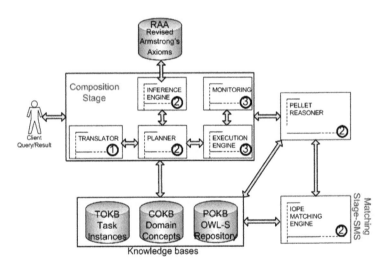

Fig. 2. The architecture of the SCA system

Initially, the **translator** performs a parsing task of knowledge bases namely, translating all atomic task instances in TOKB, all OWL-S atomic business processes of SWSs (candidates) in POKB and the goal process of client into

---

[1] An Online Book Selling domain example is in the form of TKBO,
http://cmpe.emu.edu.tr/ProcessKB/ontology/Tasks.owl
[2] An Online Book Selling domain example is in the form of TKBO,
http://cmpe.emu.edu.tr/ProcessKB/ontology/Concepts.owl
[3] The test collection (OWLS-TC3.zip) of the OWL-S atomic processes employed here can be found on SemWebCentral Website, http://www.semwebcentral.org.

I/O/P/E form. According to I/O/P/E modeling, a given set of atomic processes in the TOKB or POKB (e.g. A_i) are converted into the form $A_i \equiv I_i \rightarrow O_i$, where its inputs are $I_i \equiv I_{i1} \wedge I_{i2} \wedge \cdots \wedge I_{ik}$ and its outputs are $O_i \equiv O_{i1} \wedge O_{i2} \wedge \cdots \wedge O_{im}$. The logical expression of $A_i \equiv I_i \rightarrow O_i$ determines that O_i is obtainable only if I_i is available. If the linear implication ('$\rightarrow$') is applied, I_i is consumed and O_i is produced. Candidate SWSs and the goal process in I/O/P/E form are passed on to the planner. Next, the SCA initiates execution of the **planner, inference engine, SMS** and **Pellet OWL based reasoner** [23] jointly. Figure 3 presents the main steps of the SCA system.

In fact, the **planner** is located at the center of the SCA System. It tries to find execution sequence(s) of processes using the predefined task instances (in the TOKB) and candidate SWSs (in the POKB) while satisfying the requested complex goal process. In matching parameters, when the planner needs to find similar concepts, it calls on the SMS.

The **inference engine** performs inferencing on the task instances of TKBO using the *RAAs*. It picks one suitable rule from the RKB and then applies it to a pair of task instances. If it is able to produce (derive) a new complex process from those task instances, it sends it to the planner for checking its suitability. The inference engine and the planner are thus coordinated to work in alternating sequence in each iteration. In summary, there is a complex goal process $G \equiv I_G \rightarrow O_G$, with $I_G \equiv I_{G1} \wedge I_{G2} \wedge \cdots \wedge I_{Gj}$ inputs and $O_G \equiv O_{G1} \wedge O_{G2} \wedge \cdots \wedge O_{Gh}$ outputs. Given a path, P, planned by planner of SCA, the question is whether $P \models G$ or not. To resolve this question, we used a process derivation task that employs the RAAs. Inferencing using the RAAs is taken up in Section 6.

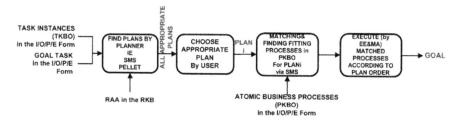

Fig. 3. The general steps of the SCA system

The **execution engine** executes the planned sequence of atomic business processes of candidate SWSs (in PKBO repository).

Finally, a **monitoring agent** monitors proper execution of the Web services processes. In the last two steps, the system initiates execution of all the components/processes in the defined plan path. After the execution, SCA serves the result to the client. The details of the semantic matching on workflows are given in the next section.

5 Semantic-Based Matching for a Composition Plan

Composition of Web services are created dynamically by using semantic descriptions of Web services to systematize them in a workflow. In most of the recent composition approaches [2 to 10 and 24], composable Web services are appended to the composition one by one. As each service is added to constructed composition workflow, a parameter matching mechanism is necessary to make sure that the service supports the I/O/P/E constraints of the workflow. A workflow can usually be described using formal or informal flow diagramming techniques, showing directed flows between processing steps (e.g. a process workflow in Figure 4).

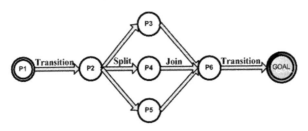

Fig. 4. A workflow scheme that is constructed from six atomic processes to attain a goal.

Individual processing steps of a workflow can basically be defined by three parameters:

- *Input: the information required to complete the step.*
- *Transformation or inferencing rules, algorithms etc...*
- *Output: the information produced by the step and provided as input to next step.*

A pair of processes can only be plugged together if the output of one is equal to the (full or partial) input requirements of the next (transitive), or, if the input of one is equal to the input requirements of the other (additive). Thus, the essential description of a process actually comprises only input and output that are described fully in terms of data types and their semantics. Planner and Inference Engine parts of the SCA System constructs a workflow for required composition. The produced workflow may contain many solution paths that satisfy the goal process. Correspondingly, the SMS step provides information about the composability by appling appropriate conversions or collecting all relational data (Synonym, Is_a etc...) between those processes. For instance, I/O concepts of a focused process (e.g. P2 in Figure 4) associate the concepts of all other possible processes (e.g. P3, P4 and P5 in Figure 4).

In addition to matching of input and output terms, preconditions and effects should also be criteria for process similarity matching for the described preconditions are essential at service provider's end and effects are needed at client's end. A precondition may be invoked and executed as a typical internal atomic process.

Depending on the outcome of a precondition process during run time, execution sequence of the composed processes in a workflow may be changed. Therefore, matching based on I/O terms alone is not adequate to find fitting processes at the current node of constructed workflow in order to attain required composition. SMS matching algorithm is based on matching of I/O terms; a similar technique can be used for preconditions and effects matching in the SMS [25 and 26]. Without the loss of generality, we will not talk about precondition and effect matching in the rest of this chapter.

Finally, process similarity assessment of SCA can be considered in three categories, such as, (1) similarity assessment of I/O concepts (namely, **rdf:ID** descriptions in the service OWL-S file), (2) of process:parameterType (***process:parameterType*** descriptions in the service OWL-S file and (3) of semantically described properties. Property-level similarity assessment is considered in SCA with two properties: **hasSynonym** and **hasIs_A**. The details of the *SMS* are given in the next section.

6 Semantic Matching Step (SMS)

Semantic matching is a well known algorithm proposed recently [27 and 28], and has been extended and cited extensively in recent proposals [29 to 33]. SMS performs the matching step on concepts of two focused processes according to their meaning, similarity, and distance of the concept relations. Consider that **Concept$_1$** is one of the concepts of a *queried (or currently focused) process* and **Concept$_2$** is *one of the concepts of another process* in a process set. Four degrees of matching are possible:

* *Exact: If Concept$_1$≡ Concept$_2$, then they are marked as **Exact** match.*
* *Plug in: If Concept$_1$⊂ Concept2, then service of the **Concept2** may fulfill the requirements of the service of the **Concept1** since service of the **Concept$_2$** serves some of the subclasses of the concept defined by service of the **Concept1**. Thus, it is a **Plugin** match.*
* *Subsumes: If Concept1 ⊃ Concept$_2$, then it is a **Subsume** match.*
* *Dissimilar: If Concept$_1$≠ Concept$_2$, then it is declared as **Dissimilar**.*

As we mentioned above, all OWL-S advertisements of the candidate SWSs (*referred to as **A** in the following*) are stored in the POKB. Then, SCA accesses POKB and also parses *process:parameterType* and *rdf:ID (I/O)* concepts of all available OWL-S advertisements. In addition, the classes, super/subclasses, and all the *is_a* and *synonym* properties corresponding to the parsed *rdf:ID (I/O)* concepts are extracted from its domain ontology during execution of the SMS. The input to the SMS is a process *currently on focus*, say *C*, in the constructed plan and the output is a set of matching *OWL-S Advertisements (of candidate SWSs)* sorted according to their similarity score. SMS focuses on these processes, which were symbolized as **C & A** in the following.

The SMS defines four different matching scores; **Dissimilar=0, Subsume=0.5, Plugin=0.75**, and **Exact=1** among two given concepts. Additionally, we describe the SMS through three sub algorithms in the following (Algorithm 1, Algorithm 2 and Algorithm 3), but first definitions of 'has synonym' (*hasSyn*) and *hasIs_a* are given. *HasSyn (Y) =Z: There exists a synonym concept of Y which is Z; such as Car's synonym is Automobile or Motorcar. HasIs_a (Y): There exists an is_a concept of Y; such as 3WheeledCar is a Car.* The two input concepts, one from C_I and one from A_I, are presented in Algorithm 1 that are being tried for matching.

Algorithm 1. *degreeOfProcessMatching(Concept C_i, Concept A_i) for calculating* $SMS_{Score(Ci,Aj)}$

1 if $((Ci{\equiv}Ai)$ **or** $(hasSyn(Ci){\equiv}Ai)$ **or** $(Ci{\equiv}hasSyn(Ai)))$ **then** return rel=**EXACT**;

2 if $((Ci{\subset}Ai)$ **or** $(hasSyn(Ci){\subset}Ai)$ **or** $(Ci{\subset}hasSyn(Ai)$) **or** $(hasIs\_a(Ci){\equiv}Ai))$ **then** return rel=**PLUGIN**;

3 if $((Ci{\supset}Ai)$ **or** $(hasSyn(Ci){\supset}Ai)$ **or** $(Ci{\supset}hasSyn(Ai))$ **or** $(Ci{\equiv}hasIs\_a(Ai)))$ **then** return rel=**SUBSUME**;

4 if $((Ci{\neq}Ai)$ **or** $(hasSyn(Ci){\neq}Ai)$ **or** $(Ci{\neq}hasSyn(Ai)))$ **then** return rel=**DISSIMILAR**;

During the matching step, *process:parameterType* property or *rdf:ID (input-output)* concepts of the processes are analyzed recovering their semantics, namely, meaning, similarities, differences, and relations (*see* Algorithm 1). Semantic distance of concepts which offer similarity information among concepts can be given by ontology developer during its development phase. If semantic distance is not scored by its developer, all direct subconcepts of a parent concept will have the same distance weight [30] according to Eq (1),

$$d_{weight_A} = \frac{1}{\#ofSubconceptsofA} \tag{1}$$

SCA finds the semantic distance weight $d_{weight(A,Z)}$ between any two concepts A and Z in particular domain ontology as in Eq (2),

$$d_{weight(A,Z)} = d_{weight(A,B)} * d_{weight(B,C)} * \dots * d_{weight(W,Y)} * d_{weight(Y,Z)} \tag{2}$$

The applied scoring method is a simple multiplicative weighting function. Given a queried process, C, with input concepts $C_I = \{C_{I1}, C_{I2} \dots C_{Im}\}$ and a service A with the input concepts $A_I = \{A_{I1}, A_{I2} \dots A_{Im}\}$ are matched and the total similarity score ($Similarity_{Score(C_I, A_I)}$) is calculated according to Eq (3),

$$Similarity_{Score(C_I, A_I)} \equiv$$

$$\frac{\sum_{j=1}^{max(n_{C_I}, n_{A_I})} \left[max_{j,i} \left\{ \prod_{i=1}^{max(n_{C_I}, n_{A_I})} \frac{SMS_{Score(Ci,Aj)} * d_{weight(Ci,Aj)}}{distance(C_i, A_j)} \right\}, delete\ C_i\ \&\ A_j \right]}{\frac{n_{C_I}}{n_{A_I}}} \tag{3}$$

$$n_{C_I} = size(C_I);\ and\ n_{A_I} = size(A_I);$$

The above equation shows only *input lists* $\{C_I, A_I\}$ or $\{Cin, Ain\}$ of the *C* and *A* processes. If sizes of the C_I and A_I are not equal then we put that much $|size(C_1)-size(A_1)|$ number of <u>null</u> values to the list for missing ones.

The *distance* (C_i, A_j) shows the number of concepts (levels) between the focused concepts; C_i and A_j in the ontology. Assuming that the *Vehicle* concept has only three sub-concepts, *Car, Ship* and *Bicycle*, additionally, if the concept of $C_i=Car$ and the concept of $A_j= Vehicle$ then the $\dfrac{SMS_{Score(C_i,A_j)}*d_{weight(C_i,A_j)}}{distance(C_i,A_j)}$ of the *(Car, Vehicle)* match will be (0.75 * 0.33)/1 = 0.2425. Because *(Car, Vehicle)* has PLUGIN relation so their subsumption score is 0.75 according to Algorithm 1. Their weight is *0.33* since Eq. (1). Also, there exists a subclass relation between *(Car, Vehicle)* thus gives their *distance (Car, Vehicle)* is 1, which is the number of levels between these concepts in the domain ontology.

As Algorithm 2 shows, *search(C)* takes the queried process (C) and iterates over every *(A)* in POKB repository in order to determine a match. SMS calculates two types of similarity scores, namely, **AdditiveMatchScore** and **TransitiveMatchScore**.

AdditiveMatchScore (Cin, Ain) is a similarity score on *C* & *A* processes that is obtained by matching only input concepts of both processes.

TransitiveMatchScore (Cout, Ain) is a similarity score on *C* & *A* processes obtained by matching on output concepts of queried process and input concepts of available candidate processes.

Algorithm 2. search(C)
1 *ResultList=null;*
2 **for** each **A** in **POKB do**
3 *AdditiveMatchScore=**getRelation(Cin,Ain);***
4 *TransitiveMatchScore=**getRelation(Cout,Ain);***
5 **if** (*AdditiveMatchScore* =**0 and** *TransitiveMatchScore* =**0**) **then goto** step2 and **take** next *A*
6 **else** *ResultList* ←*(A, TransitiveMatchScore, AdditiveMatchScore);*
7 **end if**
8 **end for**
9 *return (ResultList)*

As *Algorithm 3* shows, **getRelation** takes two lists of either Input or Output concepts of *C* and *A* for matching by SCA. The first parameter of **getRelation** is a list of *input(or output)* concepts of queried process $C_{I/O}=\{C_{I/O1},C_{I/O2},...,C_{I/Om}\}$ and the second parameter of **getRelation** is another list of only *input* concepts of a candidate service $A_{I/O}=\{A_{I/O1},A_{I/O2},...,A_{I/On}\}$. For instance, the input sets of $Cin=\{C_1,C_2,C_3\}$ and $Ain= \{A_1,A_2,A_3\}$ are executed by the *getRelation (CList, AList)* of SMS.

$$Score = (SMS_{Score} * d_{weight(C,A)}) / dis\tan ce(Ci, Aj)$$

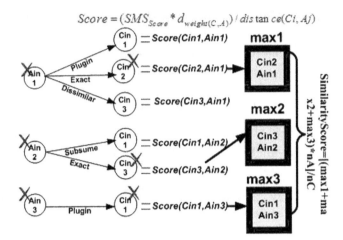

Fig. 5. A simple example for Algorithm 3.

Figure 5 shows working mechanism of the *getRelation* function through an example. In this example, the *AdditiveMatchScore (Cin, Ain)* is calculated but *TransitiveMatchScore (Cout, Ain)* is calculated similarly. The solid lines indicate the relationships inferred by the reasoner. The algorithm will first attempt to compute a max match for Ain_1. Assume that the following matches are inferred and values of the $SMS_{Score(Ci,Aj)} * d_{weight(Ci,Aj)}$ are calculated then divided by **distance** **(Ci, Aj)** value for each parameters of *Cin*:

> *[Ain₁ Subsumes of Cin₁→Plugin],*
> *[Ain₁ hasSynonym of Cin₂→Exact]*, and
> *[Ain₁ not matched with Cin₃→Dissimilar].*

Ain_1 has a max match with Cin_2. So, Cin_2 and Ain_1 are removed from their lists.

The algorithm now attempts to match the next concept: Ain_2. Assume that the following matches are inferred:

> *[Cin₁ Subsumes of Ain₂→Subsume]* and
> *[Ain₂ hasSynonym of Cin₃→Exact].*

Thus, Ain_2 is matched with Cin_3 and Cin_3 and Ain_2 are removed from their lists. The algorithm now attempts a match for Ain_3. Assume that the following match is inferred for the last parameter in *Cin*:

> *[Ain3 Subsumes of Cin₁ →Plugin].*

Algorithm 3. *getRelation (CList, AList)*
1 *List3=null; res_score=0; n_C=size(CList); n_A=size(AList);*
2 **for** *j=1 to n_A do -- for each concept Aj in AList*
3   **for** *i=1 to n_C do -- for each concept Ci in CList*
4   *score=0;*
5 *degree=degreeOfProcessMatching (Ci, Aj)*
6 **if** (*degree≠*DISSIMILAR) **then**
7   **if**(*degree=*EXACT) **then**
         score=1.00*dweight (Ci, Aj)/distance(Ci, Aj) ; **end if**
8   **if** (*degree=*PLUGIN) **then**
         score=0.75*dweight (Ci, Aj)/distance(Ci, Aj) ; **end if**
9   **if**(*degree=*SUBSUME) **then**
         score=0.5*dweight (Ci, Aj)/distance(Ci, Aj); **end if**
10 **else** *score=0;*
11 **end if**
12   *List3 ← score*
13   **end for**
14   *score← Max(List3);*
15 **if** (*score ≠ 0*) **then**
16   Find **int k**←**index(*Max(List3)*)** ;
17   *DeleteRow(C_k); DeleteRow(Aj);* -- delete concepts from CList&AList
18   *n_C←n_C-1; n_A←n_A-1;* -- decrease sizes of the AList and CList since deleted concepts
19   *i←1; j←1;* -- start the **for loops** again for the next concept
20   Calculate *res_score= res_score+score;*
21   *List3←* null; **goto step2**, take next *Aj*
22 **else**
23   Calculate *res_score= res_score+0;* --**goto step2**, for next *Aj*
24   **end if**
25 **end for**
26 **if** (*res_score= 0*) **then** no *matching* exists; **return 0;**
27   **else return**{*res_score= res_score*(size(CList)/ size(AList))*};
28 **end if**

This algorithm calculates the max values in each iteration and then finds an average $Similarity_{Score(C_I,A_I)}$ for the given *(Cin, Ain)* query. The same equations and operations are also applied to find the $Similarity_{Score(C_O,A_I)}$ of the *TransitiveMatchScore (Cout, Ain)* of C and A processes. Pseudo representation of the *getRelation* function is given above.

The algorithm finds a set of matching advertisements with their degree of match, and then returns a resulted list of processes to planner and inference engine to derive new composed processes by picking a suitable rule from the RAAs and applying it to the C and A processes. If *TransitiveMatchScore* is higher than *AdditiveMatchScore* then one of the transitive based inference rules of *RAAs* is selected to compose those processes such as *pseudo factorization, transitivity, pseudo transitivity* and *accumulation*. On the other hand, if *AdditiveMatchScore* is higher than *TransitiveMatchScore* then the *additivity* rule of RAAs is applied on those processes.

If both scores are zero then there is no suitable process that can able to append to queried process. In this state, *dissection* rule is applied to available processes that have one more than output concepts to configure suitable processes for the task.

7 Revised Armstrong's Axioms (RAAs)

Armstrong's Inference Rules (also referred to as Armstrong's Axioms-AA) are a set of axioms used to infer functional dependencies (FDs) on a relational database [10]. There are seven axioms, namely, **Reflexivity**, **Augmentation**, **Transitivity**, **Pseudo Transitivity**, **Additivity**, **Accumulation**, and **Projectivity** as shown in Table 1 (col. 1).

Table 1. Armstrong's Axioms and Revised Armstrong's Axioms.

Armstrong's Axioms	Revised Armstrong's Axioms
1. Reflexivity: A set of attributes X determines a subset Y of itself: X→Y if Y⊆X.	**1. Pseudo factorization:** If {(X∥Y→Z) ,(T∥Z→W)} and if (Z⊂T∥Z and T≡X or T≡Y) then X∥Y→W.
2. Augmentation: It allows enlarging the left-side of a FD or both side conventionally with one or more attributes. Formally, if X→Y then X∥Z→Y∥Z for any Z.	
3. Transitivity: If we have functions f : X→Y and g : Y→Z then we have a function g O f : X→Z, where (g O f)(x) = g(f (x)). Formally, If {(X → Y) and (Y→Z)},then X→Z.	**2. Transitivity:** If {(X →Y) and (Y→ Z)} then (X→Z).
4. Pseudo transitivity: is a generalization of Transitivity. It is requires the entire right hand side of a FD appears as attribute(s) of the determinant of another FD. Formally, if {(X→Y) and (Y∥Z→W)} then (X∥Z→W).	**3. Pseudo transitivity:** If {(X→Y) and (Y∥Z→W)} then (X∥Z→W).
5. Additivity: If there are two FDs with the same determinant on the left, it is possible to form a new FD that preserves the determinant and has as its right-hand side the union of the right-hand sides of the two FDs. Formally, if {(X→Y) and (X→Z)} then (X→Y∥Z).	**4. Additivity:** If {(X→Y) and (X→Z)} then (X→Y∥Z).
6. Accumulation: If there are two FDs with complementary determinant, it is possible to form a new FD that preserves the determinant X and forms its right-hand side as the union of the right-hand sides of the two FDs except the complementary determinant Z. Formally, if {(X→Y∥Z) and (Z→C∥W)} then (X→Y∥C∥W).	**5. Accumulation:** If {(X→Y∥Z) and (Z→T∥W)} then (X→Y∥T∥W).
7. Projectivity: If X determines Y and Z, then X determines Y, therefore it is possible to break each functional dependency X→Y down to X→A$_i$ for i = 1..n where Y = {A$_1$, . .A$_n$}. Formally, if (X→Y∥Z) then X→Y and X→Z.	**6. Dissection:** If (X→Y∥Z) then X→Y and X→Z.

Reflexivity and Augmentation axioms applied in sequence performs like factorization; as such, we combined the two in order to form a shorthand *Pseudo Factorization* axiom. This in turn eliminates unsuitable derivations and reduces the complexity of composition. We renamed the *Projectivity* axiom as *Dissection* for the latter better expresses the intended operation. *RAAs* were obtained by these equivalent transformations which do not alter correctness and completeness of AA. Table 1 displays AAs in the first column, corresponding RAAs in the second column.

Let us assume that all of X, Y, Z, W, and T are classes/concepts of I/O parameters of an available process set and that the right arrow is the linear implication operator. Using example $X\|Y \rightarrow Z$ in the table below, we can understand that the concept X and the concept Y are taken as concurrent inputs (that is, $X\|Y$) while execution of the process, thus their order is not important. The outcome is the concept Z after execution.

Table 2. Nine atomic processes from the OWLS-TC 3.0 with I/O/P/E parameters [34].

No	SERVICE NAME	INPUTS	OUTPUTS	DESCRIPTION
S1	Cheap_Car_Price	CheapCar<#CheapCar>	Price <#Price>	This service gives good opinion to search a cheap car.
S2	Currency_Convertor	InputAmount<#Price>, SourceCurrency <#Currency>, DestinationCurrency <#Currency>	OutputA-mount<#Price>	Converts any two currencies.
S3	CarRecommended-PriceInEuro	Car<#Car>	RecommendedPriceI-nEu-ro<#RecommendedPri ceInEuro>	Car Recommend-ed price service in Euro.
S4	CarInStock	CarModel<#Car>	ExistsInStock <#true-false>	Check the car model in stock or not.
S5	Car_Price	Car<#Car>	PriceOfCar<#Price>	Car price service.
S6	Amount-Of-Mon-ey3WheeledCar_Price	3WheeledCar <#ThreeWheeledCar>, Amount-Of-Money <#Amount-Of-Money >	Price <#Price>	This service tries to find price of a three wheeled car within given amount of money.
S7	EuroToDollarCurren-cyConverter	PriceInEuro <#RecommendedPriceI-nEuro>	PriceInDollar <#RecommendedPri-ceInDollar>	Converts Euro to Dollar.
S8	Car2PersonBicyclePri ce	2PersonBicycle <#TwoPersonBicycle>, Car<#Car>	Price <#Price>	This service re-turns price of the pair of a car and a two person bi-cycle.
S9	Car_Technology	Car<#Car>	Technology <#Technology>	Car Technology service.

A test collection (*OWLS-TC3.zip*) [34] of the OWL-S atomic processes employed for the given scenario that can be found on *SemWebCentral* Website[4]. The test collection is selected since latest version of OWLS-TC and the average quality of the descriptions is somewhat better than other test collections.

The most commonly used OWL-S Service Retrieval Test Collection version 3.0 consists of 1007 services from the following seven domains: *education, medical care, food, travel, communication, economy* and *weapon*. Table 2 presents OWL-S atomic processes for only six SWSs in this TC. Their parsed I/O and parameter types *(as an ontological reference in <#ParameterType>)* are specified in the Table 2. Additionally, Table 2 contains three supplementary atomic processes *(not in the OWLS-TC 3.0)* that are needed for the scope of next sections. Additionally, Table 2 contains two similar atomic processes, Currency_Convertor and EuroToDollarCurrencyConverter, which have same purpose but different number of concepts. Therefore, it is possible to show that the system's matching mechanism is able to differentiate and understand correlations among on focus candidate processes during the constitution of a composition plan.

A possible scenario is that a user wants to find the price of a cheap car with its technology information, however the user expects the returned price in a specific currency, say, US Dollars. Assume that the goal task called G is required by the client, **G≡Car [Car]→Technology [Technology] ‖ PriceOfCar [Price].**

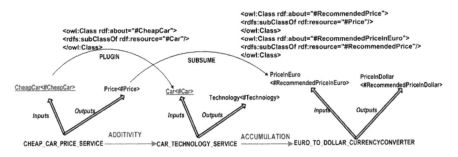

Fig. 6. A simple example of RAA.

Given above is the trace of Table 2 where Planner defines many solutions; one of them for the given G process above by synchronizing the SWSs of *Cheap_Car_Price.owl, Car_Technology.owl,* and *EuroToDollarCurrencyConverter.owl* in that order. The symbol 'O' is used to show synchronization operation when the output of a service is consumed by another service as an input, when the input of a service is the same as another service. Consequently, we need to line up suitable processes in an appropriate order to achieve the client's complex goal. If we apply an *Additivity* then an *Accumulation* rule (Figure 6), the possible chain is:

Accumulation [Additivity [Cheap_Car_Price O Car_Technology] O EuroToDollarCurrency-Convertor]].

---

[4] http://www.semwebcentral.org

Before applying the *Additivity* rule on *Service 1* and *Service 9*, SMS found a *PLUGIN* relation on the input concepts of those processes. Besides, before applying the *Accumulation* rule on the newly derived process from *(Service 1 & Service 9)* and with *Service 7*, SMS found a *SUBSUME* relation on the output concepts of the newly derived process and the inputs of the *Service 7*. The final expected new process *(named as P)* after tracing Planner and Inference Engine actions is as follows:

P≡{[CheapCar<#CheapCar>]→Technology<#Technology>]‖[PriceInDollar<#RecommendedPri

ceInDollar>]}, which gives P ⊨ G.

Service1≡Cheap_Car_Price≡[CheapCar<#CheapCar>]→ [Price <#Price>]
Service9≡Car_Technology≡[Car<#Car>]→ [Technology<#Technology>]
Service7≡EuroToDollarCurrencyConvertor≡[PriceInEuro<#RecommendedPriceInEuro>]→ [PriceInDol-
lar<#RecommendedPriceInDollar>]

ProducedProcess(P)≡[[Cheap_Car_Price.owlOCar_Technology.owl]OEuroToDollarCurrencyConvertor.owl]]

≡{[CheapCar<#CheapCar>]→ [Price <#Price>]} O {[Car<#Car>]→ [Technology<#Technology>]}
≡{[CheapCar<#CheapCar>]→ [Technology<#Technology>] ‖ [Price <#Price>]}
≡{[CheapCar<#CheapCar>]→ [Technology<#Technology>] ‖ [Price<#Price>]}
OEuroToDollarCurrencyConvertor.owl
≡{[CheapCar<#CheapCar>]→[Technology<#Technology>] ‖ [Price
<#Price>]}O{[PriceInEuro<#RecommendedPriceInEuro>] → [PriceInDollar<#RecommendedPriceInDollar>]}
≡{[CheapCar<#CheapCar>]→ [Technology<#Technology>] ‖ [PriceInDollar<#RecommendedPriceInDollar>]}

8 Inferencing in SCA: A Case Study

Details of inferencing in the SCA System is presented in this section. Semantic task contexts are supported by the OWL formal semantics; therefore, reasoning can be performed based on the task description context. In order to do this, we employ the task ontology *(Tasks.owl-TOKB)* for inferencing purpose.

OWL formal semantics is used to define the RAA as object property characteristics that will be employed during inference and planning stages: *hasTransitivity, hasPseudoFactorization, hasPseudoTransitivity, hasAdditivity, has Accumulation* and *hasDissection* are defined. These RAA rules are defined in the *Tasks ontology (TOKB)* as seen on the lines (1, 6, 11, 16, 21, and 26) in Table 3.

OWL formal semantics provide many property characteristics such as *transitive, symmetric, functional, inverseOf, inversefunctional* etc. For instance, if a property is specified as transitive then for any X, Y, and Z: $P(X,Y)$ **and** $P(Y,Z)$ **implies** $P(X,Z)$. Kang *et al.* [35] proposed that the properties are able to directly apply definitions to new properties in order to provide order among some semantic business processes for monitoring purpose. However in this study, SCA uses the RAAs as a kind of OWL properties. The transitive-based properties defined above (as shown at lines 2, 7, 12, and 22) are suitable to reason the order of several tasks while planning. The object property *hasTransitivity* is specified as transitive (line 1 and 2 in Table 3). Additionally, union-based properties (as shown at line 17) could be also suitable for additive-based operations among a set of tasks.

Table 3. Revised Armstrong's Axioms in the Tasks.owl (TOKB)

```
<!-- Object Properties of Tasks.Owl-->

1  <owl:ObjectProperty rdf:ID="hasTransitivity">    16 <owl:ObjectProperty rdf:ID="hasAdditivity">
2     <rdf:type                                     17    <rdf:type
rdf:resource="&owl;TransitiveProperty"/>            rdf:resource="&owl;UnionProperty"/>
3     <rdfs:domain rdf:resource="#Tasks"/>          18    <rdfs:domain rdf:resource="#Tasks"/>
4     <rdfs:range rdf:resource="#Tasks"/>           19    <rdfs:range rdf:resource="#Tasks"/>
5  </owl:ObjectProperty>                            20 </owl:ObjectProperty>

6  <owl:ObjectProperty                              21 <owl:ObjectProperty
rdf:ID="hasPseudoFactorization">                    rdf:ID="hasAccumulation">
7     <rdf:type                                     22    <rdf:type
rdf:resource="&owl;TransitiveProperty"/>            rdf:resource="&owl;TransitiveProperty"/>
8     <rdfs:domain rdf:resource="#Tasks"/>          23    <rdfs:domain rdf:resource="#Tasks"/>
9     <rdfs:range rdf:resource="#Tasks"/>           24    <rdfs:range rdf:resource="#Tasks"/>
10 </owl:ObjectProperty>                            25 </owl:ObjectProperty>

11 <owl:ObjectProperty                              26 <owl:ObjectProperty
rdf:ID="hasPseudoTransitivity">                     rdf:ID="hasDissection">
12    <rdf:type                                     27    <rdfs:domain rdf:resource="#Tasks"/>
rdf:resource="&owl;TransitiveProperty"/>            28    <rdfs:range rdf:resource="#Tasks"/>
13    <rdfs:domain rdf:resource="#Tasks" />         29 </owl:ObjectProperty>
14    <rdfs:range rdf:resource="#Tasks" />
15 </owl:ObjectProperty>
```

To better understand the depth of inferencing in the SCA System, let's consider another similar example of Online Car Store domain. A user wants to find price information in a desired currency of a specific car model belonging to a specific car company. Firstly, SCA searches such a task that satisfies the client's complex task. If it could not find one task that satisfies the goal task then searches a related set of tasks that contain client I/O concepts. Assume that there are four tasks found similar in TOKB: *FindCarMoldelsOfACompany, Car_Price, Currency_Converter* and *Car_Technology* tasks are given in Table 4.

In addition, if the *FindCarMoldelsOfACompany* is transitive with *Car_Price* then *FindCarMoldelsOfACompany* has to be performed before the *Car_Price* since the execution order is critical for all *Transitive* based rules to satisfying required goal process. This state is described by using semantic task contexts in the Table 4 (at line 5) which gives us an opportunity to derive a new process named *Company_Car_Price* from these two processes.

Table 4. Description of four tasks of the Online Car Store Domain in the Tasks.owl (TOKB)

```
1    <owl:Class rdf:ID="FindCarMoldelsOfACompany">
2      <rdfs:subClassOf rdf:resource="#Information_Service"/>
3        <Input_Parameter rdf:datatype="&Concept;Company">Car_Company</Input_Parameter>
4        <Output_Parameter rdf:datatype="&Vehicle;Car">Car_Models</Output_Parameter>
5        <hasTransitivity rdf:resource="&Tasks;Car_Price"/>
6    </owl:Class>

7    <owl:Class rdf:ID="Car_Price">
8      <rdfs:subClassOf rdf:resource="#Information_Service"/>
9        <Input_Parameter rdf:datatype="&Vehicle;Car">Car</Input_Parameter>
10       <Output_Parameter rdf:datatype="&Concept;Store">Car_Store</Output_Parameter>
11       <Output_Parameter rdf:datatype="&Concept;Price">Price</Output_Parameter>
12       <Output_Parameter rdf:datatype="&Concept;Currency">Result_Currency</Output_Parameter>
13       < hasPseudoTransitivity rdf:resource="&Tasks;Currency_Converter"/>
14       < hasAdditivity rdf:resource="&Tasks;Car_Technology"/>
15   </owl:Class>

16    <owl:Class rdf:ID="Currency_Converter">
17       <rdfs:subClassOf rdf:resource="#Information_Service" />
18         <Input_Parameter rdf:datatype="&Concept;Price">Input_Price</Input_Parameter>
19         <Input_Parameter rdf:datatype="&Concept;Currency">Source_Currency</Input_Parameter>
20         < Input_Parameter
rdf:datatype="&Concept;Currency">Destination_Currency</Output_Parameter>
21         < Output_Parameter rdf:datatype="&Concept;Price">Output_Price</Output_Parameter>
22    </owl:Class>
23    <owl:Class rdf:ID="Car_Technology">
24       <rdfs:subClassOf rdf:resource="#Information_Service"/>
25         <Input_Parameter rdf:datatype="&Vehicle;Car">Car</Input_Parameter>
26         <Output_Parameter rdf:datatype="&Concept;Technology">Technology</Output_Parameter>
27         <hasAdditivity rdf:resource="&Tasks;Car_Price"/>
28    </owl:Class>
```

Based on this transitive characteristic of the *hasTransitivity*, the produced new process (*Company_Car_Price*) is possible to be performed before any suitable process to continue the possible chain. Some possible reasoning operations are formalized as follows:

TRANSITIVITY1[FindCarMoldelsOfACompany(Company→Car)]O[Car_Price
(Car→Store‖Price‖Currency)]implies[Company_Car_Price(Company→Store‖Price‖Currency)]   (8.1)

PSEUDOTRANSITIVITY1[Company_Car_Price(Company→Store‖Price‖Currency)]O[Currency_
Converter (Currency‖Price‖Currency→Price)] implies [Compa-
ny_Car_Price_Currency_Converter(Company ‖ Currency → Store ‖ Price)]   (8.2)

PRODUCED NEW PROCESS (RESULTNODE): Company_Car_Price_Currency_Converter (Com-
pany‖Currency→Store ‖ Price)   (8.3)

SCA uses these task instances to find equivalent processes in the candidate SWS then creates a complex process incorporating them. The resulted composite process is implemented by using OWL-S models that are *profile, process* and *grounding*. The profile model of the resulted composition process (see above the formulation 8.3) is given below in Table 5. At the end of the composition, the resultant OWL-S based composite process is appended to TOKB as a new task instance in the form of the task ontology. This practice decreases complexity and increases speed while searching suitable processes for a constituted workflow by planner later. In Table 5, two inputs (lines at 2- 7 and 8-13) and two outputs (lines at 14-19 and 20-25) are defined with their parameter types which are some URLs pointing to the related concepts in the *Concept.owl (COKB)*.

Table 5. Profile model of the *Company_Car_Price_Currency_Converter* goal process.

```
1 < tasks:Company_Car_Price_Currency_Converter    14  <profile:hasOutput>
rdf:about="Company_Car_Price_Currency_Converter_Profi15    <process:Output rdf:ID="Store">
le">                                               16      <rdfs:label>Car_Store</rdfs:label>
2   <profile:hasInput>                            17      <process:parameterType
3    <process:Input rdf:ID="Company">             rdf:datatype="&Concept;Store/>
4      <rdfs:label>Car_Company</rdfs:label>       18    </process:Output>
5        <process:parameterType                   19  </profile:hasOutput>
rdf:datatype="&Concept;Company/>
6    </process:Input>                              20  <profile:hasOutput>
7   </profile:hasInput>                           21    <process:Output rdf:ID="Price">
                                                   22      <rdfs:label> Output_Price </rdfs:label>
8   <profile:hasInput>                            23      <process:parameterType
9    <process:Input rdf:ID="Currency">            rdf:datatype="&Concept;Price/>
10     <rdfs:label>Destination_Currency</rdfs:label>  24  </process:Output>
11       <process:parameterType                   25  </profile:hasOutput>
rdf:datatype="&Concept;Currency/>                 26</tasks:Company_Car_Price_Currency_Convert
12    </process:Input>                            er-Task>
13   </profile:hasInput>
```

In Table 6, a part of process model of the resultant composite process by SCA is given. The model contains semantic contexts of the knot points in the returned chain which are TRANSITIVITY1 (see Formula 8.1), *PSEUDOTRANSITIVITY1* (see Formula 8.2), and *RESULTNODE* (see Formula 8.3). The order is described as *TRANSITIVITY1*, *PSEUDOTRANSITIVITY1* and *RESULTNODE* which is given in the lines at 1-21 through *<process:ControlConstructList>*, *<list:first>*, *<list:rest>* and so on. Three different tasks are used for the composition task: *FindCarMoldelsOfACompany, Car_Price*, and *Currency_Converter*.

Table 6. Process model of the returned composite process *(Company_Car_Price_Currency_ Converter)*

```
1 <process:CompositeProcess                       31 <process:Process rdf:ruleID="
rdf:about="#Company_Car_Price_Currency_Con        PSEUDOTRANSITIVITY1">
verter-Process">                                  32   <process:process rdf:resource="
2    <process:ControlConstructList>               ../services/Car_Price.owl#Car_Price-Process "/>
3      <list:first>                               33   <process:hasDataFrom>
4        <process:Perform                         34     <process:InputBinding>
rdf:nodeID="TRANSITIVITY1"/>                       35       <process:toParam rdf:resource="
5      </list:first>                              ../services/Car_Price.owl#Car"/>
6      <list:rest>                                36       <process:valueSource>
7        <process:ControlConstructList>           37         <process:ValueOf>
8          <list:first>                           38           <process:theVar rdf:resource="
9            <process:Perform rdf:nodeID="        ../services/FindCarMoldelsOfACompany.owl#Car"/>
```

Table 6. (*continued*)

```
PSEUDOTRANSITIVITY1"/>
10        </list:first>
11        <list:rest>
12          <process:ControlConstructList>
13            <list:rest
rdf:resource="&list;#nil"/>
14            <list:first>
15              <process:Perform rdf:nodeID="
RESULTNODE "/>
16            </list:first>
17          </process:ControlConstructList>
18        </list:rest>
19        </process:ControlConstructList>
20      </list:rest>
21    </process:ControlConstructList>...
```

```
39        <process:fromProcess>
40          <process:Process rdf:ruleID="
TRANSITIVITY1"/>
41        </process:fromProcess>
42      </process:ValueOf>
43      </process:valueSource>
44    </process:InputBinding>
45  </process:hasDataFrom>
46 </process:Process>
```

```
22 <process:Process
rdf:ruleID="TRANSITIVITY1">
23  <process:process rdf:resource="
../services/ FindCarMoldelsOfACompany.owl#
FindCarMoldelsOfACompany -Process"/>
24  <process:hasDataFrom>
25    <process:InputBinding>
26      <process:toParam
rdf:resource="...../services/FindCarMoldelsOfAC
ompany.owl#Company"/>
27        ......
28    </process:InputBinding>
29  </process:hasDataFrom>
30 </process:Process>
```

```
47 <process:Process rdf:ruleID="RESULTNODE ">
48  <process:process
rdf:resource="..services/Currency_Converter.owl#
Currency_ConverterProcess"/>
49  <process:hasDataFrom>
50    <process:InputBinding>
51      <process:valueSource>
52        <process:ValueOf>
53          <process:fromProcess
rdf:ruleID="PSEUDOTRANSITIVITY1"/>
54          <process:theVar
rdf:resource="../services/Car_Price.owl#Price"/>
55        </process:ValueOf>
56      </process:valueSource>
57      <process:toParam
rdf:resource="../Currency_Converter.owl#Input_Price
"/>
58    </process:InputBinding>

59    <process:InputBinding>
60      <process:valueSource>
61        <process:ValueOf>
62          <process:fromProcess
rdf:ruleID="PSEUDOTRANSITIVITY1"/>
63          <process:theVar rdf:resource="
../services/Car_Price.owl#Result_Currency "/>
64        </process:ValueOf>
65      </process:valueSource>
66      <process:toParam
rdf:resource="./Currency_Converter.owl#Source_
Currency"/>
67    </process:InputBinding> ...
```

```
68 <grounding:WsdlGrounding rdf:about="#AnotherCar-ProcessGrounding">
69  <grounding:hasAtomicProcessGrounding rdf:resource="http://cmpe.emu.edu.tr/ProcessKB/services/
Currency_Converter.owl#Currency_ConverterGrounding"/>
70  <grounding:hasAtomicProcessGrounding rdf:resource="http://cmpe.emu.edu.tr/ProcessKB/services/
Car_Price.owl# Car_PriceGrounding"/>
71  <grounding:hasAtomicProcessGrounding
rdf:resource="..../services/FindCarMoldelsOfACompany.owl#FindCarMoldelsOfACompany"/>
```

The common parameter of the *FindCarMoldelsOfACompany* and *Car_Price* is 'Car'. 'Car' is produced by *FindCarMoldelsOfACompany* and consumed by *Car_Price* that is shown between the lines (31-46) in Table 6. The common parameters of the newly derived process *Company_Car_Price* and *Currency_Converter* are Price & Input_Price (50-58) and Result_Currency & SourceCurreny (59-67). These produced by *Car_Price* task and consumed by *Currency_Converter* task which is shown between the lines (47-67) in Table 6. Finally, during the *Execution* and *Monitoring* stage, SCA uses grounding information (keeps the URLs of services) of each of these atomic processes to execute. The semantic context of the grounding information is given in the lines (68-71) in Table 6.

9 Conclusion

Lack of semantic annotation parts, increasing number of Web Services on the Web, and syntactic-based search operations for current Web Services makes discovery and composition of appropriate Web Services challenging. This chapter presented an *Inference-based Semantic Composition Agent (SCA)* framework with its parts and functions that perform automatic process composition of Semantic Web Services. Starting with a scheme to parse the processes of SWSs specifications to *I/O/P/E* form, the set of *Armstrong's Axioms* were modified and applied to build a semantically-enriched planner–inference engine cycle. Besides this, the SCA framework uses its *Semantic Matching Step (SMS)* that helps to find the suitable processes for a required composition through semantic matchmaking. Utilizing *Revised Armstrong's Axioms (RAAs)* in inferring functional dependencies, SCA composes available OWL-S atomic processes. SCA System produces atomic process sequences as a workflow to achieve the required composition plan in order to satisfy user requirement as a complex task. The novelty of the SCA System is that for the first time Armstrong's Axioms are revised and used for semantic-based planning and inferencing of services.

In further study of the current SCA, optimizing the use of *RAAs* towards pure algebraic manipulation of the client's goal process and SWS description is contemplated. Incorporating more data type properties *(such as synonym, is_a, antonym, acts_on, etc.)* into the scheme to extend its capability to match even more qualified and suitable processes of services is also envisaged. Additionally, the current SCA can compose only atomic type of processes that are described in the OWL-S files of SWSs. In fact, composite type of processes is a collection of several atomic type of processes and we assumed that all available composite processes were already separated into numerous atomic processes before initiating composition. Therefore, it is possible to extend current SCA that will consider including suitable composite processes to a generated plan as a future work.

Finally, current SCA uses W3C OWL Web Ontology Language [21]. However, OWL has just been extended to OWL 2 [36]. OWL 2 contains a small but useful set of features that have been requested by users, for which effective reasoning algorithms are now available, and those OWL tool developers are willing to support. OWL 2 adds several new constructs to extend the expressivity of OWL including those for qualified cardinality restrictions, role chains, and expressive data predicates. Another future step of SCA will be modification and re-creation of the used ontology KBs (such as TKBO, DKBO and CKBO) through OWL 2 instead of the OWL. This is expected to help inferencing mechanism on SCA during the constitution of a required plan.

References

1. McIlraith, A., Son, T.C., Zeng, H.: Semantic Web Services. IEEE Intelligent Systems, 46–53 (2001)
2. Sirin, E., Parsia, B., Hendler, J.: Composition-driven Filtering and Selection of Semantic Web Services. In: AAAI Spring Symposium on Semantic Web Services (2004)
3. Sirin, E., Parsia, B., Wu, D., Hendler, J., Nau, D.: HTN planning for Web service composition using SHOP2. Journal of Web Semantics 1(4), 377–396 (2004)
4. Sirin, E., Parsia, B.: Planning for Semantic Web Services. In: Semantic Web Services Workshop at 3rd International Semantic Web Conference (2004)
5. Aydın, O., Cicekli, N.K., Cicekli, I.: Towards Automated Web Service Composition with the Abductive Event Calculus. In: Proceedings of Applications of Logic Programming in the Semantic Web and Semantic Web Services (ALPSWS 2006), Seattle, USA, pp. 103–104 (2006)
6. Aydın, O., Kesim Cicekli, N., Cicekli, I.: Automated Web Services Composition with the Event Calculus. In: Artikis, A., O'Hare, G.M.P., Stathis, K., Vouros, G.A. (eds.) ESAW 2007. LNCS (LNAI), vol. 4995, pp. 142–157. Springer, Heidelberg (2008)
7. McDermott, D.: The Planning Domain Definition Language Manual: Yale Computer Science Report 1165 (CVC Report 980003) (1998)
8. Yang, B., Qin, Z.: Composing semantic Web services with PDDL. Journal of Information Technology 9(1), 48–54 (2010)
9. Hashemian, S.V., Mavaddat, F.: Composition Algebra: Process Composition Using Algebraic Rules. In: Third International Workshop on Formal Aspects of Component Software (FACS 2006), Prague, Czech Republic (2006)
10. Redavid, D., Iannone, L., Payne, T.R., Semeraro, G.: OWL-S Atomic Services Composition with SWRL Rules. In: An, A., Matwin, S., Raś, Z.W., Ślęzak, D. (eds.) Foundations of Intelligent Systems. LNCS (LNAI), vol. 4994, pp. 605–611. Springer, Heidelberg (2008)
11. Armstrong, W.W.: Dependency Structures of Data Base Relationships. In: Information Processing, vol. 74. North Holland, Amsterdam (1974)
12. Çelik, D., Elçi, A.: Provision of Semantic Web Services through an Intelligent Semantic Web Service Finder. Multiagent and Grid Systems - An International Journal 4(3), 315–334 (2008)
13. Çelik, D., Elçi, A.: Intelligent Semantic Web Services Searcher. Turkish Informatics Foundation's Journal of Computer Science and Engineering (TBV BBMD) (2), 31–42(2006) (in Turkish); TBV Istanbul, ISSN:1305-8991
14. Çelik, D., Elçi, A.: A Semantic search agent approach: Finding appropriate Semantic Web Services based on user request term(s). In: ITI 3rd International Conference on Information and Communication Technology (ICICT 2005), Enabling Technologies for the New Knowledge Society, Cairo, Egypt, pp. 675–689. IEEE Publ., Los Alamitos (2005) ISBN: 0-7803-9270-1

15. Çelik, D., Elçi, A.: Searching Semantic Web Services: An intelligent agent approach using semantic enhancement of client terms and Matchmaker Algorithm. In: Web Technologies and Internet Commerce (IAWTIC 2005), Vienna, Austria (2005)
16. Çelik, D., Elçi, A.: A Semantic Search Agent Discovers Suitable Web Services According to an E-Learning Client Demand. In: 6th International educational Technology Conference (IETC 2006) Gazimagusa-TRNC, vol. 1, pp. 416–424 (2006)
17. Çelik, D., Elçi, A.: Discovery and Scoring of Semantic Web Services based on Client Requirement(s) through a Semantic Search Agent. In: Engineering Semantic Agent Systems (ESAS 2006), Chicago, USA, September 18-21, vol. 2, pp. 273–278 (2006)
18. SOAP versions page, http://www.w3.org/TR/soap/ (accessed November 2010)
19. Web Services Description Language (WSDL) 1.1, http://www.w3.org/TR/wsdl.html (accessed November 2010)
20. OASIS UDDI-Advancing Web Services Discovery Standard, http://www.uddi.org/ (accessed November 2010)
21. OWL Web Ontology Language Overview. OWL Web Ontology Language Overview. W3C Recommendation, February 10 (2004), http://www.w3.org/TR/owl-features/ (accessed November 2010)
22. OWL-S. Semantic markup for Web services (2004), http://www.w3.org/submission/owl-s/ (accessed November 2010)
23. Sirin, E. and Parsia, B.: PELLET: An owl dl reasoner. In: International Workshop on Description Logics (DL 2004), Whistler, Canada (2004)
24. Akkiraju, R., Srivastava, B., Ivan, A., Goodwin, R., Mahmood, T.S.: Semantic Matching to Achieve Web Service Discovery and Composition. In: Proceedings of the 8th IEEE International Conference on E-Commerce Technology and the 3rd IEEE International Conference on Enterprise Computing, E-Commerce, and E-Services (2006)
25. Bener, A., Ozadalı, V., Ilhan, E.S.: Semantic matchmaker with precondition and effect matching using SWRL. Iin Expert Systems with Applications 36(5), 9371–9377 (2009)
26. Bellur, U., Vadodaria, H.: On Extending Semantic Matchmaking to Include Preconditions and Effects. In: 2008 IEEE International Conference on Web Services (2008)
27. Paolucci, M., Kawamura, T., Payne, T.R., Sycara, K.: Semantic matching of web services capabilities. In: Horrocks, I., Hendler, J. (eds.) ISWC 2002. LNCS, vol. 2342, p. 333. Springer, Heidelberg (2002)
28. Paolucci, M., Kawamura, T., Payne, T.R., Sycara, K.: Importing the Semantic Web in UDDI. In: Bussler, C.J., McIlraith, S.A., Orlowska, M.E., Pernici, B., Yang, J. (eds.) CAiSE 2002 and WES 2002. LNCS, vol. 2512, pp. 225–236. Springer, Heidelberg (2002) ISBN:3-540-00198-0
29. Bellur, U., Vadodaria, H., Gupta, A.: Semantic Matchmaking Algorithms. In: Greedy Algorithms (2008) ISBN 978-953-7619-27-5
30. Ilhan, E.S., Akkuş, G.B., Bener, A.: SAM: Semantic advanced matchmaker. In: International Conference on Software Engineering and Knowledge Engineering, pp. 698–703 (2007)
31. Wu, J., Wu, Z.: Similarity-based Web Service Matchmaking. In: Proceedings of the 2005 IEEE International Conference on Services Computing (2005)
32. Şenvar, M., Bener, A.: Matchmaking of Semantic Web Services Using Semantic-Distance Information. In: Yakhno, T., Neuhold, E.J. (eds.) ADVIS 2006. LNCS, vol. 4243, pp. 177–186. Springer, Heidelberg (2006)

33. Kawamura, T., De Blasio, J.A., Hasegawa, T., Paolucci, M., Sycara, K.: Preliminary Report of Public Experiment of Semantic Service Matchmaker with UDDI Business Registry. In: Orlowska, M.E., Weerawarana, S., Papazoglou, M.P., Yang, J. (eds.) ICSOC 2003. LNCS, vol. 2910, pp. 208–224. Springer, Heidelberg (2003)
34. OWL-S 3, OWL-S Service Retrieval Test Collection, Version http://projects.semwebcentral.org/projects/owls-tc/ (accessed November 2010)
35. Kang, D., Lee, S., Kim, K., Lee, J.Y.: An OWL-based semantic business process monitoring framework. Journal of Expert Systems with Applications 36(4), 7576–7580 (2009) ISSN: 0957-4174
36. Grau, B.C. et. al.: OWL 2-W3C Web Ontology Language. W3C Recommendation by Oxford University (April 2008-October 2009), http://www.w3.org/TR/owl2-primer/ (accessed November 2010)

Chapter 8
Semantic Architecture for Human Robot Interaction

Sébastien Dourlens and Amar Ramdane-Chérif

Laboratoire d'Ingénierie des Systèmes de Versailles (LISV)
10/12 Avenue de l'Europe 78140 Vélizy, France
sdourlens@lisv.uvsq.fr, rca@lisv.uvsq.fr

Abstract. Robot software tend to be complex due to management of sensors and actuators in real time facing uncertainty and noise and the more complex tasks to realize in different situations like the human robot multimodal interaction task. This implies a large amount of events to exchange and to process. Robotics intelligent Architecture must be well-conceived to reduce this complexity. Information must be well organized and meaning of situation must be quickly extracted to take decision. Meaning of the situation and situation refinement require the development of a description of the current relationships among entities and events in the environment context. Extraction of meaning and ontological storage of events are very important for interpretation. Human Robot Interaction involves three main parts: awareness and acquisition context, interpretation context and execution context. They define scenarios of multimodal interaction to realize the precondition part called *fusion*, and the post condition part called *fission*. In the aim to solve the above problem, we have designed a new architecture using semantic agents and services. We propose in this chapter, simple and efficient components to any multimodal interaction architecture requirements and universal, compliant and generic architecture using a common knowledge representation language. Our framework is designed for high level data fusion, fission and components management. We don't focus on hardware parts, sensors and actuators. Semantic knowledge is expressed in domain ontologies that permit to extract the situational meaning about any entities in the environment, monitor and adapt the architecture if necessary. In this objective, we apply a narrative knowledge representation language to the memory of agents in a distributed network. We also present the structure and extension of the network for agents to act in ubiquitous environments.

1 Background

In this work, we aim to develop an intelligent robot to assist human in daily tasks. For example, by detecting possible human distress, generating alarm or phone

A. Elçi, M.T. Koné, and M.A. Orgun (Eds.): Semantic Agent Systems, SCI 344, pp. 159–185.
springerlink.com   © Springer-Verlag Berlin Heidelberg 2011

calling to support, helping to take a better posture, connecting to internet and reading vocal news and electronic mails from Internet and go and bring something to human, helping him to navigate in the city and so on. Human makes part of the environment and can interact with the robot by input modalities like gestures, body movements, voice or touch of a screen. Robot will be able to imitate, to dialog or to assist the human in his tasks, choosing the good modalities to answer and acting in the human environment with security.

It appears obvious that our intelligent robot has to manage multimodal inputs and outputs related to different contexts. Robot software need to interact with a number of sensors and actuators in real time facing uncertain, noise and complex tasks to realize in different situations. This implies to process and to exchange a large amount of events. Architecture must be well-conceived to reduce this complexity, events combination and knowledge must be well organized. Meaning of the situation must be quickly extracted to take a reactive decision. This meaning is very important to obtain a correct interpretation. Meaning of the situation and situation refinement require developing a description of the current relationships among entities and events in the environment context. The extraction of the meaning to understand what is happening as well as ontological storage of the events is very important for the interpretation. Human Robot Interaction involves acquisition and awareness context, interpretation context and execution context. These are the three main parts of the multimodal interaction. They will define scenarios of interaction to realize the precondition part called multimodal fusion, and the post condition part called multimodal fission of the cognitive process of our assistant robot. Multimodal fusion is a central question to solve and provide effective and advanced human-computer interaction by using complementary or redundant modalities. Multimodal fusion helps to provide more informative, exact, complete, reliable interpretation. The cross-modal dependency between modalities allows reciprocal disambiguation and improves recognition in the interpretation of the scene or the state of the world at a high semantic level. Multimodal fission will define the best modalities and actions to do in the environment depending of the current context and evaluation of events resulting of the fusion step. Essential requirements for multimodal fusion and fission engines are the synchronization of modalities by sending events, cognitive algorithms including a formal logic, a context representation considering all concepts and actions, evaluating the data transfer bandwidth necessary for efficient application and real-time constraints to respect in true life and robotics systems and the simulation environment to validate the architecture.

In this document, we bring intelligent and exchangeable components to fulfill these above requirements. We don't focus on hardware of robots, house and streets' sensors, or actuators' controllers. We work on a higher abstraction level managing agents' memory, meaning of the environment and communicative exchanges between intelligent software parts. Hardware parts are connected to our semantic services in the network. We have designed a new architecture using semantic agents and services components, and in a way a semantic distributed network architecture. In this chapter, we propose a universal, compliant and generic architecture using a common description language and intelligent agents. In this symbolic approach, our semantic agents required some cognitive abilities. Because of the use of a knowledge base storing all necessary contexts part of the environment, their narrative intelligence appears. The knowledge representation language will also permit to semantic agents and semantic services to directly communicate between each others. Semantic agents manage and reason on the events coming from services. Semantic services manage sensors and actuators; they are able to produce or receive events but not to reason. Cognitive abilities of agents generally need a memory that we have developed on knowledge base. It is expressed in two different ontologies that permit to extract the situational meaning about any entities acting in the environment, and to monitor and adapt the architecture if necessary.

Multimodal interaction is produced by a suitable composition of agents and services regarding to schemes templates stored in ontologies. Fusion is a part of information integration that requires particular techniques for combining these different data into one single event. We work at a higher level of the architecture where signal events have already been translated in natural language by the semantic services and according to this, events fusion is one of the processes embedded in the architecture under the form of fusion agents. Decision will result of the interpretation. In the Post condition, the scenario models permit to assume the fission of the taken decision. Then agents send several simple events to network web services.

This chapter is organized as follows. The next section provides an overview of related work. In section 3, we propose our multimodal interaction architecture design composed of semantic agents and semantic services. Components of the architecture will be described in sections 4, 5, 6, 8 with examples of our application: a human assistance robot connected to ubiquitous home network and the city network. In section 7, we present the development platform employed to realize such agents in a simulated environment. And finally, in the last section, we present the conclusion and the future directions.

2 Related Work

To decrease the late in the autonomous and assistance robotics compared to Japan advance, some military project appears in USA and lots of investment by the European Community are now in progress in several recent Robotic Human Interaction Projects: COSY[1] (cognitive Systems for cognitive assistants), JAST[2] (Joint-Action Science and Technology), ROMEO[3] and so on. As we work at a high level, the work presented in this chapter is related to different scientific domains: Architecture Modeling, Artificial Intelligence (AI), Formal Logic (FL), Knowledge Representation Languages (KRL), Knowledge Management (KM), Information Retrieval (IR), Multi Agent Systems (MAS), Multimodal Interaction (MI) and Human Robot Interaction (HRI).

Multimodal interaction refers to two important processes of interaction. They are well presented in [11] but to resume his view: Multimodal fusion starts from low level integration (signal information) to high level storage of the meaning

---

[1] http://www.cognitivesystems.org/index.asp
[2] http://www6.in.tum.de/Main/ResearchJast
[3] http://www.projetromeo.com/index_en.html

(semantic event information) by composing and correlating previous data coming from multiple sources (sensors, interaction context, software services or web services) in the case of a human-robot or human machine interactions. Therefore, information fusion refers to particular mathematical functions, algorithms, methods and procedures for data combination. Multimodal fission, in the opposite way, is the process to physically act or show any reactions to the inputs or the current situation. According to decision rules taking following the fusion process, the fission will split semantic results of the decision into single actions to be sent to the actuators. In addition, [4] proposes a review of Perceptual Anchoring, a transversal domain where it is necessary to associate a symbol to signal sensor level representation of a physical object by using context and higher level semantic information in robotics. This is important for us to have symbolic information sent by intelligent sensors. Our architecture could be complementary to this work for sensors querying symbolic concepts of the physical context.

Lots of architectures have been designed in the aim to be embodied in a robot, in a house, in the city or to simply bring an intelligent software component into a system. We focus on intelligent architecture integrating semantic agents, semantic services as structural components, ontology as knowledge base [5], inference systems and KRL as communication protocol. Ontologies offer rich representations of machine-interpretable semantics ensuring interoperability and integration [15] because of the non explicit information processed by existing complex architectures and the use of standards. The genericity of the architecture components is a key of success of designing non dedicated applications opened to several domains and mixing different technologies. MAS is a useful paradigm for distributed and ambient intelligence. It is an architectural choice permitting to model widely open, distributed and ubiquitous architecture [13, 1]. Agents are autonomous programming objects or classes with the capacity to interact with the environment because of their communication methods. Intelligent robots often need awareness model and user fusion model, using KRL in a software agents' organization permit to these robots to reason [3]. This design view is very close to our ideal robotics platform. We will adapt this work to Human Robot Interaction. We may add that agents may act socially around a common goal in groups of agents called agencies. Web services were designed by the W3C consortium in the goal to standardize software services in a distributed and interoperable way on the World Wide Web. Agents can also be seen as web services and make part of a distributed architecture. Web services are connected to the Provider agents or consumer agents who need knowledge using the Simple Object Access Protocol (SOAP[4]) for Service Oriented Architecture (SOA) in the XML format.

Semantic agents are cognitive agents implemented in a framework for programming MAS using semantic web technologies. Researches have been made on semantic agents to search, collect or index web information for users. Semantic agents have a reasoner and most of them are built on top of JADE using OWL, SWRL, PELLET and JENA libraries presented later. Ontology Web Language v2 (OWL[5]) permits to describe relationships using classes, individuals, properties in a

---

[4] http://www.w3.org/TR/soap12-part0
[5] http://www.w3.org/TR/owl-features/

hierarchical structure called a web-oriented domain ontology respecting the formal XML. It appears to be a useful storage base and may be coupled with a reasoner. Reasoners are inference systems based on description logic like KAON2[6], PELLET, JESS[7], FACT++[8], and so on, to realize the matching operation. SRWL[9] is a language that combines OWL-DL with RuleML[10] to add rules into the ontology. [18] Proposes semantic agents with the behavior making part of the knowledge base. Our memory development considers all information of the environment and of agents like states and actions. The idea behind this is to realize a distributed composition of agents respecting resources schemes available into the network. In our work, more robotics oriented, semantic agents have a semantic memory for past events sent on the network by sensors in multimodal fusion part of the architecture. They can query this knowledge base and to communicate with software and hardware services. Multimodal fission part of the architecture also called decision processes reduces complexity to achieve robotic, domotic or assistant goals. Semantic memory gives them the ability to understand what happen in the environment and then to compose or adapt the architecture managing several input and output modalities. To obtain semantic agents and services, cognitive agents require inserting a component into them to contain environment knowledge and description logic. It brings the ability to query, store or produce representative and narrative knowledge. Agent memory design permit to store the events and extract the meaning of a situation in a desired context. Recent languages appears to try to solve this issue, new Extensible Multimodal Annotation (EMMA[11]) [8], a standard language proposal from the W3C consortium is able to represent multimodal fusion and multimodal fission in a very procedural way. (Johnston 2009) presents an example of multimodal applications on iphone mobile as a proof of concepts. MurML [9], MMIL [12], MutliML [6] are pure XML and are based on a natural language processing (NLP) parser made by a combinatory categorical grammar (CCG) [17] at an intermediate level of recognition of gestures and speech utterances. It is an interesting approach to mix a grammar and semantic knowledge but it's well suited for speech recognition. But we wanted a human language grammar independent solution. [10] also propose a robot mark-up language approach based on standard XML technology.

We choose languages more suitable and powerful as KRL like Knowledge Interchange Format (KIF[12]) or Narrative Knowledge Representation Language (NKRL) [19]. NKRL is a language very close to the frames and slots idea proposed by [14] and refined by [16]. Frame is the choice we made to resolve our interpretation problem. The language we use in this chapter is very close to frames and NKRL but our ontologies and the inference system are different in terms of development, the result will be more adapted to robot architecture. It appears that the software of

---

[6] http://kaon2.semanticweb.org

[7] http://www.jessrules.com

[8] http://owl.man.ac.uk/factplusplus/

[9] http://www.w3.org/Submission/SWRL

[10] http://ruleml.org

[11] http://www.w3.org/TR/emma

[12] http://www-ksl.stanford.edu/knowledge-sharing/kif

Zarri is not free and open source so we propose another system. But more than that use of narrative languages and to respect the formal and mathematical background behind OWL (calculus completeness and consistency checking), we introduced in the memory of our "semantic agent" component (containing its ontology and the facts base) some similar meta-concepts like properties, relationships and individuals types. One important difference is the n-ary relationships implied by the use of frames and slots. This gives us the ability to import or export any OWL knowledge bases into the agent memory. In addition, the transition from frame to XML is very easy. To conclude this paragraph about languages, we will add here that the language is not so important to extract the meaning but more the structuring of the knowledge base and the relationships between models of action, models of situation linked to concepts in the representation of facts. Our ontologies contain concepts and models fully compliant with concepts package of the chapter 8 defined in the ISO/IEC 2009 CD 11179-3.2[13] standard document and value meaning of the chapter 10 of the same document. For physical agents managing our material services (sensors and controllers), we also refer to the Foundation for Intelligent Physically Agent (FIPA[14]) standards from IEEE Computer Society, in particular, the SI00091E document on frames description of FIPA-devices, the SC00001L document about the protocol schemas for our management agent in charge of the discovery, composition and adaption of the multimodal architecture.

Our KRL will be also used for the network communication between agents and services even if they may use public methods of web services to exchange information via SOAP protocol. To dialog, several techniques are available, state based, frame based and plan based. A frame is a task or a subtask at a high level of representation. NKRL is very suitable with frames because of the predicate form of models (we will detail this point later) and brings the same information on environment state coming from sensors and action done or to do. Frames of "information events" will be directly stored in the knowledge base with no specific treatment while frames of "queries events" imply to matching stored events models for example to recognize the current scenario. In this section, we have presented recent works and clarified our work area and choices. In the following section, we will explain our multimodal interaction architecture dedicated to Robotics.

3 Multimodal Interaction Architecture Design

In this section, we introduce our robotics multimodal interaction architecture as represented on figure 1. In the application of interaction, our architecture is composed of intelligent agents able to manage several multimodal services. Services are directly connected to the environment controlling sensors and actuators. They are in charge of internal of the robot or external inputs and outputs in a ubiquitous network. Semantic agents integrated into the architecture may use semantic services embedded in the robot, in the house, in the city or anywhere on Internet. Semantic agents described here are not only dedicated to robots. They may be

---

[13] http://www.jtc1sc32.org
[14] http://www.fipa.org

embedded to manage a street, a shop, a building or any Robotics or Home Automation systems. Mikes and automatic speech recognition (ASR), video camera
and gestures recognition are examples of services that will be input modalities.
Robot arms, legs or wheels, speakers and text-to-speech synthesis (TTS), TV
screens, Wireless Fidelity Coffee device will be output modalities. As you may
apprehend, this architecture will provide a kind of ambient intelligence used in
function of the objective and situation to face. We tend to bring a more generic architecture to avoid tying it to a specific problem and a specific markup language
as presented in the related work section. That's why all agents and services are
"web services" components in the network and offer discovery, composition and
orchestration in the respect of UDDI[15] and WSDL standard. Other agents will be
dedicated to composition and adaption tasks.

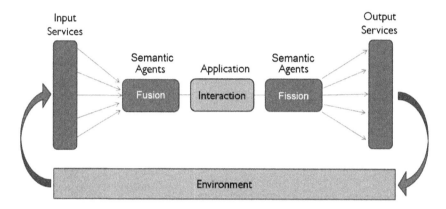

Fig. 1. Interaction Oriented Architecture

In the case of Human Robot Interaction, figure 1 shows that semantic agents
are architectural components working in the processes of fusion and fission.

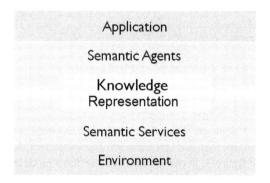

Fig. 2. Semantic Architecture

Figure 2 presents the semantic architecture where semantic agents and semantic services are components sharing the knowledge in a communication network. Knowledge can be considered like a blackboard between them but it is imported, exported or distributed into the agents (in their memory) using the network. Our software architecture is based on the three following main components:

- Semantic Services: Their role is to send any information from the environment using hardware sensors or execute a software function, or to execute orders to control actuators. Services can be seen as reactive agents with no cognitive part but enough exchangeable knowledge and code to realize the process they are designed to.
- Semantic Agents: They are cognitive or functional; they possess their own abilities and program to achieve their tasks and goals. They contain an embedded inference system able to process the matching operation. They are intelligent agent with cognitive abilities to answer queries. Scenarios or execution schemes are stored in their Memory (a knowledge base) that we will describe in the next section.
- Knowledge between agents and services: It is communicated under the form of events on the network and stored in semantic agents. Agents are specialized for certain domains of knowledge or by location or use, and filter information from the network. But, by default, events are sent to all semantic agents in a broadcast way.

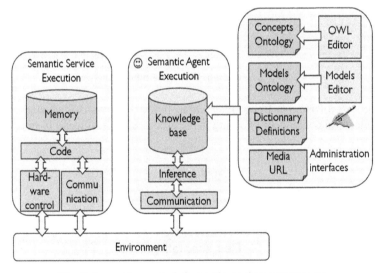

Fig. 3. Semantic agents & Semantic services components

As shown on the figure 3, in our architecture, we have modeled semantic agents and semantic services differently. Semantic services interact with the environment in two manners: the network (wireless or not) between them and, the sensors and actuators. They have code and memory (properties and methods in generic programming object model), the communication module and the hardware controller module. The hardware controller enables the service to receive information from a sensor or to drive an actuator. Semantic Agent contains its knowledge base, its inference engine and its communication module. The communication module contains the network card and its semantic functionalities to write and send the events in KRL or receive it. As we have also presented in the previous section, the figure 3 shows the agent user or developer interfaces that permit to build and import concepts, models, terms definitions and media unified related links. Media can refer to multiple pictures or video files.

OWL editor can be Protégé, SWOOP or any others OWL2 compliant editors. Thus OWL can be imported and concepts ontology in the knowledge base (agent memory) can be exported too. The models ontology editor has been developed by us to insert and modify models. It is in fact the agent memory editor. In summary, this figure shows the basic components of any complexes architectures that can be built with.

4   Semantic Agent Memory

Our semantic agents are more oriented to solve multimodal fusion, multimodal fission management and to be part of robotics architectures or human decision systems. In this section, we will present our knowledge base structure. Our architecture is fully semantic and the knowledge base is the fundamental stone of this architecture. In this work dedicated to multimodal interaction, we call our agent memory the *Multimodal Interaction Context Ontology (MICO)*. MICO is a database regrouping two ontologies (see fig. 4). Semantic agents and semantic services communicate with events. These events are written in natural language concepts (formal T-BOX of concepts) and instances of concepts (formal A-BOX of concepts) included in our first domain ontology called *Concepts Ontology*. Our second ontology is called *Models Ontology* and contains templates of events (formal T-BOX of models) under the form of predicates and instances of events called *facts* (formal A-BOX). Concepts, Events models, Query models and instances are stored in ontologies. Instances are facts, happened scenario and context knowledge. Concept ontology is fully compatible OWL2.

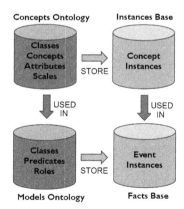

Fig. 4. Knowledge base information

In the multimodal interaction context, classes of concepts will describe, in natu-ral language (English and French languages are the only languages included for now) and in an ordered and hierarchical way, the objects of the world including sensors and actuators, entities, meta concepts to create and modify concept and all possible instances. Our concept ontology may be large and grows. We are already able to insert knowledge bases like OpenCYC[16] project and common sense projects exist, we could integrate them and work with them. But we have not the ambition to create a human brain, just an effective powerful memory for semantic agents be-cause we work by iteration, controlling the design and the safety of our approach.

Agents will contain the code and scheme of execution to realize their tasks with the help of the models of behaviors, actions or composed scenario in their mem-ory. Events instances or facts link models to concepts.

The model ontology contains predicates, roles and arguments where roles are concept classes stored in the concept ontology and arguments are concept classes for a query model or instances of concepts. Arguments must be an operation of several classes of concepts or instances by using NRKL operators as defined by Zarri [20].

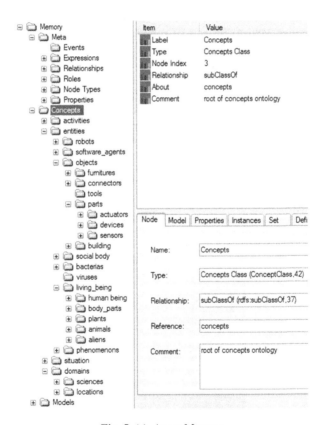

Fig. 5. (a) Agent Memory

---

[16] http://www.opencyc.org

Figure 5 shows two screenshots of the Agent Memory Editor. Meta ontology, Concepts ontology and Model ontology are in a same Memory tree. These ontologies are stored in a SQL database tables (see fig. 15). One frame query is equivalent to one SQL query sent to the database and the matching is directly done in a very fast way because of the complexity due to the storage of the ontology in database tables at the time of creation of concepts and models.

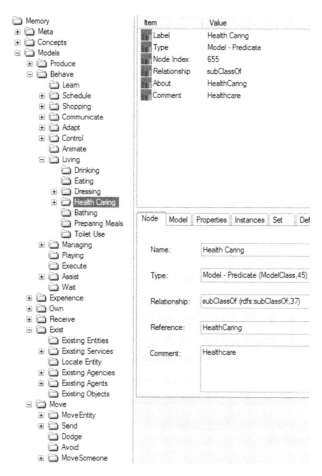

Fig. 5. (b) Agent Memory

Knowledge representation language (KRL) is a semantic formal language L that can describe events in a narrative way. The formal system is composed of the formal language based on variable arity relations in logic of predicates (event frames). It permits to realize semantic inference in order to extract the meaning of the situation. Ontologies are useful and powerful structures to store the events and extract this meaning. Inference system may use models to match the instances of the ontologies. In NKRL, frames are predicates with slots which represent pieces of information. A slot is list of roles associated to arguments. A predicate P is a semantic relationship between Roles and Arguments and represents a simple event SE or a composed event CE depending of the position of the predicate in the models ontology; it is denoted by the following formula:

$$P((R_1 A_1) \dots (R_n A_n))$$ (1)

where R_i is a role and A_i is an argument. A role is a meta concept and can be one of the possible roles of concepts in the event. Argument A_i contains one or several possible values, classes of concepts or instances of concepts in the event. And if several items are in the argument, a relationship can be used as COORD/AND (OWL intersectionOf), ALTERN/OR (OWL unionOf), ENUM (OWL oneOf), SPECIF (OWL subClassOf).

Name: <RootPredicate>:<PredicateName>
Father: <RootPredicate>:<PredicateName>
Position: <NodeTreePosition>
Natural language description: '<Predicate Description>'
<RootPredicate>   <Role1>:<Argument1>
                                   <Role2>:<Argument2>
                                   <Role3>:<Argument3>

Where Role can be OBJECTIVE, SOURCE, BENEFICIARY, MODALITY, TOPIC, CONTEXT, MODULATOR, DATE/INTERVAL or another role specified in the Meta ontology.

Fig. 6. Model of Event Structure

Figure 6 shows a sample model written with the NKRL syntax. The list of all roles is part of the Meta ontology of the agent memory. Models of events are models of predicates and instances of predicates specific to a situation, for example, "Move" is the informative term called root predicate (an event model of the events ontologies).

Name: Move:TransferOfServiceToSomeone
Father: Move:TransferToSomeone
Position: 71
Natural language description: 'Transfer or Supply a Service to Human'
MOVE OBJECTIVE   *Services*
        SOURCE   *Agents*
        BENEFICIARY   *Human being*
        MODALITY   COORD(*Composition,Execution*)
        TOPIC   *UDDI*
        CONTEXT   *Home Network Services*
     Modulators *Emergency*
        date start:
        date end:

Fig. 7. "Move: TransferOfServiceToSomeone" Predicate model

The sample model of predicate *"Move:TransferOfServiceToSomeone"* in figure 7 represents a model of possible events that can happen when a service is realized in emergency to help someone in distress. In this case, it shows a web services composition of semantic agents and semantic services where actuators can be activated by any dedicated agents to execute a task.

For example, if actuators are parts of robots arms, they may help a human to get up or to avoid falling down. If the developer needs more accuracy of this event model, he will be able to create subclasses of this event model to refine it, e.g., by replacing the "Actuators" concept by a subclass of actuators in the concept ontology and by replacing "Agents" concept by the real agent assigned to this task. And then, he has created a new sub model or subclass of the previous model. Any event facts related to this predicate will be stored under this model as instances with all roles filled by exact arguments. In this sample, in this agent memory, storage of facts is very fast because events are directly stored under its own predicate and querying past facts related to a situation will simply consist on matching predicate, roles and arguments. All situations and actions of semantic agents and services done on a time period will be quickly highlighted.

In this section, we have explicated how works the semantic agent memory. In the next sections, we will present how agents and services act with multimodal interaction in a robotics environment.

5   Multimodal Interaction Agents

In this section, we will present Fusion Agents, Management Agents and Fission Agents that are employed to resolve the multimodal interaction.

5.1   Fusion Agent

A fusion agent (FA) is a semantic agent which has the role to extract a specific meaning. Figure 8 presents the fusion agent model as a filter. The matching operation of the inference engine will extract the required meaning in the current context. Once information are extracted from the knowledge base, the fusion agent will be able to create a composed event that will be also stored in the knowledge base and eventually shared with others agents by sending it to others agents in the network. Each agent is specialized to a task or a domain to compose new events or to act in the environment using linked semantic services.

We denote fusion agent as is:

$$m = fa_x(e_1, \ldots, e_k) \tag{2}$$

where m is a meaning, fa_x the fusion agent function of the agent x and e_1 to e_k are the events or facts coming from other semantic agents. Meaning is sent under the form of composed event to other agents only if matching is true.

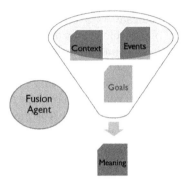

Fig. 8. Fusion Agent Model

All agents process consists to execute a loop of these five following steps:

1. Take a model of events
2. Fill roles with known or wanted arguments
3. Query the knowledge base
4. Get list of matching events
5. Use events to compose a new event (simple or composed event) by using its specialized code

```
Behave: SayHello
        SUBJECT: COORD(Gesture Arm Up,Speech Hello Robot)
SENDER: COORD(GestureDetectionVideo1,VocalRecognition1)
Date: 09/09/2009 11:34
```

Fig. 9. "Behave: SayHello" composed event instance

The following example (see fig. 9) represents a model of composed event that represents "a human saying and doing Hello". In this example, one instance of our fusion agents is in charge to merge events that happen in a same period of time. It uses a model that let it read that two events "Gesture Arm Up" and "Speech Hello Robot" sent at the index time 11:34 by the two services "GestureDectection-Video1" and "VocalRecognition1". The first service is in charge of the detection of gestures produced by human when the human is near by the camera1 sensor. The second service uses one or several mikes to recognize a speech sentence. These services have sent their basic event in parallel to any agents able to store them into their memory. As you make out, these two services are embedded in the robot or are parts of the house, no problem until they are connected to the robot agents. At a scheduled interval of time and after new events took place in its memory, our agent can compose an instance of the "Behave:SayHello" model.

Then others agents will be able to use this composed event to take decision, compose or evaluate a scenario or simply store it in their memory.

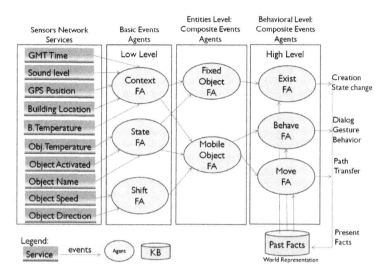

Fig. 10. Multimodal fusion agents

Figure 10 presents several examples of fusion agents (the disks) acting at different levels of the multimodal fission part of the architecture (figure 1) from the basic to the behavioral level. Basic level links fusion agents to services (the boxes) in charge of sensors. These basic agents together store their events as values or instance of concepts in their memories but not instances of models. Higher levels agents compose and store events as instances of models with the useful information coming from the previous level. In this example, the context fusion agent stores time, sound, level and location and shares these instances with agents of the next levels. Entities level agents will create simple events. At the behavioral level, agents directly store composed events under the corresponding root predicates of the memory ontology: here we have Behave, Exist and Move. We observe that fusion agent names can correspond to predicates of the agent memory. In fact, the figure shows we can use several levels of granularities of events. Fusion agents are able to query the knowledge base and produce simple or composed events.

5.2 Management Agent

A management agent is a semantic agent which has the role to manage others agents and make them work in relation with available services. All semantic agents execute the same process presented in the fusion agent section. Management agents are programmed to store events happening in the architecture, adapt and compose the architecture at a scheduled period of time. To realize their tasks, they may use the event model presented in the figure 7.

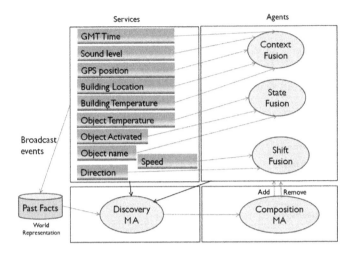

Fig. 11. Multimodal Management Agents

Figure 11 shows an example of composition of semantic agents and services realized by two management agents. The discovery management agent (DMA) is in charge of the discovery of services and agents. It uses the SOAP protocol to access UDDI server or WSDL files, to find all web services in the network and to interrogate them to know all properties and methods they have (the two arrows from the agents and services boxes). From these data, it will create semantic events about available services and agents, their location, the possible roles and their performances. Some information about us may be also directly available in its memory (represented by the Past facts disks previously read from the network as broadcast events). Then, the composition management agent (CMA) is able to compose a desired architecture with links between agents and services (arrows from services' box to agents' box). It will be capable of replacing an agent or a service by a better one more efficient or in case of failure depending on the developer or user criteria.

Three agents have a role in this multimodal fusion architecture: Context fusion agent is in charge of reading all single events from the environment context like time, sound level, location, temperature of the building. State fusion agent receives focused objects context from the environment by reading temperature of the object, activity status, name of the recognized object by the object recognition sensor. Shift fusion agent perceives basic events of the behavior of the focused object: speed and direction. The object can be any entities like the hand or the arm of the human, or any obstacles on a walking path.

5.3 Fission Agent

A fission agent is a semantic agent which has the role to manage actuators services. Fission agent acts exactly like the fusion agent except that in addition they will produce events sent to planning agent in charge to store all future jobs to execute or directly to services at a specific time by monitoring the planning agent events. So the fusion agent model presented in the figure 8 is the same for the fission agent model, only the meaning will be of different types because events will be orders or planes that we call "execution events". We denote fission agent as is:

$$a=fi_y(m_1, \ldots, m_k) \tag{3}$$

where a is an action to perform by actuator, fi_y the fission agent function of the agent y and m_1 to m_k are the meaning or composed events coming from other semantic agents. Action is sent under the form of composed event to actuators services if matching is true.

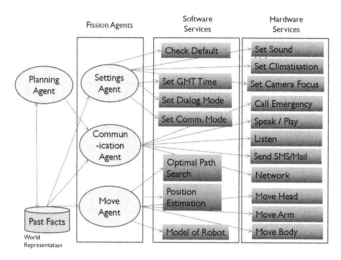

Fig. 12. Multimodal Fission Agents

Figure 12 presents an example of composition of fission agents and actuators services realized by three fission agents (settings agent, communication agent and move agent) and a planning agent. Agents are represented as disks, services as boxes and arrows represent the composition (events communications that are taken into account by the agents). Past facts are the previous events sent on the network and so available in the memories of agents. On this figure, we notice two types of services: the software services containing useful methods and the hardware services driving actuators making part of effectors network. The choice of software services and hardware services depends on the conceived robotics application.

This section clarifies the functioning of agents and services, and the interest of use of KRL in the multimodal interaction problem. In the next section, we will focus on the networking in the architecture even if we already have presented here the standard web services implied in the process of semantic agents and service composition.

6 Networking

6.1 Protocols

Our architecture is Operating System independent and built on standards to be compliant and all components are designed as web services for interoperability with external applications and agents. Simple Object Access Protocol (SOAP) is an Internet Protocol (IP) communication layer. It works with UDDI server or WSDL files to discover and integrate the web services. Some new researches bring non functional information on services like performance (execution, response time), user preferences, quality of work and availability. In a future work, it will be interesting to insert semantic markup for web services (OWL-S) to be used by management agents. Memory of these agents will keep service profile, service grounding and service model. The figure 13 presents the messages exchanged at networking level.

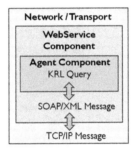

Fig. 13. Messages

Agent and Services are web services, they have an IP addresses and one or several TCP ports. Mobile agents have to change of IP addresses by moving from a network connection to another, in a wireless (WIFI) network for example. Additional security schemes may be added to manage privacy of information, services or network when the mobile agent acquires a new address or a Kerberos ticket is granted to access a service.

Our agents and services communicate by using two private methods:

- krl_send() to send events under the form of frame messages in a broadcast way to all agents and services, to an agency or to only one agent or service;
- krl_listen() to wait for any events to store or process.

These functions will use a model of events and fill the different role using values or concepts related to the service job.

6.2 Event Messages

As we have presented in the previous sections, KRL is also a communication mean used between agents and services describing happening events. It brings to a robot or any parts of the application the necessary awareness.

The knowledge is distributed on agents specialized or not (i.e using events filtering of not). Environment context comes under the form of events from the network sensors (semantic services). These events are the facts happened in the environment and in the architecture or, actions, orders or data sent by agents to other agents or to services. Events are single or composed and express the situational meaning for agents. We may call this event messaging network a semantic network. Figure 4 showed how to compose facts using two ontologies and giving two types of instances even if they are in the same knowledge base. Figure 14 is oriented to present knowledge management in the network connected to the environment and the possible actions and request so classes and instances parts of the figure 4 are presented on only two cylinders here.

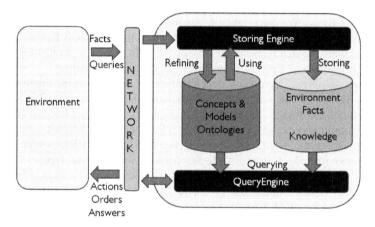

Fig. 14. Storage and Querying the agent memory

6.3 *Semantic Agencies*

In this section, for performance of using and accuracy of information, we propose to design local semantic agencies containing a set of multimodal agents specific to a location (embedded in a robot or a house) or a purpose domain (cooking, working, monitoring or health caring) using different IP networks. Depending on their goal, others agents that don't make part of the agency will be able to get a new IP address, be discovered, and eventually be used. The most important will be the possibility for it to share knowledge of the agent memories of the agency respecting a filtered or secured access by agents themselves.

Knowledge can be either an external part of the robot embedded into the autonomous agents or be elements of a nearest distributed network. Each network nodes or agencies could also be connected to other external nodes: nodes on the streets, nodes in different houses, nodes in companies, nodes on Internet servers or even virtual nodes. Virtual nodes could be incorporated into a simulation, a game or a computer. We recommend evaluating the size of the agency and the generated traffic of data in the network to be efficient, smart enough and fast to interact. With this notion of agencies and network nodes, mobile agents (embedded in a mobile robot or a mobile phone) will be able to adapt themselves to any cross environment and use local information and services to achieve its task.

7 Development Platform

Our architecture is general enough to be OS independent and developed with any programming language and any free of cost web services or multiple agent platforms. The two main development environments are Java Agent DEvelopment Framework (JADE) using JAVA and Microsoft .NET, CCR, DSS using C, C#, C++, Visual Basic. We have taken C for most of the code and VB.NET for the Memory Editor interfaces to be easily integrated in the Microsoft Robotics Studio simulation platform that respects the physics of earth gravity and physical parts of the robot (dimensions, weights, boundaries, degrees of freedom).

Our two ontologies have been developed using our own agent memory interface (see fig. 5). For the concepts ontology, our interface is compatible with any OWL editors like Protégé (Java), OWLVE (Eclipse) or Swoop (Java) as our interface may import and export the concepts tree. For the models ontology, we use our editor to respect the NKRL description language as no NKRL source code is available and free. As a future work, we think to develop this type of ontology with the Protégé-frames[17] free framework. It is an integrated Protégé version giving the ability to add frames (our predicates) and slots (our roles and arguments). To store and access quickly ontologies, some people will prefer to write XML text files, we preferred to build our ontologies on a database engine. XML text files are very quick to write and read but the difficulty comes to parse and recursively match events with models. Our database (see fig. 15) is fully compatible with MySQL, PostgreSQL and Microsoft SQL Server. The SQL query engine is much more efficient on a huge amount of data because of the indexing system. Researches are done in the field of domain ontologies to make hybrid systems linking ontologies and database but we think a choice must be made to avoid managing two environments. Oracle developers have also recently added semantic SQL procedures to manage OWL ontologies and match instances. Our ontologies are stored in a database with different tables containing links between concepts, roles and arguments of events models and queries models to accelerate the matching/unification process by using a simples SQL request. Links represents meaning by associating concepts and events nodes. Table named "RA" stores facts under the form of predicates (see fig. 6) and links all roles and arguments of facts (propositions in terms of KRL). Unification is done very quickly because it is a $O(n)$ comparison between a query model and n facts. As instance of a previous work, [2] describe a smart house platform using TAOM4E, JADE and the SOCAM ontology.

---

[17] http://protege.stanford.edu/overview/protege-frames.html

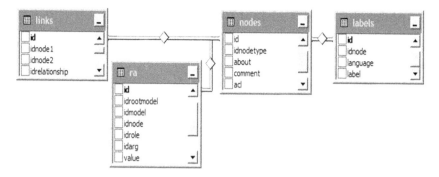

Fig. 15. Ontologies stored in database with Role and Arguments (ra) fully linked ontologies

Our framework and previously presented components have been developed to be integrated in "online" real-time software embedded in robots, intelligent houses or mobile phones. Our framework is also designed to continuously insert new instances to the agent memory. Then we have integrated to a simulation robotics platform fully compatible with the robot hardware.

Knowledge representation languages used are NKRL and OWL v2 but other description languages like KIF may be also included. The only constraint is that the knowledge base must have a conversion procedure to store events as predicates instances. We can export our database to XML format with one direct SQL query. Concerning natural language integration, as our concepts ontology contains words of natural language, it is also possible to consider verb as predicate, subject and others as instances of arguments. Searching for the closest predicates model containing arguments related to these instances. The procedure will take little more time but should work. In addition, to insure interoperability, the network layer is made of standard TCP/IP and SOAP protocols. Web services may also be developed with AXIS, ANT or Tomcat. We have developed this architecture with Microsoft DSS and CCR (see fig. 16) to guarantee the scalability. Graphics simulations will be realized with Microsoft Robotics Studio and DirectX.

Layer	Description
Agents	Software components of the application
Services	Software and hardware services
Knowledge Representation Language	Events encoded into Predicates
Microsoft CCR and DSS	Concurrent Agents & Web Service Management
SOAP / XML	Standard Web Services Protocol
TCP-IP Network	Standard TCP/IP protocol

Fig. 16. Application Layers

8 Application to an Assistant Robot

As presented in the introduction, we want to develop an intelligent robot to assist human in daily tasks at home or in the city.

8.1 Robot Composition

First our robot may know its possibilities of actions and initiate its own architecture. In the software, we have implemented our semantic agents presented in the figures 10, 11 and 12. The management agents will discover all services using the SOAP webservices discovery and using the two models of Exist root predicate in the models ontology: "Exist:Existing Services" and "Exist:Existing Agents". To link the services to the agents, CMA will use the sub classes of "Behave:Adapt" model. Our two agents repeatedly adapt the architecture depending of the current context: multimodal interaction context (too much noise to use vocal recognition, too much noise to produce speech or sounds, , don't use keyboard input as light is too dark, because and so on), user context (user is sleeping in front of television, user doesn't want robot to make noise, user is talking to someone else, dumb user can't speak so robot will read on lips or read signs, and so on), location context (don't make sound because it's forbidden in this location, outside noise is too high, robot must avoid people and obstacles, global positioning system is activated outside but not inside house, and so on), awareness context (speech, move or sound detection) and all other desired situational contexts. Once composition is done, all agents are able to read events messages on the network and get all information from the services like real possibilities of actuators (for instance, minimum and maximum angles of a arm rotation) and list of functions they can execute (rotate arm to right, grasp something with left hand). As agents are specialized to a given task, they will keep all or part of knowledge available on the network in their memory and use it to make composed events or act. Before starting the robot, we have given to agents and to services the maximum of information about environment and jobs to realize in their memories by filling concepts and models ontologies.

8.2 Robot at Home

At home, lots of services are available, detectors in the doors, coffee maker in the kitchen, drink distributor in the saloon, human detectors in the rooms and corridors. Once connected to the home area network (HAN), our robot is able to use all services according to their availability. Management agents of the robot discover services and compose a new architecture depending on the application or robot objective. All actuators, devices and sensors are fully described in the concept ontology of the agents under the "entities:parts" class. It is the same for the walls position and furniture in the house. To know where the human, animal or any entities are, agents may use the following models to query their memory: "Exist:Existing Entities" and "Exist:Locate Entity". To achieve scheduled tasks, robot agents will read planning in their memory and compose or adapt the architecture with available resources of the robot then acts. Triggered tasks are also taken into account by receiving an emergency event from the fusion agents following some priorities.

Name: Move: Someone Falls Down
Father: Move:Move Someone
Natural language description: 'James is falling down'
MOVE
      SUBJECT: *"fall" (possible moves detected by the camera)*

      SENDER: *MoveDetectionVideo5 (instance of sensors)*
        SOURCE: *James (instance of human_being)*
        MODALITY: *COORD(Behave:Adapt, Move, Behave:Catch Someone)*
        TOPIC: *Monitoring*
        CONTEXT: *House Activities*
        LOCATION: *Living room (instance of Building)*
      Modulators: *Emergency*
      date start: 10/10/2010 10:21:59

Fig. 17. People falls down detection implies robot assistance

Figures 17 and 18 represent the situation where someone is falling down in the living room. Management agents delay all current tasks and adapt the architecture for the robot to hold, retain or catch the one if possible. Else it will alert the rescue from a phone call.

Name: Behave: Catch Someone
Father: Behave:Assist
Natural language description: 'Robot is catching James'
MOVE
        SOURCE: *Robot (instance of robots)*
        BENEFICIARY: *James (instance of human_being)*
        MODALITY: *COORD(Left Arm,Left Hand,Right Arm,Right Hand)*

        TOPIC: *Assistance*
      Modulators: *Emergency*
      date start: 10/10/2010 10:22:01

Fig. 18. Robot has helped James

Robot was doing something else, agents receive the "Move:Someone Falls Down" event coming from the *MoveDetectionVideo5* service located in the corridor1 (see fig. 19).

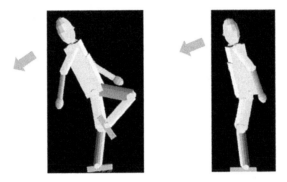

Fig. 19. Body motion detector

As it's a case of emergency and the modality "Behave:Adapt" is defined in this event, management agents organize the change by ending events to planning agent and compose architecture to move the robot near the human being called James and to catch him before James hurts the ground. All input modalities stay activated but output modalities are chosen. Fission agents in charge of the coordination of moves control the robot wheels and the rotation actuators to put arms and hands in the right position. During the fall, agents will modify the trajectory of robot arms. To obtain the right position of the body of the human and the speed of fall to cal-culate the appropriate position of the robot, fission agents can send queries to *Mo-veDetectionVideo5* service. After the reply; fission agents control the actuators to achieve the goal. The scenario will end when *MoveDetectionVideo5* service de-tects a secure position of the human being. Lots of scenario can be integrated by the addition of our predicate models: More simple scenarios like to recognize a vocal order from the human and executing it, to send an event to actuators to close the shutters at night or call the police on intrusion detection. Or more complex scenarios like to dialog with the human, to learn gestures or actions by imitating human, and realize complex behaviors.

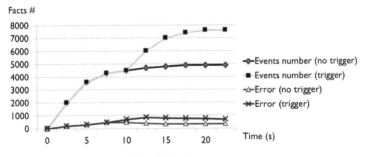

Fig. 20. Performances

We have built and simulated the house and robot sensors and MAS architecture composed with 10 different sensors services sending a lot of events and 20 agents working in parallel. We trigger the falling of James at 10 seconds. We study the performance of the agents by analyzing the number of facts stored in their memory and the well composed events resulting of multimodal fusion and fission agents. The figure 20 shows these results. We observe that errors in output events are less than 10% with no trigger and less than 20% with trigger depending of the number of events and the events processing. These results are due to simulation; in real, much more uncertainty occurs. In a future work, we think to improve this by a quality management of events and a model checking.

8.3 Robot in the City

We showed our robot acting in the house but why not use agencies of a simulated city and extend the robot into the city. Semantic agencies presented in section 6.3 are integrated in several locations. Each agency has a network of agents with specialized roles customized to exchange information about their accomplished jobs.

Figure 21 presents a simple simulation for city streets map scenarios (streets are in white color and buildings in grey color) where a robot helping a human at home, to navigate in car or by walking to shop, work and post office. Some agencies (dark disks) are accessible from home location like police and Hospital for the robot to call them in case of emergency. Some agencies like Street1 and Street2 will be accessible to navigate from home to work, to inform robot about the traffic road and even to give the current color of the traffic lights at the crossroads. Baker and Shop gives information about the opening hours, the available quantity of bread or the new fashion clothe. Same for the post office which could inform about a packages delivery bought on Internet. At home, the network may inform the robot about the current health state of the human, daily living tasks in progress or unusual situations. Importation of knowledge from other agencies is interesting for an agent to learn experience and to change job in a composition where agents are not too busy. Many applications are possible.

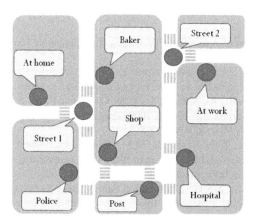

Fig. 21. Semantic Agencies City streets map

9   Conclusion and Future Work

To conclude this chapter, after a background review, we have presented a new architecture based on semantic components which offer a good framework to design complex and intelligent applications like our implementation in a human robot interaction platform. Semantic agents and services work in parallel and are distributed in a network. Knowledge representation languages and ontologies conceived to build the agents memories give the ability to reason about the environment very quickly. The use of frames and matching operations is a promising way of developing artificial intelligence. Agents are able to continuously adapt the robot and the modalities depending on changes. In addition, all current standards are respected and may be integrated.

In future work, we will work on semantic pervasive architecture and ambient intelligence applied to learning and training of robots. One important work to realize is the measure of performance of the architecture. We have not yet measured all performances and compared to OWL QL parsers for now but we are confident about the space of complexity and we think to integrate later models of quality evaluation of the system in term of networking and distributing (quality of service) and decision taken (scenario analysis).

One main problem is the limitation of the network bandwidth. In this case, it should be possible to reduce the number of events on the network by using private channels between agents like human communication. In the work done here, all agents store all events and use it wisely by filtering unnecessary events.

Concepts ontology of the agent memory contains enough knowledge for our robotics application but we think to accumulate more concepts and definitions by automatically importing CYC common sense ontology, English and French thesaurus by linking them directly to our existing concepts.

References

1. Allan R.J.: Survey of Agent Based Modeling and Simulation Tools. Technical Report. STFC (2008-2009),
   http://epubs.cclrc.ac.uk/work-details?w=50398,
   http://epubs.cclrc.ac.uk/bitstream/3637/ABMS.pdf
2. Benta, K.-L., Hoszu, A., Vacariu, L., Cret, O.: Agent Based Smart House Platform with Affective Control. In: EATIS 2009, Prague, CZ, June 3-5, Art. 18. ACM, New York (2009) ISBN:987-1-60558-398-3
3. Erik, B., Ivan, K., John, S., Kokar, M.M., Subrata, D., Powell, G.M., Orkill, D.D., Ruspini, E.H.: Issues and Challenges in Situation Assessment (Level 2 Fusion). Journal Of Advances In Information Fusion 1(2), 122–139 (2006)
4. Silvia, C., Amy, L.: A review of Past and Future Trends in Perceptual Anchoring. In: Fritze, P. (ed.) Tools in Artificial Intelligence. I-Tech Education and Publishing, Vienna (2008)
5. Guarino, N.: Formal ontology, conceptual analysis and knowledge representation. Human-Computer Studies 43(5/6), 625–640 (1995)

6. Manuel, G., Alois, K.: MultiML - A General Purpose Representation Language for Multimodal Human Utterances. In: ICMI 2008, Chania, Crete, Greece (2008)
7. Michael, J.: Building Multimodal Applications with EMMA. In: ICMI-MLMI 2009, Cambridge, MA, USA, November 2-4, pp. 47–54. ACM, New York (2009) ISBN: 978-1-60558-772-1/09/11
8. Michael, J., Paolo, B., Burnett Daniel, C., Jerry, C., Dahl Deborah, A., Gerry, M., Dave, R.: EMMA: Extensible MultiModal Annotation markup language. W3C Recommendation (February 2009)
9. Kranstedt, A., Kopp, S., Wachsmuth, I.: Murml: A multimodal utterance representation markup language for conversational agents. In: Proceedings of the AAMAS, Workshop on Embodied Conversational Agents - Let's Specify and Evaluate them!, Bologna, Italy, July 16 (2002)
10. Jun-young, K., Young, Y.J., Shinn Richard, H.: An Intelligent Robot Architecture based on Robot Mark-up Languages. In: Proceedings of IEEE International Conference in Engineering of Intelligent Systems, pp. 1–6 (2006)
11. Frédéric, L.: Physical, semantic and pragmatics levels for multimodal fusion and fission. In: Seventh International Workshop on Computational Semantic (IWCS 2007), Tilburg, The Netherlands, pp. 346–350 (2007)
12. Frédéric, L., Denis A., Ricci A., Romary L.: Multimodal meaning representation for generic dialogue systems architectures. In: Proceedings on Language Resources and Evaluation (LREC 2004), pp. 521–524 (2004)
13. Macal Charles, M., North Michael, J.: Tutorial on agent-based modeling and simulation part 2: How to model with agents. In: Perrone, L.F., Wieland, F.P., Liu, J., Lawson, B.G., Nicol, D.M., Fujimoto, R.M. (eds.) Proceedings of the 2006 Winter Simulation Conference, Monterey, CA, USA, December 3, pp. 73–83 (2006)
14. Marvin, M.: Matter, Mind and Models. In: Proceedings of IFIP Congress, May 1965, pp. 45–49. Spartan Books, Washington D.C (1965); Reprinted in Semantic Information Processing
15. Leo, O.: Ontologies for semantically Interoperable Systems. In: Proceedings of the twelfth International Conference on Information and Knowledge Management, New Orleans. LA, USA, pp. 366–369. ACM Press, New York (2003) ISBN: 1-58113-723-0
16. Ross, Q.: Semantic memory. Ph.D. thesis, Carnegie Intstitute of Technology (1966); In: Minsky, M. (ed.) Semantic Information Processing, pp. 227–270. MIT Press, Cambridge (1968)
17. Mark, S., Jason, B.: Combinatory Categorial Grammar. In: Borsley, R., Borjars, K. (eds.) Non-Transformational Syntax. Blackwell, Malden (2005)
18. Subercaze, J., Maret, P.: SAM: Semantic Agent Model for SWRL rule-based agents. In: Proceedings of the International Conference on Agents and Artificial Intelligence, Valencia, Spain. Agents, vol. 2, pp. 244–248. INSTICC Press (2010) ISBN 978-989-674-022-1
19. Piero, Z.G.: Representation and Processing of Complex Events. In: Association for the Advancement of Artificial Intelligence AAAI Spring Symposium (2009a)
20. Piero, Z.G.: Representation and Management of Narrative Information: Theorical Principles and Implementation. In: Jain, L., Wu, X. (eds.) vol. 1, pp. 978–971. Springer, Heidelberg (2009b) ISBN:978-1.84800-078-0_1

Part III

Applications of Semantic Agent Systems

Chapter 9
A Semantic Agent Framework for Cyber-Physical Systems

Jing Lin, Sahra Sedigh, and Ann Miller

Department of Electrical and Computer Engineering
Missouri University of Science and Technology
Rolla, MO, USA, 65409
{jlpg2,sedighs,milleran}@mst.edu

Abstract. The development of accurate models for cyber-physical systems (CPSs) is hampered by the complexity of these systems, fundamental differences in the operation of cyber and physical components, and significant interdependencies among these components. Agent-based modeling shows promise in overcoming these challenges, due to the flexibility of software agents as autonomous and intelligent decision-making components. Semantic agent systems are even more capable, as the structure they provide facilitates the extraction of meaningful content from the data provided to the software agents. In this book chapter, we present a multi-agent model for a CPS, where the semantic capabilities are underpinned by sensor networks that provide information about the physical operation to the cyber infrastructure. As a specific example of the semantic interpretation of raw sensor data streams, we present a failure detection ontology for an intelligent water distribution network as a model CPS. The ontology represents physical entities in the CPS, as well as the information extraction, analysis and processing that takes place in relation to these entities. The chapter concludes with introduction of a semantic agent framework for CPS, and presentation of a sample implementation of the framework using C++.

Keywords: cyber-physical systems, agent-based modeling, semantic capabilities, fault detection, multi-agent system, sensor networks, intelligent water distribution.

1 Introduction

The synergy between agent-based modeling and semantic technologies holds promise for the resolution of challenges posed by a broad range of complex systems, in particular cyber-physical systems (CPSs), where embedded computing and communication capabilities are used to streamline and fortify the operation of a physical system [1]. In CPSs, sensors collect information about the physical operation of the system, and communicate this information in real-time to the

A. Elçi, M.T. Koné, and M.A. Orgun (Eds.): Semantic Agent Systems, SCI 344, pp. 189–213.
springerlink.com   © Springer-Verlag Berlin Heidelberg 2011

computers and embedded systems used for intelligent control. These cyber components use computational intelligence to process the information and determine appropriate control settings for physical components of the system, such as devices used to control the flow of a physical commodity, e.g., water or electric power, on a line.

A fundamental challenge in research related to CPSs is accurate modeling and representation of these systems, especially as related to reliability. Simplistic models that assume components fail independently are rendered unusable for the majority of CPSs, due to significant interdependencies within the cyber and physical infrastructures, respectively, and across the cyber-physical boundary. In other words, modeling of any CPS is hampered by the need to model both the cyber (software, communication network, computing hardware) and the physical infrastructure (physical components and their interactions). Furthermore, the application of graph-theoretic models is complicated by heterogeneity in the notion of "flow" in CPSs. "Information" is the flow on the cyber infrastructure that provides communication and computing capabilities. The flow on the physical infrastructure is domain-specific, e.g., power for an electric power grid or vehicles for a ground transportation system. Both types of flow need to be represented accurately, such that effects of any event are reflected in either or both networks. Thirdly, existing explicit communication protocols used to impart information between the cyber and physical infrastructures do not fully capture the semantics of the interaction between the two. The vision of using distributed computing resources in the cyber networks to manage the distributed resources in the physical infrastructure further complicates modeling of CPSs.

Among existing techniques, agent-based modeling holds promise in surmounting the aforementioned challenge, due to its capability of encapsulating diverse attributes within one agent, as well as its emphasis on the interaction among autonomous, heterogeneous agents, which share a common goal achieved in a distributed fashion. Sensors are the key to this approach, as they provide situational awareness to the agents and enable them to function based on the semantics of their mission and the specifics of their environment. The research presented in this book chapter aims to accurately model a CPS as a multi-agent system, where each agent is an independent entity that manages resources within its local scope. In the proposed model, information from the sensor networks is dynamically integrated with semantic services to support real-time decision support in the information-rich environment of a CPS.

The CPS domain used as a case study for an application of this model is intelligent *water distribution networks* (WDNs). In a WDN, physical components, e.g., valves, pipes, and reservoirs, are coupled with the hardware and software that supports intelligent water allocation. Fig. 1 depicts a sample WDN.

The primary goal of WDNs is to provide a dependable source of potable water to the public. Information such as demand patterns, water quantity (flow and pressure head), and water quality (contaminants and minerals) is critical in achieving this goal, and beneficial in guiding maintenance efforts and identifying vulnerable areas requiring fortification and/or monitoring. Sensors dispersed in the physical

Tank

Pump

Valve

Reservoirs (storage tanks)

Junction

Customer (industrial,

commercial, residential...)

Transmission line

Operator workstations

Field data interface devices (RTU, remote data unit)

MUX

Server

Communication line (arrow shows the direction)

Command control line (arrow shows the direction)

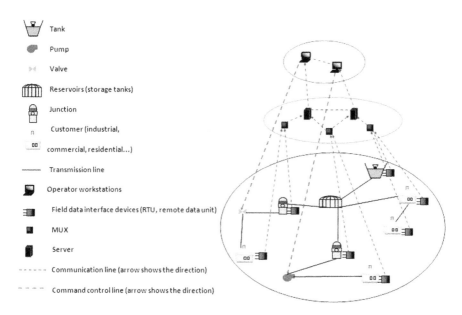

Fig. 1. Cyber and physical components of an intelligent WDN

infrastructure collect this information, which is summed by multiplexers and servers for hierarchical semantics interpretation. The processed and reasoned sensor data is then fed to distributed algorithms running on the cyber networks. These algorithms provide decision support to hardware controllers that are used to manage the allocation (quantity) and chemical composition (quality) of the water. The algorithms are implemented through software executing on multiple distributed computing devices. This software is represented by the agents in our model, each of which is capable of perceiving its environment, acting on that perception, communicating with other agents and exhibiting behavior that fits its goal.

This book chapter is an extension of our previous work, where we first articulated the use of semantic agents in modeling CPSs [2]. The extended content includes: a more detailed discussion on the agent-based modeling technique and its use in addressing design challenges in complex CPS, a more comprehensive presentation of our work on construction of semantic agent framework, and an introduction to data type processing.

The remainder of this book chapter is organized as follows. Section 2 presents an overview of related literature. In Section 3, we present tools and procedures to construct an agent-based model, the method for defining an agent, and a UML multi-agent model that captures the static structure and dynamic behavior of a WDN. The semantic interpretation service is elaborated upon in Section 4, where the sensor information ontology and associated semantic service model are defined. In Section 4, we also propose a semantic agent framework for interpretation of the semantics of raw data streams, describe data type processing of the raw data stream, and provide an overview of implementing semantic interpretation capabilities through C++ on Matlab. Section 5 concludes the book chapter and describes future research directions.

2 Background Work

CPSs are an emerging research area, and the body of related literature is limited. A considerable fraction of related work examines critical infrastructure systems, which are prime examples of CPSs. Salient studies, e.g., [3, 4, 5, 6] are on interdependencies among different components of critical infrastructure systems, and [3], which provides a relatively comprehensive summary of modeling and simulation techniques for critical infrastructure systems. System complexity has been identified as the main challenge in characterizing interdependencies in CPSs [4]. Other challenges include the low probability of occurrence of critical events, differences in time scales and geographical locations, and the difficulty of gathering the accurate data needed for modeling. These challenges are clearly articulated in the literature, but solutions are very scarce.

The need to use agent-based modeling for distributed complex system has been investigated in [7]. The work in [8] adopts a distributed multi-agent architecture to analyze the observed information in real-time to adapt the multi-agent system to the evolution of its environment. To address the dependability issue in multi-agent system, [9] improves the capability of calculating how critical an agent is to the system through its interactions with other agents and provides a framework that uses this information to ensure availability and reliability. A multi-agent system (MAS) in [10] represents a powerful model to solve distributed computation problems. A particularly relevant study is presented in [11], where agent-based modeling is used to estimate residential water demand. An agent community is assigned to behave as water consumers, and econometric and social models are incorporated for estimating their water consumption. However, this study considers the WDN as a purely physical system with no cyber control.

As a formal specification language with precise semantics, UML 2.0 has been adopted to model multi-agent systems with precise semantics. A detailed demonstration of how UML 2.0 can be used for the specification of an agent-based system has been presented in [12]. UML 2.0 has been adopted during the analysis and design phase in [13], to model the physical and social contexts for embedded multi-agent systems. The specification of Action Semantics (AS) in [14] shows how the applicability of AS to the UML meta-model paves the way for powerful meta-programming for model transformation.

Semantic agent technologies are typically closely associated with sensor networks, and several prototype systems or software architectures have been proposed based on the combination of the two. A prototype for battlefield information systems has been described in [15], where the stated goal is to dynamically integrate sensor networks with information fusion processes to support real-time sensing, interpretation, and decision-making in an information-rich tactical environment. In [16], an architecture and programming model for a semantic service-oriented sensor information platform has been presented. In contrast to [16] our work expands the semantic service model to a semantic agent framework, whereas [16] focuses on how to use the semantic model to query the system for high-level events without processing raw sensed signals. The use of autonomous semantic agents in developing new software architecture for distributed processing environments has been proposed in [17]. The discussion in [17] involves software architecture in general, and utilizes semantic web technologies; whereas our work is tailored to the specific requirements of CPSs. Due to the stringent security requirements of critical infrastructure and the vulnerabilities of web technologies, we do utilize them at this stage of our research.

The complexity of CPSs, as well as the necessity of capturing embedded computing and communication capabilities motivates the use of distributed agents and semantic services for representing the relationship between the cyber and physical infrastructures. In our work, the distributed semantic agent model augments the data acquisition of sensors in the CPS with ontological decision-making intelligence. The proposed model not only captures the complexity of the CPS in a clear and understandable way, but also takes accurate semantic interpretation into consideration. To our knowledge, our work is the first study to apply semantic agents to modeling of CPSs.

3   Agent-Based Modeling Technology

As a visual modeling language for representing object-oriented systems, UML is an intuitive choice for supporting agent-based modeling, in both the design and the communication phases. UML consists of several types of structured diagrams and graphical elements that are assembled to represent a model. The high level of abstraction is independent of the implementation of the model, especially when an object-oriented programming language is used.

Generally, by our definition, the agent is a piece of software code with intelligent decision-making functionality. An agent can be considered a self-directed object with the capability to autonomously choose actions based on its situation, and therefore the object-oriented paradigm is a useful basis for agent modeling. Object classes can be used as agent templates, and object methods can represent agent behaviors. The data-driven, rather than process-driven perspective of object-oriented modeling also makes it well-suited to agent-based modeling.

The construction of an agent-based model can be broken down into the following steps, each of which is described in one of the subsections that follow.

1)   Defining the agents in the context of the system; and identifying attributes of the agent; and other classes, along with their attributes.
2)   Defining the environment where the agents reside, and the objects with which the agents interact.
3)   Designing the methods by which agent attributes will be updated in response to agent-to-agent interactions or agent interactions with the environment.
4)   Implementing the designed agent model in modeling software.

Our work specifically defines agent as software code and differentiates agents from the other devices, such as sensors and actuators. In contrast, architectures proposed in a number of other studies place the agents in the context of embedded devices. For example, in [18], the agent construction model is composed of components that are the basic building blocks for an agent; and the generic functionalities of these components are further divided into information collection (sensors), information storage (infostores), decision-making (controllers), and affecting change in the environment (actuators). In this book chapter, our focus is on the software aspects of agents, particularly on the implementation of semantic interpretation. Related work takes a more application-specific approach, e.g., the study in [19] discussed software aspects of information agents in a pervasive computing environment.

3.1 Definition of the Agents

In our defined context of WDNs, the agent is the software code embedded in one or more computing devices on the cyberinfrastructure, with the goal of exerting control over components of the physical infrastructure. Other than this software, all other system components or subsystems are mechanical or hardware parts, including all mechanical water facilities, e.g., pumps, valves, reservoirs, water consumption junctions; communication links and sensors; or even more intelligent field-programmable gate array (FPGA) or programmable logic controllers (PLC) devices.

Regardless of their specific task, agents share the following characteristics.

1) An agent is an identifiable and discrete individual, as each segment of software code is located on distributed PCs to control a local water area. The subprogram code inherits the attributes and methods of the main program and develops its unique attributes or operations to manage water resources within its scope. Therefore, it is constrained by rules governing its behavior, and in possession of decision-making capability.

2) An agent is situated in an environment where it interacts with other agents. In our model, each agent is in charge of its local scope, but they collectively interact for information sharing, data transmission and parallel computing.

3) An agent is goal-directed. Its major tasks include managing the raw data, using real-time data to quantify the overall reliability of the CPS, making a decision to take appropriate action if risk is anticipated in the near future, and sending control commands to actuators to meet the broader system objective or prevent potential damage. Approaches adopted for decision-making include game theory, which can be used to allocate water resources; the Leontief model [20], which can be applied to quantify the effect of a failure in one scope on operation of another scope; and Markovian models, which can estimate the likelihood of a transition from the current state to a given future state.

4) An agent is flexible, due to its nature as a segment of code. It can learn and adapt its behavior to the environment, based on new information, which includes data from sensors or from peer agents; and experience, such as data retrieved from a history database.

5)   An agent is responsible for its intelligent semantic inference. After receiving raw data from the sensors, each agent should firstly check the integrity of the data, specifically, whether the data is deemed legitimate per scientific hydraulic relationships among its various physical parameters; and whether the data is reasonable, compared with history data and that of surrounding nodes. A large number of failures can be screened out through this procedure, and the redundancy of data can be greatly reduced through semantic aggregation.

6)   An agent requires some form of memory, either on the computing device or in a separate database, to store the data of various water attributes for a period of time.

3.2 Construction of an Agent-Based Model

In this section, we present an agent-based model for an intelligent WDN, as a case study of CPSs. We use the various types of UML diagram to gain insights to the system functionality, property and behavior of system components, software architecture, and the dynamics aspects of the complex system.

- **Use Case Diagram**

Creating a use case diagram is the first step for system analysis. A use case captures the interaction of a number of external actors with the system towards accomplishment of a goal. Fig. 2 shows the actor and the use cases involved in the intelligent WDN. The use case diagram presented here can be generalized to other CPSs whose main goal is management of a physical commodity. Examples

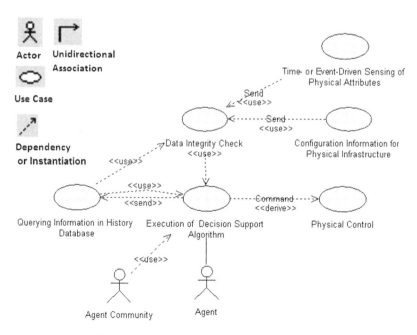

Fig. 2. Use case diagram of an intelligent WDN.

include power grids and intelligent transportation systems. As described in Section 1, the primary goal of WDNs is to provide a dependable source of potable water to the public. The specific role of the agents in the system is to intelligently guide water allocation, per the algorithm programmed in the agents.

The CPS agent is the actor in the use case diagram, and associated with the decision support algorithm. For simplicity, only use cases associated with one agent are shown in Fig. 2; all other agents have similar use cases associated with them. As shown in the use case diagram, sensors collect information about the physical operation of the system on a time- or event-triggered basis. The collected information is aggregated by a multiplexer and sent to the *Data Integrity Check* for intelligent semantic inference. The *Data Integrity Check* use case uses three main data streams, specifically, raw data from the corresponding sensor, real-time data from nearby sensors for the same or related physical attributes, and the data from a history database. The second and third data streams mentioned are used for corroboration of the first, by checking for discrepancies in the values, whether in variation or in conformance to physical (hydraulic) laws that govern the physical operation of the WDN. If no data is available from nearby sensors, as would be the case if all nearby sensors are in sleep mode, the history database will serve as a source of data for corroboration. As indicated in Fig. 3, the values of physical attributes, such as water quantity, of nearby nodes, should not be significantly different from each other for the same time period. For instance, adjacent water nodes should have similar water temperature and similar water pressure value. If significant discrepancy exists, the use case can conclude that the collected data may not be a legitimate group of data and should not be used for further information processing.

Fig. 3. Flow among nearby nodes

The decision support algorithm uses three data streams, one data stream from the *Data Integrity Check*, another from the history database, and a third data stream from other agents. The decision support algorithm is an advanced algorithm implemented through software code for intelligent management of physical commodities. The algorithm can make use of legitimate (corroborated) data whose integrity has been checked, and can also resort to history data for adjustment (rectification) of the calculated values in determining an appropriate strategy for resource allocation. Meanwhile, the local agent interacts and negotiates with other agents by sharing real-time information that provides global perspective of resources in the system, and adjusts its own strategy accordingly. For instance, one adjacent agent reports that pipe bursting have been detected and more water is needed from neighboring areas to guarantee regular water consumption before restoration. In this case, the well-being agents will adjust the strategy to maintain the local water consumption and support extra quantity of water to its neighbor. Various algorithms can be the candidate for the decision support algorithm, and the game theory holds the greatest promise.

- **Class Diagram**

Based on the use cases and interconnections defined in Fig. 2, Fig. 4 provides an overview of different classes in the intelligent WDN, along with the specified attributes and the corresponding methods for each class. Fig. 4 also depicts and how the classes interrelate. Other information provided in Fig. 4 includes the data types of the attributes and the main constraints used in the decision making algorithm. The attributes of the water facility classes have been chosen to be most representative of both static (elevation) and dynamic aspects (head loss) of water.

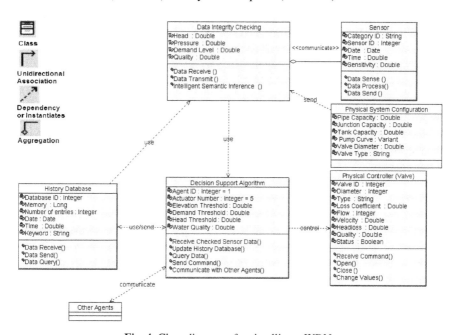

Fig. 4. Class diagram of an intelligent WDN.

The *Data Integrity Checking* class takes three data streams, from *Sensor, Physical System Configuration* and *History Database*, respectively. Data collected by the sensors is aggregated by the multiplexor (representing by the small diamond) and sent for data integrity checking. The *Physical System Configuration* block specifies the basic configuration and topology physical water infrastructure. This configuration data is sent to *Data Integrity Checking* to assist in evaluating physical constraints, e.g., judging whether a newly requested water value (such as quantity) will exceed the capacity of a pipe. History data can be queried by the *Data Integrity Checking* for comparing abnormal real-time data with historical values. Various types of semantic analysis are carried out through *Intelligent Semantic Inference*, including the aforementioned evaluation of physical constraints and corroboration with historical data or data from nearby nodes.

The purpose of this semantic inference is to screen out illegitimate or corrupted data (based on the preliminary judging criteria), to ensure that only legitimate data is sent to the decision making algorithm. A domain ontology for more advanced semantic interpretation and system failure detection based on semantic interpretation will be introduced in Section 4.

The agent has varied types of association with other classes: it receives the data after semantic processing, stores the data in the history database or queries data from the database to assist in decision making (bidirectional), negotiates resource allocation with other agents, and exerts control over actuators (valves and pumps).

- **Component Diagram**

In Fig. 5, the main program that implements water allocation executes on the cyberinfrastructure. The physical location of the main program is immaterial. The main program is directly dependent on the code specification, which is the head file of the agent class. It includes prototype information for the class function. The remainder of the script is the package body, which exhibits functionality similar to that of the main program and executes in distributed fashion within its autonomous management scope. If the script is written in C++, the package body is a .cpp file. An independent database is attached to each script, meaning that the script can only retrieve data from or store data to the database for management purposes within its own scope. All the data sent to the script for advanced semantic analysis or decision making has been checked its integrity, as described earlier in this chapter.

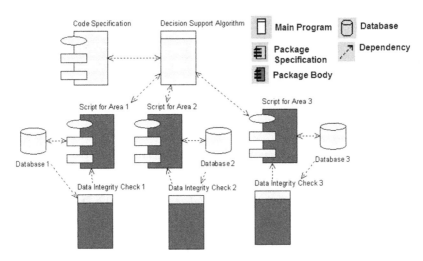

Fig. 5. Component diagram of an intelligent WDN.

- **State Transition Diagram**

Fig. 6 depicts the state transitions of data in one period, which is the time span from the point that data is collected at preset time (start state) until control has been exerted on the water consumption entity (end state). As agent-based modeling is a data-driven modeling method, it is vitally important to track each state transition of the data. The condition that can trigger entry to or exit from a particular state has been specified. The history state (encircled 'H') records the state of the system immediately before query of the history database. Once the agent has finished data retrieval, the state reverts back to the original state before data storage, and the agent begins processing based on the combination of retrieved historical data and the originally collected data. The flow of the decision making procedure, whose goal is to allocate water (quantity), has been specified in the figure with two decision blocks (encircled diamonds).

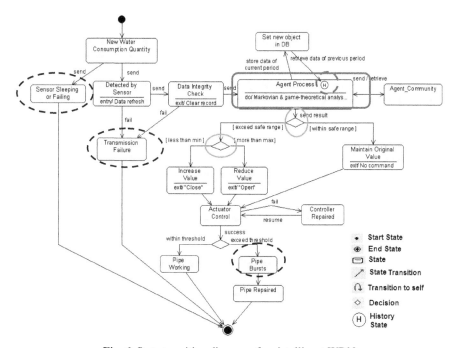

Fig. 6. State transition diagram of an intelligent WDN.

The *Agent Process* (in a solid rectangle) is the critical state within the context, as it provides a precise numeric value to guide the control over actuators. Markov and game-theoretic analysis are included in the state as instances. Since the state of the data within the system has already been identified through the agent-based model, we can use a vector to numerically represent these states. Failure is a state transition from functional to non-functional, such as the Sensor *Sleeping or Failing, Transmission Failure,* and *Pipe Bursts* states (all in dotted ellipses) shown in the figure. We can define the normal functioning state to be 1 and failure state to be 0. Therefore, vectors formed by 0 and 1 can precisely represent the state of the system. At this point, to identify the functional states, a Markov reliability model can be built to estimate the probability that the next state of the system is an operational state.

Game-theoretic analysis can be applied to calculate the equilibrium state of the water allocation among the agents. In the context of a city, the water quantity allocated to each sub-area can be determined, subject to the constraints of the physical facility. Base on the threshold values of the constraints, the optimal water allocation scheme can be obtained as well.

- **Activity Diagram**

In Fig. 7, which depicts the activity diagram for an intelligent WDN, three entities are involved, including the physical networks; agent 1, acting as the main agent; and agent 2 as the agent interacting with agent 1.

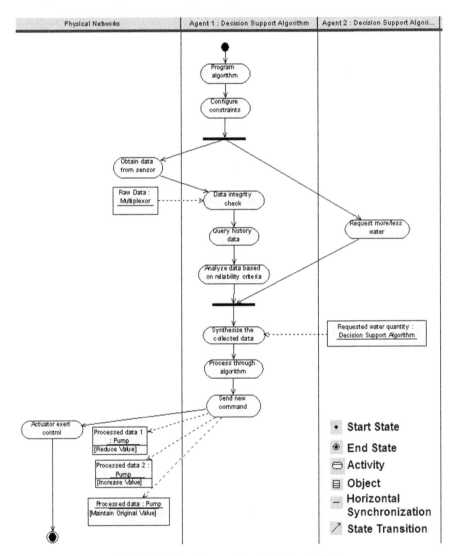

Fig. 7. Activity diagram of an intelligent WDN.

The activity diagram reflects how an agent interacts with the environment, and how the values in the associated object change after date integrity checking and data processing. For instance, the raw data is changed into semantically-processed data for control, and the requested water quantity of one agent may affect another agent's water consumption quantity.

- **Sequence Diagram**

Fig. 8 depicts the sequence of messages exchanged among different entities in the intelligent WDN. The message on the line shows the method adopted by the receiver (class defined in the class diagram) upon receiving the message. The figure shows the sequence of data received by the data integrity checking object and the decision support algorithm object of agent. For the former object, it directly receives and checks the raw data from the sensors (collected by multiplexor) and then if it needs to compare the real-time data with previous history data, it will receive data from its local database to make sure the result of checking is based on a reliable history record. The water consumer object and the adjacent agent object are eliminated after they send the return message, which means that no message from these two objects will be accepted outside of particular periods. The decision support algorithm of the agent first receives checked sensor data first, queries data from the history database, and finally communicates with the adjacent agent. Such a sequence is from the physical infrastructure to the cyber infrastructure (bottom-up). After the decision has been made, the calculated result will be sent to the community agent first, then a command will be sent to actuator to exert real-time control over the physical commodity, and finally the calculated data is recorded as history data in the database. Such a sequence is from the cyber network to the physical infrastructure (top-down), culminating in data recording.

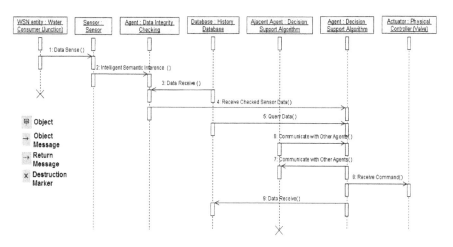

Fig. 8. Sequence diagram of an intelligent WDN.

A clear and correct sequence diagram of the agent-based WDN is the prerequisite for resolving challenges related to timing in CPS modeling. As the CPS is a two-layered system, the intelligent agents make decisions based on the collected data, but when the control command is sent back the actuator, the previous data for computation has already changed. Therefore, how to select an appropriate cycle period and how to process the changing data are open problems for further research.

4 Semantic Interpretation Services

4.1 Sensor Information Ontology

Semantic interpretation is carried out on semantic streams, each of which is defined in a domain-specific ontology associated with the agent. The specific domain in this book chapter is intelligent water distribution. Generally, an ontology is a description, e.g., a formal specification of a program, of the concepts and relationships that can exist for an agent or a community of agents. The notion of ontology utilized in this book chapter is a model that describes semantic relations among components of the physical and cyber infrastructures, respectively, as well as the interdependencies across the cyber-physical boundary. Each component in the ontology model is a unique class in terms of programming implementation, with properties and parameters described in the class definition. The relations define how classes can be related to one another. Semantic interpretation is implemented through distributed software with capabilities of extraction, analysis, and processing of the semantic stream. The definition of ontology for the WDN domain helps unify information presentation and permits software and information reuse, so as to reduce information redundancy during the process of semantic interpretation in the agents.

The use case diagram in Fig. 2 depicts intelligent control of the physical infrastructure by the cyberinfrastructure of an intelligent WDN. To achieve the goal of intelligent management and control, a number of tasks are involved to implement various functionalities (use cases), such as the pre-processing of the raw data from the sensors, coordinating the time sequence to query data from the history database and communicate with peer agents, converting the logical command to physical control over actuators, and so on. As ontology has advantage over other information representations in terms of capturing the structure and meaning of information, we use ontology to represent the failure detection procedure, which is an important component of the intelligent information reasoning functionality in CPSs. Similar ontologies can be identified for other functionalities or use cases of a CPS.

Fig. 9 shows the information hierarchy for failure detection through the semantic interpretation process. In the UML class diagram, each block represents one type of semantic stream in the intelligent WDN. The attributes of each class have been omitted in the interest of figure clarity. Details of the attributes are presented in Fig. 14, which shows pseudo code for the semantic service.

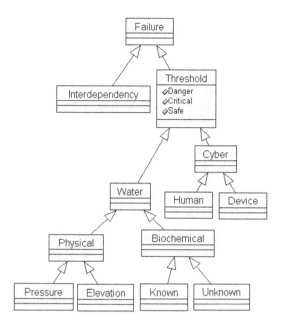

Fig. 9. UML representation of failure detection ontology in CPS for an intelligent WDN

Fig. 9 shows that a failure in the WDN can be detected by the agent in the event of device or information failure, the latter of which occurs when data falls outside a pre-defined safety range. Failures in the physical infrastructure of a WDN are of two main two types, physical failure due to excessive values of pressure and elevation, or biochemical failure due to excessive quantities of a biochemical substance or discovery of unknown biochemical materials. Cyber failures can be caused by human error or device malfunction. Each class identifies one type of semantic stream that can lead to failure in the CPS, and the ultimate determination of failure (or the overall interpretation) is carried out by the corresponding agent, which is in charge of all sensors deployed within its administrative scope.

The main reason that the attributes of the class have not been defined here is for simplifying reasoning procedure on information. For instance, the danger threshold can be triggered by both excessive water quantity or cyber malfunction, but the excessive quantity of one single attribute of the water class is sufficient to diagnose the source of failure is from physical networks. Besides, the undefined attributes can help to reduce the semantic redundancy in terms of automatic semantic conversion, which not every property field of a class needs to be filled or met before performing detection. For example, to identify biochemical attack from the excessive biochemical quantity (such as excessive bacteria), the agent can just check if the detected biochemical element falls into the database of known elements. As long as one type of elements is unknown, even other co-existing elements fall into the knowledgeable scope, the agent can immediately determine that a failure can be caused by the unknown biochemical element.

The sensor information ontology captures the semantic entities (classes in the UML diagram) and the relations of events and objects, deriving a reasoning procedure beyond what sensors can directly provide through detection. The ontology proposed in Fig. 9 is specific to the WDN domain, but can be readily adapted to other CPSs, such as smart power grids.

4.2 Model for Semantic Services

Based on the sensor information ontology proposed, we can develop components that convert semantics between classes in the information processing hierarchy, by extracting new semantic information from existing data streams. In other words, the components encapsulate the semantic service into a ``black-box" containing the execution method, which takes as input information (defined as precondition [21]) corresponding to events detected by sensors and generates as output a number of meaningful new events (defined as postcondition [21]). The process is depicted in Fig. 10.

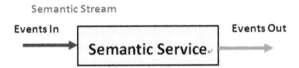

Fig. 10. Semantic service on the semantic stream

We propose a semantic service model that overlays the ontology defined in Fig. 9. The semantic service model allows the agents (users) to annotate semantics of data transmitted between the services of each entity on the ontology, and can check and automatically convert between data semantic whenever possible. As the services are event-driven, the events passing between services not only carry their value information, but also serve as triggers for service execution. The semantic services can be categorized into two types, i.e., a) the service that supplements input events with additional semantic annotation, and b) the service that produces new semantic streams.

The first type of service can only identify additional properties carried by the input event. For example, a sensor has detected that the water pressure in a certain area has exceeded the safety threshold and reports this event to its semantic service component, which can be a superior sensor or multiplexer. The semantic service model associated with this component will add the geographical location as an additional identifier to distinguish this event from events reported from other areas. Such functionality is particularly useful for distributed control and management in the context of CPS, where a service may not correspond to a centralized component that physically exists on one device; it can be physically implemented on several distributed devices, but logically exists as a single service.

The second type of service automatically terminates the input semantic stream, and uses the generated output semantic stream as the new stream propagating on the ontology. The essence of this type of service is semantic transformation, where the input and output events are different classes in the ontology. One typical semantic transformation is generalization. For example, in Fig. 9, an excessive pressure quantity will be interpreted as physical failure due to an abnormal pressure value. Later on, the semantic stream of physical failure will be propagated to a higher level for ultimate decision making, instead of the semantic stream of abnormal pressure quantity, which no longer exists. This case will be illustrated by the code in Fig. 14. Another example can be the derivation of danger event by passing the threshold component, which abstracts the possible sources of dangers in terms of water failure and cyber failure. Such case falls into the category of generalization.

The benefits of proposing such a semantic service model on information ontology include the reduction of information redundancy, pre-processing and abstraction of data for the agent, and the facilitation of semantic query by a user. A user can issue a query that requests that a certain data stream with desired semantics be provided to a certain component device to diagnose whether failure exists on the queried level.

4.3 Semantic Agent Framework

Fig. 11 illustrates how the agents use the information detected by sensor networks and the interpreted semantics through components based on the defined ontology. Raw data is obtained from sensor networks, and since each agent is an independent entity in charge of a particular geographical area, the sensors located in distributed areas are managed by different agents (with possible overlap). For a semantic service component, the input semantic events are preconditions of the service. The postconditions, i.e., the processed output semantics, are provided to the agents for further computing.

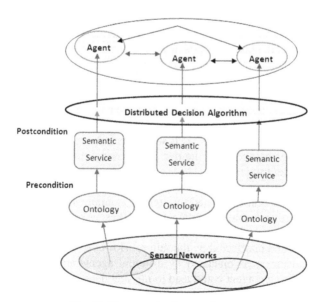

Fig. 11. Framework of semantic agents

The agents must host uncoordinated tasks, which are events triggered by different physical events and sensor data categorized in the ontology at unpredictable times. The data collected and process carried out strongly depends on the agents' surroundings. It is very likely that redundant information will exist among the concurrent tasks of different agents. For example, a pipe burst can impact several agents corresponding to nearby areas. Therefore, parts of the information for these tasks can be shared. To reduce the redundancy and use the computing resources more economically, we adopt a distributed decision algorithm, such as Maxflow, executing in parallel on multiple agents [22]. In the distributed decision algorithm, each agent uses only a portion of the computing resources to process the data within its own administrative scope, and the agents circulate the calculated results among each other to share information that may be helpful to the local decision making strategy.

A common application case is water allocation, and the algorithm we have adopted in the model is game theory. The details of our work have been presented in [23]. The sensors located in different areas collect water quantity information and send the data to the game theory algorithm. The water allocation strategy aims to achieve Nash equilibrium among the participating players - agents in the model. The equilibrium strategy seeks to achieve more efficient and fair water allocation, as compared with the low-level self-regulation that would occur if physical and hydraulic rules alone are left to govern the water flow. The proposed work can be further refined by introducing additional constraints, such as different pipeline thresholds and the maximum/minimum water quantity requirements of different water consumers.

However, some limitations exist in the model. The distributed algorithm is adopted by the agents due to limited computing resources, and the agent is responsible for the final decision used to exert control over the physical network. Several factors will impact this final decision. As shown in the use case diagram (Fig. 2), the execution of the decision support algorithm takes into consideration the data after integrity checking and comparison with historical values, and will also be affected by data from nearby agents. If the decision algorithm waits for data from all three sources to become available, the timeliness of the decision cannot be guaranteed, particularly if the speed of water flow is high and water attributes change rapidly. If the decision algorithm does not wait for data from all sources to be ready, its implementation may suffer the risk that the final decision does not meet the requirements or violates constraints, as it will already be outdated by the time it is made. Designating the agents as the decision making authority should take such factors into consideration.

Another shortcoming of the hierarchical organization is information asymmetry. This problem is best illustrated through an example scenario, where the program segment used to manage a certain area crashes. As the computing resources and the algorithm are distributed, the code running on other computers should remain operational and unaffected by the failure. However, water flow is a dynamic commodity, and the areas managed by different agents are interconnected. The failure of even a single agent could lead to missing data, which in turn can lead to flawed decision-making by other agents, and potentially a cascading failure of a significant portion of the system. Such a scenario serves as a cautionary tale of the vulnerabilities introduced by the use of cyber control.

The fault tolerance of the model is limited by several factors. A number of them has are apparent from the state transition diagram of Fig. 6. Data collection by the sensors may suffer the risk of sensor failure or sleep, and information loss can be caused by transmission failure. Lost, delayed or incomplete data can directly affect the functionality of higher-level components. Besides, the cyber components suffer risks from computer crash, disconnect of communication links, and internal design issues of the decision algorithm, such as interference among the agents. Increasing the robustness of the system by addressing these issues is an open research topic.

Before the input events are processed in the semantic service model, the event stream will undergo data type processing, including data type definition, data type checking, and data type conversion. A number of issues related to timing synchronization and sampling remain to be resolved for the data type processing as well.

4.4 Data Type Processing

The main purpose of data type processing is to reduce runtime redundancies, based on event semantics. The events in the event-driven model serve two roles: carrying values and triggering further services defined in the failure detection ontology. It is crucial for agents to identify the maximum sensing overlap and to reduce runtime redundancy, which is an intermediate information reuse and summarization problem. In light of the ontology defined in Fig. 9 and the semantic agent framework of Fig. 11, the intermediate data processing should carry out three functions: a) identification of overlapping information from multiple sensor nodes, including those that collect physical water data, others that monitor communication links, and yet others that supervise the cyber infrastructure; b) suppress parts of the data that are useless for failure detection, keeping only the critical information active and sending it to the higher-level entity; and c) sharing intermediate data with its peer entities. The second and third attributes can be realized in the service-oriented architecture proposed in Fig. 11, and will be elaborated upon in Section 4.4 and Section 4.5. To maximally identify redundant sensing is the task of data type processing.

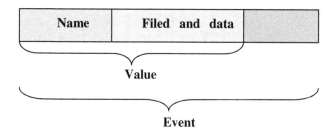

Fig. 12. Visual representation of an event

Based on inspiration from [21] and [24], we define an event as a tuple with two elements: a *value* and a *tag*. In contrast to [24], we define a triggering event as a signal consisting of its respective value and tag types. The composition of an event can be represented as Fig. 12.

The value of an event is represented in the following form:

$$V = (name, \{n_1,p_1\},\{n_2,p_2,...,(n_k,p_k)\}) , \tag{1}$$

where *name* represents the name of the signal, and (n_i,p_i) represents the field and the data type. Such a representation of *value* can help parameterize the event and facilitate implementation of the service in the subsequent stage. For example, the *value* of an event that corresponds to measurement of physical attributes of water can be represented as:

water sensing = (water_sensor, {geoID, int}, {width, float}, {length, float},
$$\{height, float\}) \tag{2}$$

Representation of events with a *value* also facilitates information aggregation. Based on record types in OCL programming languages [25], a pure specification language for expression, if two events have identical names, but the field type of one value is a subclass of the other, then the event value with a more generic field type can subsume the other event value to reduce data redundancy, while keeping the unique field type of the subclass. The following instance can be aggregated by (2).

water sensing = (water_sensor, {geoID, int}, {width, float}, {length, float},
$$\{height, float\}, \{biochemical, float\}) \tag{3}$$

Aggregation of subclass data type is depicted in Fig. 13 (a), where the larger area denotes the common field type, and the smaller area denotes the unique field type of the subclass. After aggregation, the event record with the subclass field type does not exist. Information reuse, as an extension of information aggregation, keeps the common field type and the unique field type of the subclass separate, as shown as Fig. 13 (b).

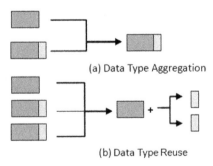

(a) Data Type Aggregation

(b) Data Type Reuse

Fig. 13. Data type processing

```
Sensor{
water_sensor, geoID,[width,length,height], /* properties of sensors:
        water detection, geographical ID, location*/
Outputs(pressure, elevation, biochemical, location, T1);  /*the pa-
                    rameters can be detected by sensor at time T1* /
  }

Pressure component:
Pressure_Service{
service(pressure),   /*service indicates execution method and the
                        parameter is pressure*/
Inputs (sensor(water_sensor, geoID, [width,length,height], T1);
If (pressure > normal range)
Outputs (pressure_normal (false), detected (pressure,geoID, T2));
/*add the judgment result and time T2 when make judgment*/
  }

Physical component:
Physical_Service{
service(physical_failure),
Inputs(pressure_service(pressure_normal(false), detected (pres-
sure,geoID,T2)));
If ((elevation < normal range) && (pressure_normal = false))
             /*guarantees pressure is the unique reason*/
Outputs(physical_normal(false), de-
tected(physical_failure,pressure,geoID,T2)); /*inherit inferior
attribute*/
  }

Water component:
Water_Service{
service(water_failure),
Inputs(physical_service(normal(false),
(physical_failure,pressure,geoID,T2)));
If ((biochemical_normal = true) && (physical_normal = false)) /*same
as above*/
Outputs(water_normal(false), detected(water_failure,physical
_failure,pressure,geoID,T2));
  }

Threshold component:
Threshold_Service{
service(failure),
Inputs(physical_service(normal(false),
detected(physical_failure,pressure,geoID,T2)));
Switch(detected(pressure))
  {
Case (within range for safe): service terminates;
Case (within range for critical): send (pressure,geoID,T2) to
database;
Case (within range for safe): output system failure alert;
Default: service terminates;
  }
Outputs(system_failure_alert, detected( water_failure,physical
_failure,pressure,geoID,T2));
  }
```

Fig. 14. Pseudocode for semantic service

Tags are utilized to represent timing and ordering relations among events. Sophisticated timing issues include the sampling frequency of sensors, difference and conversion between discrete time signals and continuous time signals, interpolation to merge different timing signals, and so on. More detailed information can be found in [24]. In this chapter, we simply use time, T, to represent the tag for each event.

Upon defining the value and tag, we can easily check and if necessary, convert the data type. In checking the data type, the focus is on checking the field type of the data, with the premise that the data has passed the integrity check. Data type checking can facilitate reuse of services from existing tasks, by comparing newly injected information about an event with existing information, including matching the names of the event, checking for the existence of subset relations in the field type, and checking the data types in the field. The focus of data type conversion is mainly on reconciling the tag value of the triggering event with the tag value of the subsequently triggered event. As the tags dictate the timing and ordering relations among events, synchronization issues needs to be resolved, and the solution method depends on whether the triggered event is periodic or aperiodic. Interpolation is one solution method for periodic signals [24].

4.5 Implementation in C++

The choice of C++ for implementation of the semantic service was motivated by several factors. Firstly, a service component is an information-rich component that needs to define the semantic service for execution, extract information from the input, and produce new semantics at the output. Such logical analysis is best implemented through a high-level programming language such as C++ or JAVA, rather than a computing tool. As the modeling approach adopts UML, it is also natural to use the Object-Constrain Language [26] to specify the pre- and post-conditions and the actions in C++ or JAVA. Secondly, a class in C++ is a good fit for our definition of the service component; the declaration of service properties and the execution method of the service can be encapsulated into one class. Thirdly, Matlab2008b can integrate C++ and support parallel computing, and the integration of the semantic service and computation of the algorithm in Matlab will make simulation of the CPS more compact and faster.

To implement the service in C++, the properties of the service are parameterized, and the execution method of the service becomes the corresponding method in the service class. To illustrate the method, we choose the branch of *Pressure* to *Failure* in Fig. 9 as an example. Sensors are treated as services with only output semantics, which are parameterized into data that can be used by superior service components. Each component has been specified with a service name and a parameter associated with the service. Each service takes the outputs of an inferior component as the input to its execution method, and inherits the parameters to ensure that attributes of a potential failure source (such as pressure, failure time, or location) are not lost during information propagation on the ontology. The pseudo-script for C++ implementation is shown in Fig. 14.

5 Conclusions

CPSs are the topic of emerging research, but existing tools and techniques for modeling them are still limited. A number of related challenges were discussed in this book chapter, with focus on the importance of capturing interdependencies and flow heterogeneity, and streamlining semantic interpretation between the cyber and physical infrastructures. The use of agent-based modeling was proposed, and related methods and tools were introduced. An intelligent WDN was presented as a case study for demonstrating the ability of the technique to capture various facets of the operation of a CPS. A semantic service model based on the definition of ontology was presented, with the goal of reducing information redundancy and simplifying the data interpretation procedure of the agents. The data processing carried out for parameterizing and aggregating the raw data streams was described, as was the implementation of the semantic service model in C++. The proposed model reflects the semantics of intelligent water distribution, but can be modified for use in other CPS domains.

The modeling work presented in this book chapter is a preliminary step that will facilitate the broader goal of modeling CPSs. Future extensions to this work will incorporate sophisticated decision support algorithms, e.g., game theory, for the agents. The semantic service model implemented through C++ will be integrated with Matlab to facilitate the complex computation required. Provision of the semantic service in C++ to the decision support algorithm in Matlab will create an advanced simulation environment for CPSs, which can be invaluable to gaining a more profound understanding of the operation of CPSs.

References

1. Lee, E.: Cyber physical systems: Design challenges. In: Proc. of the 11th IEEE International Symposium on Object Oriented Real-Time Distributed Computing, pp. 363–369 (May 2008)
2. Lin, J., Sedigh, S., Miller, A.: Modeling cyber-physical systems with semantic agents. In: The 5th IEEE Workshop on Engineering Semantic Agent Systems in Conjunction with Proc. of the 34th IEEE International Computer Software and Applications Conference, Seoul, South Korea (July 2010)
3. Rinaldi, S.M.: Modeling and simulating critical infrastructures and their interdependencies. In: Proc. of the 37th Hawaii International Conference on System Sciences (2004)
4. Pederson, P.: Critical infrastructure interdependency modeling: The survey of U.S. and international research (August 2006)

5. Svendsen, N.K., Wolthusen, S.D.: Analysis and statistical properties of critical infrastructure interdependency multiflow models. In: Proc. of the IEEE Information Assurance and Security Workshop, pp. 247–254 (June 2007)
6. Lin, J., Sedigh, S., Miller, A.: Towards integrated simulation of cyber physical systems: A case study on intelligent water distribution. In: The 8th International Conference on Pervasive Intelligence and Computing (2009)
7. Macal, M.C., North, J.M.: Tutorial on agent-based modeling and simulation Part 2: How to model with agents. In: Proc. of the 38th Winter Simulation Conference, pp. 73–83 (2006)
8. Guessoum, Z., Faci, N., Briot, P.J.: Adaptive replication of large scale multi-agent systems - towards a fault-tolerant multi-agent platform. In: Proc. of the 4th International Workshop on Software Engineering for Large-Scale Multi-Agent Systems. ACM, New York (2005)
9. de C Gatti, M.A., de Lucena, C.J., Briot, J.: On fault tolerance in law governed multi-agent systems. In: Proc. of the 5th International Workshop on Software Engineering for Large-Scale Multi-Agent Systems. ACM, New York (2006)
10. Poslad, S.: Specifying protocols for multi-agent systems interaction. ACM Transactions on Autonomous and Adaptive Systems 2(4) (November 2007)
11. Athanasiadis, I.N., Mentes, A.K., et al.: A hybrid agent based model for estimating residential water demand. Simulation 81(3) (March 2005)
12. Bauer, B., Odell, J.: UML 2.0 and agents: How to build agent based systems with the new UML standard. Engineering Applications of Artificial Intelligence 18(2) (2005)
13. Klein, F., Giese, H.: Analysis and design of physical and social contexts in multi agent systems using UML. In: Proc. of the 4th International Workshop on Software Engineering for Large-Scale Multi-Agent Systems. ACM, New York (2005)
14. Sunye, G., Le Guennec, A., Jezequel, J.: Using UML action semantics for model execution and transformation. Information Systems 27, 445–457 (2002)
15. Jiang, G., Chung, W., Cybenko, G.: Semantic agent technologies for tactical sensor networks. In: Proceedings of the SPIE, pp. 311–320 (2003)
16. Liu, J., Zhao, F.: Towards semantic services for sensor-rich information systems. In: 2nd International Conference on Broadband Networks, pp. 44–51 (2005)
17. Elci, A., Rahnama, B.: Consideration on a new software architecture for distributed environments using autonomous semantic agents. In: Proc. of the 29th Annual International Computer Software and Applications Conference (2005)
18. Finin, T., Joshi, A., Kagal, L., Ratsimore, O., Korolev, V., Chen, H.: Information agents for mobile and embedded devices. In: Klusch, M., Zambonelli, F. (eds.) CIA 2001. LNCS (LNAI), vol. 2182, pp. 264–286. Springer, Heidelberg (2001)
19. Ashri, R., Luck, M.: An agent construction model for ubiquitous computing devices. In: Odell, J.J., Giorgini, P., Müller, J.P. (eds.) AOSE 2004. LNCS, vol. 3382, pp. 158–173. Springer, Heidelberg (2005)
20. Haimes, Y.Y., Jiang, P.: Leontief-based model of risk in complex interconnected infrastructures. Journal of Infrastructure Systems 7(1), 1–12 (2001)
21. Lee, E.A., Vincentelli, A.S.: A framework for comparing models of computation. IEEE Transactions on CAD 17(12), 1217–1229 (1998)
22. Armbruster, A., Gosnell, M., et al.: Power Transmission Control Using Distributed Max-Flow. Proc. of the 29th International Computers, Software, and Applications Conference (2005)

23. Lin, J., Sedigh, S., Miller, A.: A Game-Theoretic Approach to Decision Support for Intelligent Water Distribution. In: Hawaii International Conference on System Sciences (January 2011)
24. Liu, J., Cheong, E., Zhao, F.: Semantics-based optimization across uncoordinated tasks in networked embedded systems. In: The International Conference on Embedded Software (September 2005)
25. Mitchell, C.J.: Foundations for Programming Languages. MIT Press, Cambridge (1996)
26. Object Constraint Language Specification Version 2.0, OCL,
   http://www.omg.org/technology/documents/formal/ocl.htm
   (accessed February 20, 2010)

Chapter 10
A Layered Manufacturing System Architecture Supported with Semantic Agent Capabilities

Munir Merdan[1], Mathieu Vallée[2], Thomas Moser[3], and Stefan Biffl[3]

[1] Automation and Control Institute (ACIN), Vienna University of Technology,
Vienna, Austria
merdan@acin.tuwien.ac.at
[2] Institute of Computer Technology, Vienna University of Technology, Vienna, Austria
vallee@ict.tuwien.ac.at
[3] Christian Doppler Lab Software Engineering Integration for Flexible Automation Systems
Vienna University of Technology, Vienna, Austria
{thomas.moser,stefan.biffl}@tuwien.ac.at

Abstract. Manufacturing control systems are a mission-critical application domain for semantic agents systems. While multi-agent systems have been explored in the manufacturing systems domain, there is very little work on semantically enabled agent systems. This chapter introduces a layered architecture for manufacturing systems based on agent systems and discusses relevant capabilities of the semantic agents based on real-world use cases.

1 Introduction

The manufacturing sector, faced with growth in the variety of products and at the same time with a decreasing product life cycle duration, is forced by global competition to produce customized products in a short time at low price. Manufacturers have to be capable to effectively react to sudden changes in customer demands, as well as to cope with unpredictable events such as failures and disruptions. The aspects of complexity and flexibility of mission-critical manufacturing systems make this domain interesting as test bed for semantic agents systems.

A manufacturing system is defined as "*a collection or arrangement of operations and processes [...] to make (a) desired product(s) or component(s)*" [9]. Such a system consists of interrelated elements (people, equipment, sub-systems, etc.) introduced to cooperatively achieve the overall objective of transforming raw material into commercial products effectively, efficiently, and robust against failures.

The control systems, which are currently applied in practice, usually consist of heterogonous units, which use different types of data and data structures, and are not capable to ensure the uninterrupted flow of information between and sometimes through the controlled levels [32]. The applied methodologies in these systems are based on disconnected ordering, scheduling and execution processes, and lack the

A. Elçi, M.T. Koné, and M.A. Orgun (Eds.): Semantic Agent Systems, SCI 344, pp. 215–242.
springerlink.com

agility needed for enterprise-wide integration. Process planning is usually separated from scheduling as well as control activities and unnecessary information gaps between implicated systems are created, even though the outputs and data from one application could be fluently used as inputs for another application [50] .

Multi-agent system (MAS) technologies offer a convenient way to cope with dynamics in large complex systems, using distributed control of the system, thereby reducing the complexity, increasing flexibility and enhancing fault tolerance [22]. This approach replaces a centralized database and control computer by a network of agents, each endowed with a local view of its environment and the ability and authority to respond locally to that environment.

Agents communicate and negotiate with each other in order to perform the operations based on the available local information or to solve possible conflicts. In order to ensure correct understanding of the exchanged messages, agents must have the same presentation of the environment, or at least that part of the shared environment about which they are exchanging information with each other. Ontologies are of vital importance for enabling knowledge interoperations between agents and at the same time a fluent flow the different data from different entities [29]. The ontology can be a description of the concepts and relationships that can exist within a multi-agent system [53].

In this chapter we report on the engineering of the semantic agent architecture in four layers (management, planning, scheduling, and execution) and the related ontology for each particular layer in the application domain, the manufacturing system, and provide lessons learned based on real-world use cases.

2 State of the Art

This section summarizes related work on manufacturing system control, multi-agent systems and on semantic systems.

2.1 Centralized Manufacturing System Control

The factory control is defined *"… as the actuation of a manufacturing plant to make products, using the present and past observed state of the manufacturing plant, and demand from the market"*. It is the fundamental system of a factory, because *"it coordinates the use of the factory's resources, giving the system its purpose and meaning"* [7]. Manufacturing control can be divided into low-level and high-level control [14]. The high-level control (HLC) of the factory is responsible for the coordination of the manufacturing resources and government of the production including the ERP (Enterprise Resource Planning) as well as the MES (Manufacturing Execution System) levels. The low-level control (LLC) is focused on the control of the individual manufacturing resources and their reliable function during the execution of operations organized by the HLC.

In current manufacturing systems, the centralized and hierarchical structures are the most commonly used control architectures. However, due to their rigid character and limited adaptation capabilities such systems respond weakly to frequently changing customer demands in terms of performing necessary changes in the manufacturing environment itself [23, 40, 49]. Additionally, the construction of a centralized system, due to large complexity and the necessity to centralize all logic for sensing, actuating and control into a single entity, usually requires a huge investment, long lead times, and in turn, results in generating a rigid control system [16]. The central controller, as it needs to have the accurate information about each unit in the system in order to make right decisions, can be seen as a single point of failure and its breakdown could stop the whole system [25]. Scheduling, in centralized and hierarchical control structures, is established such that each level creates the scheduling for its subordinate levels having a weak feedback from lower levels and almost without any consultation and coordination with higher layers of neighboring units. Such an approach works well only if everything goes as expected; otherwise it could completely fail when unpredictable disturbances occur [11]. Recently, the holonic manufacturing paradigm was introduced, with the idea to combine hierarchical and decentralised control. A holon, the core component of this architecture, can be part of another holon, e.g., a holon can be broken into several others holons, which in turn can be broken into further holons, which allows the reduction of the problem complexity [18]. Leitao and Restivo presented the multi-enterprise model in a structure similar to the general holonic architecture [31]. The most important and relevant approaches in this area are presented by Babiceanu and Chen [5].

2.2 Multi-Agent Systems as Foundation for Decentralized Control

The application of decentralized control architectures based on autonomous and co-operative units is considered as a promising approach for overcoming the weaknesses of centralized manufacturing control. The multi-agent systems (MAS) approach has been widely recognized as enabling technology for designing and implementing the next-generation of distributed and intelligent manufacturing systems [11, 42]. Making the control of the system decentralized, intelligent agents offer a convenient way of modeling processes and systems that are distributed over space and time, thereby reducing the complexity, increasing flexibility and enhancing fault tolerance [22]. MAS can be defined as a network of autonomous, intelligent entities – agents – where each agent has individual goals and capabilities as well as individual problem-solving behaviors. Due to their lack of a global system objective and overview, agents have to cooperate and communicate with each other in order to achieve common aims, which are beyond the individual capabilities and knowledge possessed by each agent. There are two architecture approaches for agent encapsulation in agent-based manufacturing systems: the functional decomposition approach and the physical decomposition approach [49]. In the functional decomposition approach, agents are used to encapsulate modules assigned to functions such as an order, task, etc. In the physical decomposition approach, agents are used to represent entities in the physical world, such as a robot, conveyor, or pallet.

2.3 *Agent Systems Facilitated by Semantic Technologies*

As agents are applied in a distributed and heterogeneous environment and have by themselves only partial representations of the environment, these agents have to communicate with each other to coordinate their activities. Semantic technologies, such as ontologies, have been developed and investigated in the areas of artificial intelligence and natural language processing to facilitate knowledge sharing and reuse [26]. In a general sense, semantic technologies aim at bridging the gap between human-accessible meaning and machine-processed data [30]. As such, they cover several aspects such as knowledge representation, extraction of meaning from syntax and context-based interpretation of meaning. For agent systems, semantic technologies are vital for enabling knowledge interoperations between agents and, at the same time, a fluent flow of heterogeneous data from a range of entities. Ontologies allow the explicit specification of a domain of discourse, increase the level of specification of knowledge by adding semantics to the data, and promote knowledge exchange in an explicitly understandable form.

An ontology is defined as an explicit specification of conceptualization [20], where conceptualization means the shared view of environment representation. From the viewpoint of inter-agent interactions, the explicitly defined and commonly accepted ontology is an indispensable tool for ensuring interoperability between agents in the sense of providing a formally defined specification of the meaning of those terms which are used during the inter-agent communication. Ontologies can also capture actions and events in a uniform and processable way so that they can be recorded in time and further analyzed. The usage of ontologies for knowledge representation, sharing and high-level reasoning could be seen as a major step ahead in the area of agent-based control solutions [37].

Nevertheless, ontologies have been rarely used with software agents and most of the existing MAS are not aware of ontologies at all: the information processing and reasoning are hard coded in the agents' behaviors. Although important standardization work has been done by introducing the message transport service for sending FIPA-ACL Messages [19] by defining message types, protocols, etc., the agents are not able to semantically interpret the domain-specific content of the exchanged messages or the knowledge held by other agents [44].

3 Research Issues

Multi-agent system (MAS) technologies offer a convenient way to cope with dynamics in large complex systems, in our case manufacturing control systems. This approach replaces a centralized database and control computer by a network of agents, each endowed with a local view of its environment and the ability and authority to respond locally to that environment. Each agent is a representation of a manufacturing component and serves as an artificial "brain" of the real-world physical component. The agents are supposed to supervise the physical components of the system and to cooperate with other components. The MAS approach has proven to successfully handle complexity and dynamics in a number of comparable systems.

Agents communicate and negotiate with each other in order to perform the operations based on the available local information or to solve possible conflicts. Inter-agent communication capability provides the essential means for agent collaboration. In order to ensure correct understanding of the exchanged messages, agents must have the same presentation of the environment, or at least that part of the shared environment about which they are exchanging information with each other. Ontologies are of vital importance for enabling knowledge interoperations between agents and at the same time a fluent flow of the different data between different entities. Ontology is defined as an explicit specification of conceptualization. Explicit specification of conceptualization means that ontology is a description of the concepts and relationships that can exist within a multi-agent system.

A key research question is how to reduce the complexity of a manufacturing system. Therefore, we propose to divide the control of a manufacturing system into "hierarchically" ordered layers. The layered system structure enables the functional decomposition into subsystems, which can then be easily further decomposed but also integrated and managed in bigger systems as well. In the layered structure specification, besides the fact that particular subsystems logically symbolize some layers (e.g., planning and planning control layer), we consider that particular decisions have to be made in a real time requiring observation of a limited environment in contrast to decisions that require a global view without a special time limitation. The introduction of layers limits their responsibility and planning perspective, improving system performances and simplifying the concept.

In the context of a research project platform based on a real-world industrial use case we describe the layers, their contributions, and report lessons learned from the research work.

4   A Layered Manufacturing System Architecture

The layers of control in a manufacturing system correspond to planning and decision tasks in the domain and can, in general be divided between high-level control and low-level control. In more detail we specified four layers: management, planning, scheduling and executive layer (see Fig. 1) [34]. These layers structure agent-based systems control and provide the context for deriving the necessary semantics. The functions and responsibilities of the four layers are defined as follows:

- The *Management Layer* is responsible for entire system stability and functionality. It supports production and resource initialization as well as their determination. It is also concerned with the communication with the external environment and provides solutions for complex problems related to the global environment. It accepts production orders on a routine basis.
- *The Planning Layer* links process planning with product design. It is basically concerned with the sequencing of process steps, identification of product types and quantities to be produced. It defines equipment and resources that could be used and ensures that the parts or components required for the production are available and the final product delivery dates not exceeded. The shop floor layout is also defined on this layer.

- *The Scheduling Layer* is concerned with the synchronization of production needs with available resource capacities. The goal is to reach the internal deadlines that are set on the planning level. This layer is responsible for negotiating with the resources, the tasks as well as parts, tools, and product allocation between resources.
- *The Execution Layer* is related to the physical job shop equipment. On this layer, the production tasks are executed considering the resources' constraints and abilities, their performances measured and if a failure or disruption is diagnosed, the scheduling as well as management layer is informed. Also specific activities related to the execution of particular actions (i.e., pallet routing, removing, fixing, etc.) are coordinated on this layer.

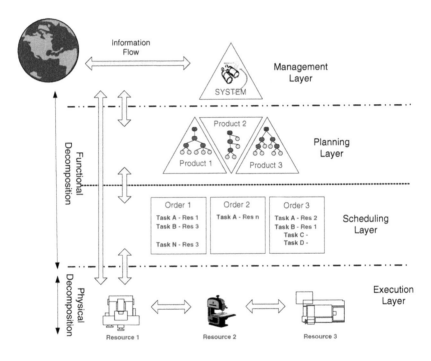

Fig. 1. The layers of manufacturing system control.

In this chapter we will present an approach where the manufacturing layers have been "agentified" to implement behaviors that represent the specific layer or the component of this layer as well as their objectives and functionalities, with each agent being responsible for carrying out different specific functionalities. We used the functional decomposition approach to create agents responsible for system support, process modeling and task scheduling (see Fig. 1), applying this approach to each of the three upper layers and generating particular agent types for each layer. We also applied the physical decomposition to create the related agents that have physical representation. In the rest of the chapter we will present the multi-agent system that governs specified manufacturing layers. Considering the role of the ontologies to enable data and information interoperability in extensive heterogeneous and content rich environment, we present the approach that integrates the related agent types at each manufacturing layer with ontologies. In the remainder of the chapter we will present the resulting multi-agent architecture.

5   The Management Layer

This section describes the management layer. First, a short introduction to Enterprise Resource Planning (ERP) and Virtual Enterprises is given. Then, the used layers and agents are introduced, followed by a description of the production process cycle used throughout this chapter.

5.1   Enterprise Resource Planning (ERP) and Virtual Enterprises

One of the most important trends in current IT development for companies is the trend to support and optimize the daily business in all areas of the business. Primary goal when deploying and using an ERP-System in a company is to increase the efficiency of established processes by supporting them through information technology. If the adaption of this system is a success and it supports the daily to-dos of the employees, it is over all a big opportunity to save time and money for the company. Of course, the first problem of establishing the new system and major investments has to be carefully planned [21].

ERP-software involves the business task to plan the resources of a company like money, personal or equipment persistently, carefully and in an efficient form during the whole production process. Therefore companies need more and more IT infrastructure which gets even more complex and so harder to maintain. The simulation of the production flow by multi agent system tools also belongs to this package of planning software. Furthermore it can be an essential part of it and complete the functionality of standard ERP-software by providing further information about the production process and add important data into the central available data source. In this case, the ERP-software acts as media, which transports this information towards all roles that need the data.

A virtual enterprise (VE) [26] is seen as an integrated network of regular companies that join their core services and resources in order to respond to unexpected business opportunities collaborating on an ad hoc basis. Such a network includes also suppliers, distributors, retailers and consumers requiring from involved companies to gather and share data and information about markets, customers and internal competences [1]. The capability of companies to form virtual enterprises and cooperate with partners is an important factor for keeping a competitive position on the market.

A new approach for virtual enterprise modeling as well as the fulfillment and consideration of several research challenges, such as improved knowledge exchanging and sharing, fast reaction to customer demand, re-organization capability, and integration of heterogeneous entities, are required [47]. The introduction of tools, techniques and methodologies that will support interoperability, information search and selection, contract bidding and negotiation, process management and monitoring, etc., is also highly required [12]. In this context, the information and knowledge exchange between partners plays a critical role for the success of such networks. This is particularly due to the extreme heterogeneity of the VE environment, in which it is usually not transparent to the partners, which knowledge is available at whose partner's site or even if so, then in most cases the knowledge is not understandable due to the usage of different formats and tools.

Ontologies have been developed and investigated for quite a while in artificial intelligence and natural language processing to facilitate knowledge sharing and reuse [26]. They are of vital importance for enabling knowledge interoperations between partners and, at the same time, a fluent flow of different data from diverse domains. Ontologies allow the explicit specification of a domain of discourse, increasing the level of specification of knowledge by incorporating semantics into the data, and promote its exchange in an explicitly understandable form [51]. Semantic means in this context that all relevant concepts important for partners will be modeled in an ontology by capturing the associations between the domains ensuring at the same time the understanding of exchanged knowledge during the inter- as well as inside-company communication. This allows business partners to build open communities that define and share the semantics of the information exchanged in their domain.

5.2 Layers and Agents

As shown in Fig. 2, each of the layers introduced in the previous section involves a certain type of agent, which will be described in the following section. For the management layer, this agent type is the so-called *Contact Agent (CA)*.

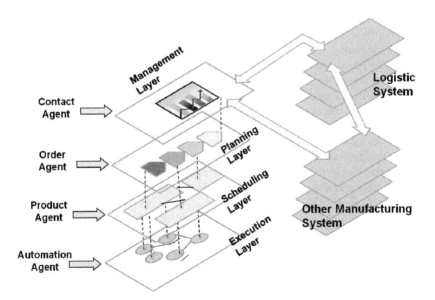

Fig. 2. Overview Agents and Layers.

The *Contact Agent* (CA) is related to the Management Layer and accordingly it has responsibilities that encompass organizational and supervisor functions. The CA is created at the start-up of the system and it is always active. It is concerned with the system stability and in the case that one part of the system collapses; this agent considers its influence on the system performance and, if significant, undertakes particular steps in order to bring the system back into the optimal state. Its further responsibilities are to receive a customer order and create one Order and Supply agent for each related product order. This agent also creates an agent for each new resource introduced in the system. After the order was accomplished or particular resource removed from the system, the CA determinates the related agent. However, having only one instance of this agent for the whole system and considering it as a possible single point of failure, the replication technique [33] and replication service provided with used agent platform [8] can be applied to enhance agent's failure tolerance level.

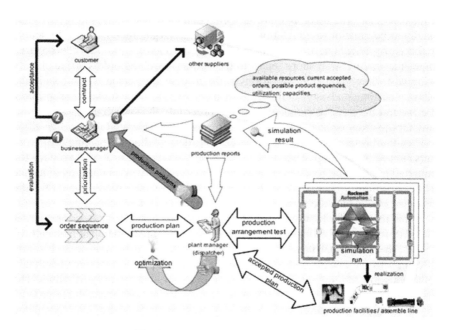

Fig. 3. Production Process Cycle [36].

5.3 Production Process Cycle

The following description shows a possible execution of incoming orders of customers for a company. This designed process is the economic background and core of the project focused in this chapter to reproduce a manufacturing assemble line to simulate actions on a flexible production unit. The optimization focus of the job-shop-scheduling is realized over simulation runs with different parameter settings to maximize the production system output. The explanation describes the involved roles on the one side and necessary information flows on the other side.

As outlined in Fig. 3, the normal treatment of incoming orders from customers involves more than one enterprise-internal role in a coordinated and balanced task sequence. The described draft of the process shows the close dependencies between all involved acting units/agents. In addition the draft expresses the possibility, how appearing production complications like resource bottlenecks, machine or conveyor failures or capacity overruns are handled within a production planning and control system. To use the available capacity optimally, the reactions to solve such unexpected problems on the first level are focused directly at the executive layer by the involved role. Only if this is not possible an agent on a higher level receives an exception ticket to avert the danger. The role/agent/execution unit on the higher level eventually can use other possibilities to solve the problem because of more available information or more granted authority (e.g., it is not possible to fulfil the order with the actual time and resource restrictions – they have to be extended or manufacturing of the product has to be outsourced).

Starting point for all production planning actions is the incoming detailed order of the customer. The assumption of the order can take place on several ways: personally at negotiations (completion of contracts at large orders) written by letters or forms, by telephone or on new media like the internet (online orders via E-commerce-trading, web-shops, etc.). In the optimal case, the orders arrive already in digital form on prepared web forms at the responsible person which acts in the role of a customer adviser. In this manner media breaks are eliminated as a source of error and the passing-on/takeover of the order into the system used in the company for the order management as well as the PPC is accelerated immensely.

Then the business manager takes over the order winding by making a first optimisation as a next step. Therefore he lines up the received orders to produce the requested quantity of goods in time according to certain criteria (such as sequence of the incoming orders, dates of delivery, urgency of the ordered goods for the customer, etc.) which are actually decisive for the enterprise success.

In this manner the first order sequence arises considering certain priority criteria and thus serves as an input for the production planning of the plant manager.

The job of the plant managers is to distribute the order list transmitted to him efficiently to the available capacities and resources. However, besides he should be anxious to keep the predefined sequence to avoid unnecessary difficulties. His concrete job is to compile an efficiently allocation plan for the available production lines under the current time, capacity and resources restrictions. Available simulation tools can be used by the plant manager to complete this job. By defining the available orders by configuration of input parameters (e.g., number and kind of products) as well as different settings of factors which influence the production process (e.g., speed of conveyor belts, available free pallets, shift duration, assemble line arrangement) the expected production process can be predicted to find out the best production sequence. This simulated and tested production order can be transmitted to the physically existent production units as an optimized production plan.

The assemble lines should preferably act automated to carry out the transmitted production instructions to fulfil their manufacturing task. During the whole production process data of the current state (order, product sequence, number of pieces, order status, utilisation, …) is collected and made available on the information system to the different roles of the working on unities and employees.

If failures occur during the production, the units/agents of the assemble line automatically react to the resulted failure to compensate (compensation at operation layer) without further intervention. Only if this is not possible any more, i.e., the entire processing by resource lack, machine failure, overload, etc. cannot be executed in the planned time, a suitable announcement about the production report is done.

6 The Planning and Scheduling Layers

The production planning and scheduling issues are of essential importance for the manufacturing domain today, especially due to the dynamic and competitive nature of the nowadays global market that needs enterprises to be adaptive, flexible, robust and collaborative. The process planning is usually separated from scheduling and unnecessary breaks between these systems are created, even though the outputs and data from one application could be fluently used as inputs for another one. In this section we present an approach to integrate these layers using agent technology and ontologies.

6.1 Planning

Process planning (PP) plays a very important role in the product life cycle by linking the product design with the manufacturing phase. Process planning resolves between what and how products will be produced. The process planning phase has to consider the product requirements (price, quantity, geometry, tolerance, material, etc.) as well as the production constraints (machine capacity, tool characteristics, etc.).

Due to the lack of intelligent capabilities, current process planning systems have difficulties to automatically adapt plans according to the availability of resources, or share knowledge among the various planning related functional modules [10]. Additionally, a low frequency of planning runs and difficulties in coping with new organizational forms of manufacturing such as product oriented or customer driven production, require new skills and new approaches capable to handle the shortcomings mentioned above [3].

6.2 Application of Agents in Process Planning

As a possible way to overcome these shortcomings, a decentralized architecture that spreads the planning process between several entities/agents, each capable to create, control, and observe the execution of its own plans, is suggested. The agents cooperate and coordinate their actions in order to effectively accomplish their plans as well as to reach the commons system goals. Such organization brings time improvements, since the complex problems are partitioned between the entities by giving each entity a part of the problem to solve instead of dispatching the whole problem to the central unit, what could cause difficulties especially when the data is voluminous and changes frequently. The distributed agent-based approach allows proactive data processing at the place of its origin and data exchanges are only those necessary for effective system functioning [43]. Moreover, large and complex problem structures become more simple and the possible failures easier to track [48]. An extensive literature reviews related to agent-based collaborative process planning is provided by Zhang et al. [55].

6.3 Production Scheduling

The task of scheduling is the allocation of jobs and activities to available resources over time considering relevant constraints and requirements [46]. Its main objectives are the minimization of the production time of jobs, production costs, increased resource utilization, etc. Most of the developed scheduling systems are based on centralized structures, which make manufacturing systems scheduling even more complicated [50]. The application of agent technology can significantly improve the efficiency and performance of the entire system [36]. Extensive surveys of dynamic scheduling in the manufacturing environment considering also agent-based systems were done by Babiceanu and Chen [6] as well as by Ouelhadj and Petrovic [39].

6.4 Integration of Process Planning and Scheduling

Process planning and scheduling are highly related, because when the planning ends the scheduling phase starts. The process plan restrictions and shop floor constraints have to be considered in the scheduling phase, which could become a very complicated and time consuming process, if applied in a dynamic environment. In order to make more realistic and applicable plans, the integration of the planning and scheduling phase is necessary. Nevertheless, the traditional approaches execute these processes separately, mostly ignoring the condition of resources on the shop floor (e.g., machine workloads, etc). That leads to the under- or over-utilization of certain resources or even that some of the process plans perhaps cannot be executed requiring alterations or replanning [27]. A lot of work has been done in the past to optimize and integrate process planning and scheduling in the area of manufacturing [50].

However, one of the main shortcomings of the presented architectures is the lack of interoperability, since the applied methodologies separate planning activities (e.g., process planning) from executing activities (e.g., production control and scheduling), creating a gap between the involved systems. The problem in current distributed systems is that they are still tightly coupled from the point of view of automated gathering and integration of data, information and knowledge, being programmed with the focus on performing particular tasks rather than on interoperability and openness [38]. Shen et al. defined the integration of process planning, manufacturing scheduling, and control as one challenging research topic where much more attention has to be set on the complexity analysis and formal modeling of such integration [50]. The assimilation of different knowledge sources is considered as an important problem that has to be solved being marked as not easy task due to different representations, foundations, and levels of abstraction of various knowledge sources [10]. Being mostly applied in heterogeneous environment, an agent has to understand it as well as the knowledge of related agents in order to reason about it, prior to making decisions. Moreover, considering that future distributed manufacturing systems will need to handle a great diversity of autonomous agents and mechatronic devices interacting intensively, there is as strong need that all components understand the exchanged information and know how to communicate [15]. According to Finin et al. [17] for software agents to interact and interoperate effectively three fundamental and distinct components: (i) a common language; (ii) a common understanding of the exchanged knowledge; and (iii) the ability to exchange whatever is included in the previous two; are required. The usage of machine-interpretable semantics (ontologies) to describe the components of manufacturing systems enables other intelligent components (agents) to perform reasoning and infer sufficient knowledge to interact as well as to overcome current interoperability barriers [28].

6.5 Planning and Scheduling in the Assembly Domain

In order to automate the process planning generation, the product model representation has to be made in a way that enables understanding the designer's intention, offering information about specific features (connections, definitions, constraints, etc.) that could be used for the selection of appropriate equipment as well as tool set-up definitions. At the same time, knowledge about the capabilities of the equipment could facilitate the product design and ensure its manufacturability. The ability to present the product model in a same way as production process and production equipment can support easier mapping between these three key manufacturing elements and enable easier optimization of both planning and scheduling process.

Assembly is much more than a process where two or more parts are connected, since the whole process is accompanied with preceding as well as following actions (supply, transportation, inspection, handling, delivery, etc.). "Assembly model" or models must be capable of capturing a diverse set of information needed to describe the entities and activities associated with assemblies and assembling so that designers of products, assembly systems, logistic systems, supplier relations, field support, and finally disassembly and recycling, can have access to the information they need [54]. However, the lack of ways to standardize and describe the assembly domain knowledge is an obstacle to achieve an easy flow of information. This is the reason why we select the assembly domain and its automation as test case for our knowledge-based multi-agent concept.

Traditionally, the product model is based on geometry and provides incomplete product definitions. Besides the assembly geometry, the understanding of its physical effects as well as the design intentions (e.g., joint type) is required. The meaningful representation of product data is necessary to enable semantic interoperability across different application domains [41]. An ontology-based assembly model was presented in [24] and serves as a formal, explicit specification of the

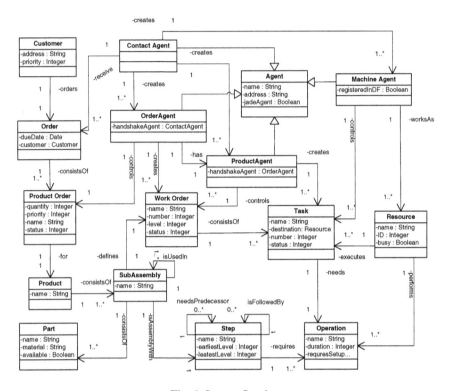

Fig. 4. System Ontology.

assembly design so that it makes assembly knowledge both machine-interpretable and shareable at the same time. We use the same concept to link product designs, assembly planning processes and required assembly equipment together. The ontology based product model is used to extract the production/assembly operations from the product design and link particular tasks, which have to be performed for the production/assembly of a *product*, to particular *resources*. Each *task* consists of a series of actions that can be executed by one or more *resources*. The connection to the *product order* is made through the type of ordered product, quantity, which defines the number of parts that have to be available to start the assembly process, and due date that defines the priority of the order.

In this context, a *product* is presented as a hierarchy of subassemblies and parts together with all their properties and relationship between them. *Parts* are defined as components, described by a set of attributes, properties, constraints and relations to other parts. A *subassembly* is a non-empty subset of parts that either has only one component or is such that every part has at least one surface contact with another part in the subset [45]. The relationship between parts within a subassembly defines operations that have to be done to connect these parts and represents how these subassemblies should be put together to complete the product. An *operation* is defined as a discrete set of actions which leads to a certain change of state in or on the part.

Each product type is described through its own process plan. A *process plan* (assembly tree) specifies the sequence of manufacturing or assembly operations which have to be performed in order to make a product. Each operation could be performed by different resources and consequently can be accompanied by related transport operations, making all together *work orders*. The ontology-based concept of the product production/assembly described with *Steps* ensures the exact decomposition of the product orders to related work orders and further associated tasks and their correct indexing. This is particularly supported with the integrated planning relationships *needsPredecessor* (Step) and *isFollowedBy* (Step) that enables an agent to reason when and why to start particular task allocations, which are known as scheduling activities. A *resource* is a physical component able to perform a certain action. However, since this component embodies agent as its control part, we consider also the agent concept as integral resource element.

In order to reach the goal of modern assembly planning systems to create activity sequences that are not only feasible but also optimized according to one or more parameters, such as makespan, machine or tool utilization, the agents that are supervising a particular resource or process planning system can use the accurate information stored in the ontology to reason about available resources and utilize appropriate optimization heuristics [28]. We are using this ontology to provide semantic understanding among software agents. The agent interaction with the ontology in the background ensures that when an agent extracts relevant information from a message it understands the meaning of the terms in the message and the way this terms are combined in the statement. The presented concept distribution and ontological representation of a production process improves the way components communicate and exchange information in the manufacturing environment. Our ontology covers the environment structure, characteristics, states and components interrelationships enabling the related agents to interpret their environment, reason about it and make right decisions.

We used the functional decomposition approach to create agents responsible for process modeling and task scheduling (see Fig. 1), applying this approach to each planning and scheduling layer and generating particular agent types for each layer.

The **Order Agent** (OA) captures the goals and tasks of the Planning Layer. The OA is responsible for accomplishment of one product order, respecting due dates and the like; and handling customer requests for modifying or cancelling their orders. The essential information for an order agent is: type of product, the production deadline, quantity, and the priority of the client. Having the knowledge about all products corresponding to a single order, this agent combines the ontology-based model for a particular product together with other information, sequences this into work orders and sends it to the supply agent. Based on this knowledge and contacting the storage, the OA checks if all parts and materials required for execution of a single order are available. During the production, this agent collects also information concerning the status of current product orders or the system's performance. The OA is responsible for loading products into the system when a product order reached the system and for unloading products from the system when all of their processes are finished.

The **Product agent** (PA) maps the functions of the scheduling layer. The PA is in charge for coordinating the production execution in order to achieve the best possible production results, including on-time delivery, cost minimization, and so forth. It also manages the movement of related product order's subassemblies and materials across the job shop. After the OA decomposes the product order into work orders, they are forwarded to the PA. Using the ontology and taxonomic relations specified in the product definition (see Fig. 4), the PA extracts tasks from work orders and schedules the ones that have to be completed at first. After that, the PA initially sends requests for bids to all machine agents that have the capability to process the first task. The interested machine agents respond with their bids. Each bid contains an estimated queuing time and finishing time for the requested operation. After collecting the bids, the PA evaluates the bids and selects the best one. When the related machine is identified, the agent negotiates with transport agents to route the task there. Whenever a current task is completed, this agent sends bid requests for the next operation. This bidding procedure continues until all the requested features of a job are finished. When the last task in the production process is finished, the agent sends the notification to the OA.

The important advantage of the introduced ontology-based approach is the achievement of the preconditions for easy assembly and disassembly of the product. Our knowledge-based system does not need to be told, how a problem has to be resolved (i.e., which and when particular tasks have to be done), but the concept and the goal is described instead. The system decides on its own how to achieve the goal.

7 The Execution Layer

Within the Layered Manufacturing System, the execution layer is responsible for the execution of production tasks. More specifically, the main role of the execution layer is to integrate the basic functionalities provided by physical resources (e.g., drilling, transporting pallets) into the Layered Manufacturing System.

In this section, we first recall the general requirements on the execution layer. We then detail the design of the execution layer as a system of semantic agents. Finally, we discuss some lessons learned in using semantic agent technologies in the execution layer.

7.1 Requirements of the Execution Layer

In order to correctly play its roles in the Layered Manufacturing System, the execution layer needs to satisfy a number of requirements. We could categorize these requirements into three groups: requirements regarding the *execution of production tasks*, requirements regarding the *robustness of machine control*, and requirements regarding the *diagnosis of disturbances and failures*.

Providing the Layered Manufacturing System with means of executing production tasks is the main requirement of the execution layer. More specifically, this can be decomposed into two aspects: identifying the appropriate resources for executing a production task and controlling the execution of a particular task. To do so, the execution layer relies on a set of machines (mechatronic components), each capable of performing specific tasks. Because of the heterogeneity of these machines, the identification of appropriate resources is not straightforward. Therefore, the execution layer must be able to match the requested tasks with available resources. In addition, the execution layer should consider the runtime constraints of the machines (load, temporary unavailability, etc...), in order to obtain the most suitable solutions.

Providing a robust control of production machines is a major requirement for the execution layer. To do so, the execution layer should keep the various control algorithms and components as independent and loosely-coupled as possible, in order to avoid local failures to hinder the whole system. Decomposing control into smaller, well-defined control units also makes it easier to design and to test, thus enabling a better quality.

Being able to diagnose disturbances and failures is an important requirement. In such as physical system, it is not possible to assume that all machines will function seamlessly and always respond as expected. On the contrary, disturbances and failures can happen at any time, and their occurrence is likely to impact the production process negatively if they are not tackled properly. Therefore, the execution layer must cope with unexpected disturbances and failures. In particular, it should be able to detect disturbances and failures, to determine their impact of the production tasks it is in charge of, to report problems to the scheduling layer, and possibly to elaborate recovery solutions when possible.

7.2 Semantic Agents for the Execution Layer

This section presents an overview on semantic agents used in the execution layer. We call them automation agents [52], as these semantic agents are more specifically dedicated to the control of mechatronic components.

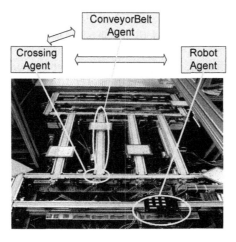

Fig. 5. Overview of some automation agents in the execution layer.

The execution layer consists in various types of automation agents, corresponding to the various mechatronic components and machines (resources) available in the factory. Fig. 5 presents an overview of some automation agents of the execution layer. Each agent is responsible for the control of its dedicated resource, and can only influence the behavior of another resource by sending a request to the corresponding automation agent.

The Automation Agent Architecture

In order to facilitate the design of agents for specific mechatronic components, we define a generic architecture for automation agents [52]. This architecture distinguishes between a high-level control (HLC), a low-level control part (LLC) and the mechatronic component itself (see Fig. 6).

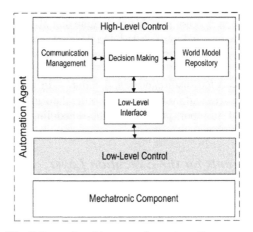

Fig. 6. Internal architecture of an automation agent.

The LLC is built on top of the physical mechatronic component and is directly responsible for managing the underlying physical system. It consists in "pure" control software and comprises a limited set of reactive behaviors to directly control the physical component, collect and process the information from sensors and based on the result perform particular actions. The LLC is particularly designed to perform in real-time. It can also diagnose certain types of failures (e.g. a conveyor stock) and informs the upper layer about it.

The HLC is composed of four main modules:

- The world model repository contains a world model, i.e., a symbolic representation of the world of the agent. The world model repository provides facilities for updating and querying the world model.
- The low-level interface enables the agent to use functionalities provided by the LLC. It especially provides facilities for receiving event notifications about the current operations of the LLC and for requesting particular operations from the LLC. Because various kinds of LLC may exist in a system, we use a unified low-level interface as described in [32].

- The communication manager provides facilities for managing communication with other agents. The communication between agents depends on the knowledge they have to exchange and on the tasks they have to achieve collectively.
- The decision-making component is in charge of reasoning about the states of the world and deciding what to do (e.g., communicate with other machines, request an operation from the LLC, issue notifications to an operator). Event notifications generated by the low-level, by communication with other agents or by the world model trigger the decision-making procedures. These procedures then update the world model, request operations from the LLC and communicate with other agents.

Semantic Technologies for Automation Agents

Within an automation agent, semantic technologies are mainly used for dealing with the representation of activities. In our context, an activity can be defined as "a process occurring in the world in which the agent is participating". This definition not only covers actions directly performed by the agent, but also the occurrence of events that the agent only observes. In the first case, the agent is an actor in the activity, while in the second case it is only an observer. In the automation domain, both aspects are useful, as a change in the environment often happen without an agent being directly responsible for the change. Using this general notion of activity enables representing several important concepts related to an automation agent, such its capabilities (what activities it can perform), its goals (which activities it intends to perform), its actions (what activities it is actually performing) or its interdictions (what activities it does not have the right to perform).

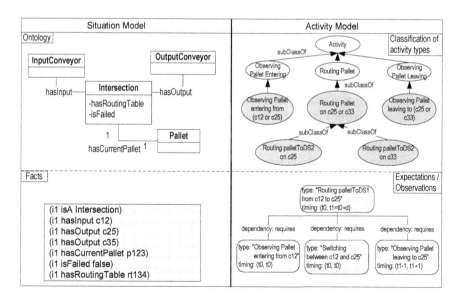

Fig. 7. World Model of an Automation Agent controlling an intersection.

This representation is contained in the world model of the automation agent. Fig. 7 depicts the world model of an automation agent for an intersection. It consists of two parts: the situation model and the activity model. The situation model holds knowledge about the agent's situation. The situation of an agent consists both of its own characteristics and its relations to other entities in the world. The activity model holds knowledge about activities of the agent. The situation model is composed of a domain ontology and a set of facts:

- The *domain ontology* (top) is a model of the type of entities in the domain of the agent [13]. It defines relevant classes of entities as well as relations between entities. It also serves as a vocabulary for referencing these classes and relations, thus ensuring the consistency of the world model as well as the interoperability between different agents. For our example, the ontology defines that an intersection can have input conveyors, output conveyors, and one current pallet inside it. Such concepts and relations can be extracted from existing ontologies, such as [35].
- The *facts* (bottom) express the current knowledge about the world. Facts are expressed using the vocabulary defined by the ontology. For our example, facts express that i1 is an intersection, which has one input conveyor (c12) and two output conveyors (c25 and c33). It is important to note that facts expressed in the situation model do not intend to represent completely the world of the agent. They rather represent an abstraction of some meaningful aspects of the world, which can be used for realizing high-level control tasks. Although more complete and dynamic information about the world may be available at the low-level control, it is only processed in this layer, usually under real-time constraints.

The activity model is composed of a classification of activity types and a model of expectations and observations:

- The *classification of activity types* (top) models the types of activities in which the agent can be involved. Types are defined formally using description logics formulas and they are organized hierarchically based on the subsumption relationship (noted subClassOf) [4]. Primitive types are defined as direct subclasses of Activity. Derived types are defined by restriction of the primitive types to take into account the actual world of the agent. For instance, the generic type "Routing Pallet" is refined to the more specific type "Routing pallet on c25 or c33" corresponding to the present situation.
- The *expectations and observations* (bottom) is a model of the activities that are expected and observed by the agent. Expectations and observations are defined by the specification of a type (based on the classification of activity types) and timing. Timing is expressed using time intervals [2]. While observations can express a precise timing, expectations rather express constraints on their timing. Expectations are linked by dependencies, indicating how observations on one expectation can have consequences on other expectations. For instance, Fig. 8 depicts that the expectation that ``Routing palletToDS2 on c25" should occur

starting at time t0 and ending at time t1=t0+d (where d is the time for going through the intersection). This in turns implies that the activity ``Observing palletToDS2 entering from c12" should occur at t0, the activity ``Switching between c12 and c25" should also occur at t0, and the activity ``Observing palletToDS2 leaving to c25" should occur at t1.

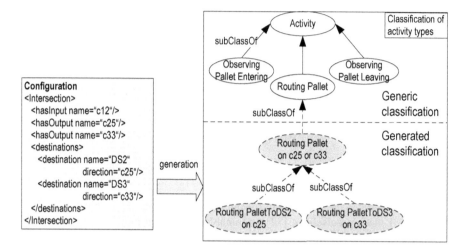

Fig. 8. Mechanism for initial configuration.

The world model using semantic technologies plays a key role in three mechanisms of automation agents: the *initial configuration* (and eventual reconfiguration), the *response to production task requests* and the *interaction with the LLC* for executing and monitoring production tasks.

Fig. 8 illustrates the mechanism for initial configuration of an automation agent. On the right is a declarative description of the configuration, indicating the input and output conveyors of the intersection. Based on this description, the generic activity type "Routing Pallet" is refined for this particular intersection.

A more specific class describing the capability of this intersection is defined as "Routing Pallet on c25 or c33". Even more specific classes can be defined based on the routing table contained in the configuration. They indicate that pallets with destination DS2 are routed on c25 and pallets with destination DS3 are routed on c35.

Fig. 9 illustrates the mechanism for responding to production requests. When a message containing a request for routing a pallet arrives, it is translated in terms of an activity type to be performed. This request is matched with known types from the activity classification, in order to evaluate if the request can be honored. A request can be honored only if it some activity type can be both more specific than the request and than one capability of the agent. This is the case here for "Routing Pallet124 on c25".

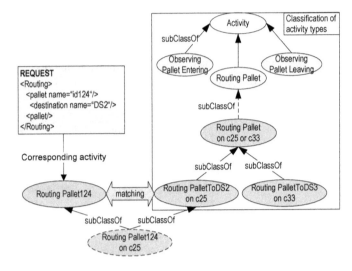

Fig. 9. Mechanism for responding to production requests.

Fig. 10 illustrates the mechanism for interaction with the LLC. This mechanism is described more precisely in [32] and enables the integration of heterogeneous types of LLC. The HLC expresses a request using a description of a type of activity, which is independent of a particular LLC. Depending on the technology employed for the LLC, this request is translated into a specific message. To do so, we use a semantic description of the LLC functionality, which relates the LLC-independent concepts used by the LLC and the LLC specific types of message.

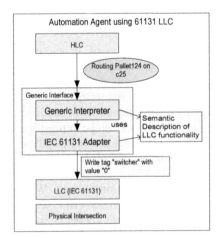 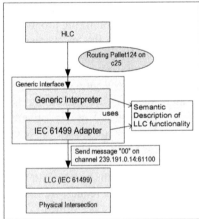

Fig. 10. Mechanism for interaction with the LLC.

7.3 Lessons Learned – Practical Use of Semantic Agent Technologies

Designing and using the automation agent architecture described above enables us to discuss more precisely the practical use of semantic agent technologies in a Layered Manufacturing System.

Automatic reasoning and interoperability in an open environment are often claimed a major advantage of semantic technologies. This approach provides the most flexibility, as the agent can operate in an open environment and cope with unexpected cases. However, this requires a completely consistent, formal definition of the ontologies in use. In particular, various ontologies used within the system have to be properly aligned in order to provide the expected results. This is often unpractical, mainly due to the complexity of the domains involved and the lack of experts in ontological modeling. Moreover, solutions based on runtime reasoning from basic principles (e.g., the axioms of Description Logics underlying OWL) can be computationally expensive for a large ontology.

Using a rule-based engine with domain-specific rules operating on a knowledge base can be a more tractable solution to enable interoperability among agents through translation between the ontologies in use. This facilitates the partitioning of the domain ontologies into several parts, reconciled with appropriate rules. In our example, reasoning about activity type for the intersection can be performed in a restricted domain, without considering a general ontology of all types of activities likely to appear at the level of the whole factory.

Model-based generation of procedural code for managing interactions is another practical approach. In particular, an efficient handling of ontologies can be obtained by taking into account that many entities in the domain of a production system are fixed, at least for some reasonable time scale. Although changes of the production process, of the types of products or even of the types of machines may occur, this only happens at a reduced rate. Therefore, it is practical to transform the knowledge expressed in ontologies into a fixed set of useful data structures and constants, easier to manipulate at runtime. This can be done at deployment time, when the system starts to operate under a fixed configuration (with a fixed set of machines and products). Parts of the code can be re-generated when new conditions appear, thus providing sufficient flexibility. In that sense, ontologies are used as a tool for humans to design and understand the system, rather than as a runtime artifact.

In the execution layer of the Layered Manufacturing System, we employ these three ways of dealing with ontologies. In particular, we found model-based generation of procedural code to be particularly suitable for dealing with ontologies when interacting with the LLC. As only few changes in the LLC can be expected after deployment, it is possible to obtain a specific set of adapters for transforming all possible requests of the HLC into relevant LLC messages. A rule-based engine is useful for interpreting declarative configuration and messages and translating them into the corresponding activity types. Especially, implicit references to various entities are resolved by the use of a knowledge base which keeps track of the successively available pieces of information. Automated reasoning tools are best used for matching requests with capabilities, as they provide the most flexibility. Complexity can be tackled by considering only a small ontology of activity types, formally defined only in the context of a particular agent.

8 Conclusion and Further Work

Manufacturing control systems are a mission-critical application domain for semantic agents systems. While multi-agent systems have been explored in the manufacturing systems domain, there is very little work on semantically enabled agent systems.

In this chapter we reported on the engineering of the semantic agent architecture in four layers (management, planning, scheduling, and execution) and the related ontology for each particular layer in the application domain, the manufacturing system. Lessons learned based on real-world use cases in the context of the Layered Manufacturing System are:

- The semantic agent architecture enables related layers to communicate directly avoiding unnecessary "stage by stage" procedures, since the agents from each layer are able to communicate and understand agents from any other layer. For sure, this direction communication is based on the ontology used in the background for understanding and/or translating specific messages exchanged between agents. However, the perceived system complexity of the single agents is lower, and the complexity regarding translation between layers is centrally stored in a single point, namely in the ontology.
- The presented system assures clear definitions of each layer role in the system as well as associated tasks that have to be done in order to achieve common goals. This further enables the smooth creation of related agent classes, e.g., bv using Model-Driven Development (MDD) paradigms, and mapping of ultimate system goals to these agents.
- Within each layer, agents have the ability to maintain an accurate internal representation of pertinent information about the environment in which they operate. The world model of an automation agent contains a symbolic representation of the manufacturing domain, more precisely in terms of situations and activities. For this representation, we make use of ontologies.
- The application of ontology based representation enables solving the interoperability problem in all the layers (including control, planning and scheduling) which are managed by agents.

We can summarize that in order to make agent-based system more reliable and more attractive the importance of symbiosis of this technology with other emergent technologies such as semantic web is essential, since it brings advances in a way of treating and dealing with information which is the most important factor in present information age.

Future Work. As one can see major steps for introduction of semantic technologies in manufacturing domain have been already done and the pioneer work for its fusion with multi-agent technology exists. Nevertheless, the related works that will be concerned with ontology life cycle in terms of maintaining and engineering of ontology models in distributed manufacturing environment is still missing and could be seen as next logical step.

Acknowledgments. This work has been supported by the Christian Doppler Forschungsgesellschaft and the BMWFJ, Austria, as well as by the FIT–IT: Semantic Systems program, an initiative of the Austrian federal ministry of transport, innovation, and technology (bm:vit) under contract FFG 815132. In addition, this work has been partially funded by the Vienna University of Technology, in the Complex Systems Design & Engineering Lab.

References

1. Aerts, A.T.M., Szirbik, N.B., Goossenaerts, J.B.M.: A flexible, agent-based ICT architecture for virtual enterprises. Computers in Industry 49, 311–327 (2002)
2. Allen, J.F.: Maintaining Knowledge about Temporal Intervals. Communications of the ACM 26, 832–843 (1983)
3. Azevedo, A.L., Sousa, J.P.: A component-based approach to support order planning in a distributed manufacturing enterprise. Journal of Materials Processing Technology 107, 431–438 (2000)
4. Baader, F., Horrock, I., Sattler, U.: Description Logics. In: Handbook on Ontologies, pp. 3–28. Springer, Heidelberg (2004)
5. Babiceanu, R., Chen, F.: Development and Applications of Holonic Manufacturing Systems: A Survey. Journal of Intelligent Manufacturing 17, 111–131 (2006)
6. Babiceanu, R., Chen, F.: Development and Applications of Holonic Manufacturing Systems: A Survey. Journal of Intelligent Manufacturing 17, 131, 111 (2006)
7. Baker, A.D.: A survey of factory control algorithms that can be implemented in a multi-agent heterarchy: Dispatching, scheduling, and pull. Journal of Manufacturing Systems 17, 297–320 (1998)
8. Bellifemine, F., Caire, G., Greenwood, D.: Developing multi-agent systems with JADE. Springer, Heidelberg (2008)
9. Black, J.T.: The Design of the Factory With a Future. McGraw-Hill Companies, New York (1991)
10. Bose, U.: A Cooperative Problem Solving Framework for Computer-Aided Process Planning. In: Proceedings of the Thirty-second Annual Hawaii International Conference on System Sciences, vol. 8, pp. 8015–8024. IEEE Computer Society, Los Alamitos (1999)
11. Bussmann, S., Jennings, N.R., Wooldridge, M.: Multiagent systems for manufacturing control: a design methodology. Springer, Berlin (2004)
12. Camarinha-Matos, L.M.: Virtual organizations in manufacturing: trends and challenges. In: 2th International Conference on Flexible Automation and Intelligent Manufacturing (2002)
13. Chandrasekaran, B., Josephson, J.R., Benjamins, V.R.: What are ontologies, and why do we need them? IEEE Intelligent Systems and Their Applications 14, 020–026 (1999)
14. Christensen, J.H.: HMS/FB architecture and its implementation. In: Deen, V.S.M. (ed.) Agent-based Manufacturing, pp. 53–87 (2003)

15. Christo, C., Cardeira, C.: Trends in Intelligent Manufacturing Systems. In: Conference Trends in Intelligent Manufacturing Systems, pp. 3209–3214 (2010)
16. Colombo, A.W., Schoop, R., Neubert, R.: An agent-based intelligent control platform for industrial holonic manufacturing systems. IEEE Transactions on Industrial Electronics 53, 322–337 (2005)
17. Finin, T., Labrou, Y., Mayfield, J.: KQML as an agent communication language. In: Software Agents, pp. 291–316. MIT Press, Cambridge (1997)
18. Fisher, K.: Agent-based design of holonic manufacturing systems. Robotics and Autonomous Systems 27, 3–13 (1999)
19. Foundation: Foundation for Intelligent Physical Agents. FIPA Communicative Act Library Specification (2003), http://www.fipa.org/specs/fipa00037/ and FIPA ACL Message Structure Specification, http://www.fipa.org/specs/fipa00061
20. Gruber, T.R.: A translation approach to portable ontology specifications. Knowl. Acquis. 5, 199–220 (1993)
21. Hansen, H.R., Neumann, G.: Wirtschaftsinformatik. Fischer (1982)
22. Jennings, N.R., Bussmann, S.: Agent-based control systems: Why are they suited to engineering complex systems? IEEE Control Systems Magazine 23, 61–73 (2003)
23. Jones, A.T., McLean, C.R.: A proposed hierarchical control model for automated manufacturing systems. Journal of Manufacturing Systems 5, 15–25 (1986)
24. Kim, K.-Y., Manley, D.G., Yang, H.: Ontology-based assembly design and information sharing for collaborative product development. Computer-Aided Design 38, 1233–1250 (2006)
25. Krothapalli, N.K.C., Deshmukh, A.V.: Design of Negotiation Protocols for Multi-Agent Manufacturing Systems. Int. Journal of Production Research 37, 1601–1624 (1999)
26. Kulvatunyou, B., Cho, H., Son, Y.J.: A semantic web service framework to support intelligent distributed manufacturing. Int. J. Know.-Based Intell. Eng. Syst. 9, 107–127 (2005)
27. Kumar, M., Rajotia, S.: Integration of process planning and scheduling in a job shop environment. The International Journal of Advanced Manufacturing Technology 28, 109–116 (2006)
28. Lastra, J.L.M., Delamer, M.: Semantic web services in factory automation: fundamental insights and research roadmap. IEEE Transactions on Industrial Informatics 2, 1–11 (2006)
29. Lastra, J.M., Delamer, I.: Ontologies for Production Automation. In: Advances in Web Semantics I, pp. 276–289 (2009)
30. Berners-Lee, T., Hendler, J., Lassila, O.: The semantic web. Scientific American (2001)
31. Leitão, P., Restivo, F.: A layered approach to distributed manufacturing. In: ASI 1999 International Conference, pp. 21–23 (1999)
32. Lepuschitz, W., Vallée, M., Merdan, M., Vrba, P., Resch, J.: Integration of a Heterogeneous Low Level Control in a Multi-Agent System for the Manufacturing Domain. In: 14th IEEE International Conference on Emerging Technologies and Factory Automation (ETFA 2009) (2009)
33. Mellouli, S.: FATMAS: A Methodology to Design Fault-tolerant Multi-agent Systems, PhD. Université Laval (2005)
34. Merdan, M.: Knowledge-based Multi-Agent Architecture Applied in the Assembly Domain. Vienna University of Technology (2009)

35. Merdan, M., Koppensteiner, G., Hegny, I., Favre-Bulle, B.: Application of an Ontology in a Transport Domain. In: IEEE International Conference on Industrial Technology (IEEE ICIT 2008). Sichuan University, Chengdu (2008)
36. Merdan, M., Moser, T., Wahyudin, D., Biffl, S.: Performance evaluation of workflow scheduling strategies considering transportation times and conveyor failures. In: IEEE International Conference on Industrial Engineering and Engineering Management, pp. 389–394 (2008)
37. Obitko, M., Mařík, V.: Ontologies for Multi-Agent Systems in Manufacturing Domain. In: Proceedings of the 13th International Workshop on Database and Expert Systems Applications, pp. 597–602. IEEE Computer Society, Los Alamitos (2002)
38. Obitko, M., Vrba, P., Mařík, V., Radakovic, M.: Semantics in Industrial Distributed Systems. In: The 17th IFAC World Congress, Seoul, Korea (2008)
39. Ouelhadj, D., Petrovic, S.: A survey of dynamic scheduling in manufacturing systems. Journal of Scheduling 12, 417–431 (2009)
40. Parunak, H.V.D.: Applications of distributed artificial intelligence in industry. In: Foundations of Distributed Artificial Intelligence, pp. 139–164. John Wiley & Sons, Inc., Chichester (1996)
41. Patil, L., Dutta, D., Sriram, R.: Ontology-based exchange of product data semantics. IEEE Transactions on Automation Science and Engineering 2, 213–225 (2005)
42. Pěchouček, M., Mařík, V.: Industrial deployment of multi-agent technologies: review and selected case studies. Autonomous Agents and Multi-Agent Systems 17, 397–431 (2008)
43. Pěchouček, M., Vokrinek, J., Becvar, P.: ExPlanTech: multiagent support for manufacturing decision making. IEEE Intelligent Systems 20, 67–74 (2005)
44. Qiu, X.: Agent interaction in a Semantic Web environment: A state-of-the-art survey and prospects in knowledge mobilization. In: Information Systems Research in Scandinavia IRISCA28, Kristiansand, Norway (2005)
45. Rabemanantsoa, M.: Knowledge-based system for assembly process-planning. In: Proceedings of Software Engineering Standards Symposium, 1993, pp. 267–272 (1993)
46. Rajpathak, D.G., Motta, E., Zdrahal, Z., Roy, R.: A generic library of problem solving methods for scheduling applications. IEEE Transactions on Knowledge and Data Engineering 18, 815–828 (2006)
47. Roche, C., Fitouri, S., Glardon, R., Pouly, M.: The Potential of Multi-Agent Systems in Virtual Manufacturing Enterprises. In: Proceedings of the 9th International Workshop on Database and Expert Systems Applications. IEEE Computer Society, Los Alamitos (1998)
48. Seilonen, I., Pirttioja, T., Appelqvist, P., Halme, A., Koskinen, K.: Distributed planning agents for intelligent process automation. In: IEEE International Symposium on Computational Intelligence in Robotics and Automation, vol. 2, pp. 614–619 (2003)
49. Shen, W., Norrie, D.H.: Agent-based Systems for Intelligent Manufacturing: A State of-the-Art Survey. An International Journal of Knowledge and Information Systems 1(2), 129–156 (1999)
50. Shen, W., Wang, L., Hao, Q.: Agent-based distributed manufacturing process planning and scheduling: a state-of-the-art survey. IEEE Transactions on Systems, Man, and Cybernetics, Part C: Applications and Reviews 36, 563–577 (2006)
51. Silva, N., Rocha, J.: Ontology mapping for interoperability in semantic web. In: IADIS International Conference WWW/Internet (2003)

52. Vallée, M., Kaindl, H., Merdan, M., Lepuschitz, W., Arnautovic, E., Vrba, P.: An Automation Agent Architecture with A Reflective World Model in Manufacturing Systems. In: IEEE International Conference on Systems, Man, and Cybernetics (SMC 2009), San Antonio, Texas, USA (2009)
53. Vrba, P., Radakovič, M., Obitko, M., Mařík, V.: Semantic extension of agent-based control: The packing cell case study. In: Mařík, V., Strasser, T., Zoitl, A. (eds.) HoloMAS 2009. LNCS, vol. 5696, pp. 47–60. Springer, Heidelberg (2009)
54. Whitney, D.E.: The potential for assembly modeling in product development and manufacturing. MIT Press, Cambridge (1996)
55. Zhang, W., Xie, S.: Agent technology for collaborative process planning: a review. The International Journal of Advanced Manufacturing Technology 32, 315–325 (2007)

Chapter 11
Semantic Multi-Agent mLearning System

Stanimir Stoyanov[1], Ivan Ganchev[2], Máirtín O'Droma[2], Hussein Zedan[3],
Damien Meere[2], and Veselina Valkanova[1]

[1] Dept. of Computer Systems, Plovdiv University "Paisij Hilendarski", Bulgaria
stani@uni-plovdiv.bg, veselinaviva9@gmail.com
[2] Telecommunications Research Centre, University of Limerick, Ireland
{Ivan.Ganchev,Mairtin.ODroma,Damien.Meere}@ul.ie
[3] Software Technology Research Laboratory, De Montfort University, UK
hussein.zedan@googlemail.com

Abstract. Within this chapter, an agent-oriented middleware created to support the delivery of context-aware mLearning services provision is presented. This middleware architecture, based on the concept of InfoStations and developed within a University campus domain, is described in detail. Concepts for the control and management of service sessions and communications scenarios are also presented. The multi-agent approach adopted for the implementation of this system and indeed the system entity interactions involved in service delivery are discussed. The harvesting and utilisation of semantic information in order to facilitate the contextualisation and personalisation of mLearning services is also detailed.

1 Introduction

A distinguishable feature of contemporary mobile eLearning (mLearning) systems is the anywhere-anytime-anyhow aspect of delivery of electronic content, which is personalised and customised to suit a particular mobile user [1, 2]. In addition, mobile service content is expected to be delivered to users always in the best possible way through the most appropriate connection type according to the always best connected and best served (ABC&S) communication paradigm [3, 4]. In the light of these trends, the goal is to develop a context-aware and agent-oriented middleware system. The middleware described within this chapter uses an InfoStation-based communication environment with distributed control [5, 6]. The InfoStation paradigm is an extension of the wireless Internet, where mobile clients interact directly with Web service providers (i.e. InfoStations). The users request services (from their mobile devices) from the nearest InfoStation utilizing Bluetooth or WiFi wireless communication. Of course in future, the incorporation of technology such as WiMAX will greatly aid the deployment of this architecture and the delivery of services to a much wider community. In our approach, each application utilizing the InfoStation infrastructure consists of two components:

A. Elçi, M.T. Koné, and M.A. Orgun (Eds.): Semantic Agent Systems, SCI 344, pp. 243–272.
springerlink.com

(i) a standardized middleware, which identifies and locates the changes in the environment in order to prepare the adequate delivery of requested services; and (ii) a set of electronic services (in this case mLearning services), which are adapted and controlled by this middleware.

In the architecture presented in this chapter, the context is processed by the middleware components deployed on different InfoStations. The determination of a particular context is accomplished within the framework of predefined scenarios. In certain cases, it is necessary to identify different local circumstances at different moments and in different network nodes (InfoStations). What is required in these cases is a strict synchronisation between the middleware components deployed on different InfoStations.

The rest of the chapter is organized as follows. Section 2 presents a state of the art of various related works. Section 3 presents the InfoStation communication infrastructure. Section 4 describes our concept for context-aware service provision. Section 5 presents briefly the layered system architecture developed to support the context-aware service provision. Section 6 describes the agent-oriented middleware architecture of an InfoStation. Section 7 details the use of the Ontology Web Language for services (OWL-S) in describing the properties and capabilities of the various mLearning services, essentially providing the mechanism to advertise the services to agents operating within the architecture. Section 8 is concerned with issues related to the context-aware management of service sessions. Section 9 details the contextualisation and adaptation of presented services based on information conveyed through the *Resource Description Framework* (RDF)-based user-centric profiles. Section 10 provides an outline of the various entity interactions and semantic information transactions involved in the provision of a contextualised mLearning services, in this case the mTest service. Section 11 considers relevant implementation issues. Finally section 12 concludes the chapter.

2 Related Works

The efforts to support electronic services for different application domains/areas, on one hand, and for the provision of heterogeneous access to them through different types of networks (Bluetooth, WiFi, WiMAX, sensor networks, etc.), on the other, have in recent years led to the development of hybrid software architectures. Within such architectures, intelligent agents operate jointly with electronic services. Advantages of this combined use of agents and services are presented in detail in [7]. In the last few years, there has been a growing trend in realisation of this support for context-aware and adaptable middleware. A variety of such systems are described in the specialised literature, which are focused on various aspects of this support, i.e. from systems ensuring efficient heterogeneous access to middleware systems oriented to the specifics of the application domain.

A large portion of middleware deals with the extraction and support of context information. A context-aware architecture composed of clients (moving nodes), a context-server and middleware (fixed nodes connected through TCP/IP with the context-server) is described in [8]. Within that architecture, the middleware plays important role both in identifying the clients using the Bluetooth technology and in finding a suitable executable module in accordance with the requisite context acquired from the context-server. Another architecture described in [9] uses a Context Engine for context-aware delivery of Web services, whereby a rule-based approach based on first-order logic is adopted for the centralised processing of context. In [10] a context broker architecture based on ontologies for context representation is presented. A context-aware analog to a telephone service provider is described in [11]. More common context-aware architectures are presented in [12, 13]. The context within these architectures is stored and processed in a centralised manner whereby the middleware is used mainly as a mediator.

Various specialized literature sources also describe interesting applications of agent-oriented middleware architectures targeting different application domains. For instance, [14] presents a system for controlling the transport of valuable cultural-historical objects, which also provides opportunities for users to receive information about these objects on their mobile phones, according to their personal preferences. The system described in [15] details a multi-agent approach for air traffic control. Agent-oriented middleware to detect abnormalities in the traffic of computer networks is presented in [16]. An agent-oriented hybrid system in the field of bioinformatics is presented in [17].

Many of the aforementioned sources treat the context-aware middleware components and the adaptation to heterogeneous networks separately. However, in recent years, more complex middleware systems, containing the necessary resources for context-dependent and personalised delivery of electronic services, have become increasingly prevalent. An architecture is presented in [18], where these two topics are considered jointly. There, a middleware for multimedia services in heterogeneous networks is described, which includes two frameworks: (i) an adaptive service-provisioning middleware framework ensures the delivery of services to mobile users, at any time, any place and for any context, by interacting with existing heterogeneous networks, (ii) a context-aware multimedia middleware framework supports content filtering, adaptation, aggregation of context, drawing of conclusions and training. The architecture described in [19] illustrates an agent-based and service-oriented, adaptive and context-dependent architecture. The architecture supports semantic match-making, composition, implementation and control of services. The offered functionality is divided in the following three abstract layers: an agent-based peer service composition agent layer, an agent-based wrapper service layer, and a resource and asset layer.

The main requirements, which an agent-oriented middleware for adaptable systems must satisfy, are defined in [20]. Meeting some of these requirements is demonstrated by the presented there middleware architecture. The middleware is implemented as a web of agents, operating as mediators between the operating system and the user applications.

3 InfoStation-Based Network Architecture

The continuing evolution in the capabilities and resources available within modern mobile devices has precipitated an evolution in the realm of eLearning. The architecture presented here attempts to harness the communicative potential of these devices in order to present learners with a more pervasive learning experience, which can be dynamically altered and tailored to suit them. The following network architecture enables mobile users to access various mLearning services, via a set of intelligent wireless access points, or InfoStations, deployed in key points across the University Campus. However, these services are not only optimized to suit the operating environment onboard the mobile device, but also customised according to the learners' own personal details. In order to accomplish this, and indeed, due to the inherent mobility supported by the system, it is necessary for intelligent agents to operate throughout every level of the system. Indeed, both the InfoStation and the InfoStation Centre, each take on the appearance of a multi-agent system in itself, as is outlined within Section 6. These intelligent agents gather information and accomplish tasks without the requirement of human interaction and coordinate themselves in order to complete various network management tasks. Throughout this chapter we will outline how these agents interact with each other, and detail the semantic information transactions necessary for the delivery of highly contextualised learning content to learners. The system architecture consists of the following three tiers as shown in Fig. 1:

- 1st tier – encompassing the user mobile devices (cell phones, laptops, PDAs), equipped with intelligent agents acting as Personal Assistants to users. The Personal Assistant gathers information about the operating environment onboard the mobile device, as well as soliciting information about the user. Supplied with this information, the InfoStation can make better decisions on applicable services and content to deliver to the Personal Assistant;

- 2nd tier – consisting of InfoStations, satisfying the users' requests for services through Bluetooth and/or WiFi wireless mobile connections. The InfoStations maintain connections with mobile devices, create and manage user sessions, provide interface to global services offered by the InfoStation Centre, and host local services. The implementation of these local services is an important aspect of this system. By implementing particular services within specific localised regions throughout the University campus, we can enrich the service users experience within these localities. A prime example of how this type of local service can enrich a learners experience, is the deployment of library-based services [21]. Within the library domain, library users experience can be greatly enhanced through the facilitation of services offering resource location capabilities or indeed account notifications. The division of global and local services allows for a reduction of the workload placed on the InfoStation Centre. In the original InfoStation architecture [5], the InfoStations operated only as mediators between the user mobile devices and a centre, on which a variety of electronic services are deployed and executed. The InfoStations within this architecture do not only occupy the role of mediators, they also act as the primary service providing nodes;

3rd Tier: *InfoStations Centre*
(with Profile Managers and Global Services' Content Repository)

2nd Tier: *InfoStations*
(with cached copies of recently used user/service profiles, and Local Services' Content Repository)

1st Tier: *Mobile Devices*
(with Intelligent Agents acting as Personal Assistants for mobile users)

Fig. 1. The 3-tier InfoStation-based network architecture

- 3rd *tier* – this is the InfoStation Centre – the core of the overall architecture – concerned with controlling the InfoStations, and overall updating and synchronisation of information across the system. The InfoStation Centre also acts as the host for global services.

We have developed a set of mLearning service prototypes that use the resources distributed throughout this architecture in an efficient manner. For instance, the user requests for local services are satisfied directly by the InfoStations without a need to access the InfoStation Centre's resources. These local services are provided only to mobile devices currently operating within range of a particular InfoStation. On the other hand, user requests for global services require redirection to the InfoStation Centre in order to be satisfied. In some cases/scenarios a local service could be the initial and/or final phase of some global service provision. For instance, a typical action concluding a global service provision is to forward the final result to the mobile device that has requested it. Also, the initial execution of some service requests could be handled by one InfoStation and later completed by another one, due to mobility supported by the system, e.g. when during the service provision the user moves into the range (service area) of another InfoStation, as detailed in [22]. Middleware software, with specific functions utilized to support this architecture, ensures sufficient flexibility, adaptability and autonomy of the system architecture's nodes. In addition, the autonomic software components must be able to communicate at a higher semantic level with regard to the context, business-logic of the provided service, and the individual characteristics of users. For this reason, an agent-oriented approach has been adopted for the implementation of the required middleware.

4 Context-Aware Service Provision

Any system that ensures a context-aware service provision could be built as an integration of two components [23]:

- A standardized <u>middleware</u>, which is able to detect the dynamic changes in the environment (c.f. communication scenarios below) during the processing of user requests for services (*contex-awareness*) and correspondingly to ensure their efficient and non-problematic execution (*adaptability*);
- A set of <u>electronic services</u> realizing the functionality of the application area (*business functionality*), which could be activated and controlled by the middleware. In our case, the application area is the eLearning.

As the middleware described further in here is concerned with the context-awareness and adaptability aspects, it is important to first clarify these concepts. Within this development approach, Dey's definition [24] was adopted, according to which "context is any information that can be used to characterize the situation at an entity". An entity could be a person, place, or object that is considered relevant to the interaction between a user and an application, including the user and applications themselves. Context could be of different type, e.g. location, identity, activity, time.

Dey's definition is utilized here as a basis for further discussions. In order to elaborate on this definition a working one for the creation of the desired middleware architecture, we first solidify the definition as presented further in the chapter. We want clearly to differentiate context-awareness from the adaptability. Context-awareness is the middleware's ability to identify the changes in the environment/context as regards:

- Mobile device's location (*device mobility*) – in some cases this change leads to changing the serving InfoStation. This is especially important due to the inherent mobility within the system, as users move throughout the University campus. This information has a bearing on the local services deployed within a particular area i.e. within the University Library;
- User device (*user mobility*) – this change offers different options for the delivery of the service request's results back to the user. What is important here is to know the capabilities of the new device activated by the user, so as to adapt the service content accordingly;
- Communication type – depending on the current prevailing wireless network conditions/constraints, the user may avail of different communications possibilities (e.g. Bluetooth or WiFi);
- User preferences – service personalisation may be needed as to reflect the changes made by users in their preferences, e.g., the way the service content is visualised to them, etc.;
- Goal-driven sequencing of tasks engaged in by the user;
- Environmental context issues such as classmates and/or learner/educator interactions.

The goal of adaptability is to ensure trouble-free, transparent and adequate fulfilment of user requests for services by taking into account the various aspects of the context mentioned above. In other words, after identifying a particular change in the service environment, the middleware must be able to take compensating actions (counter-measures) such as handover of user service sessions from one InfoStation to another, re-formatting/transcoding of service content due to a change of mobile device (varying device capabilities), service personalisation, etc.

To ensure adequate support for user mobility and device mobility (the first two aspects of the context change), the following four main communications scenarios are identified for support in our middleware architecture [22]:

- *'No change'* – a mLearning service is provided within the range of the same InfoStation and without changing the user mobile device;
- *'Change of user mobile device'* – due to the inherent mobility, it is entirely possible that during an mLearning service session, the user may shift to another mobile device, e.g. with greater capabilities, in order to experience a much richer service environment and utilize a wider range of resources;
- *'Change of InfoStation'* – within the InfoStation paradigm, the connection between the InfoStations themselves and the user mobile devices is by definition geographically intermittent. With a number of InfoStations positioned around a University campus, the users may pass through a number of InfoStation serving areas during the service session. This transition between InfoStation areas must be completely transparent to the user, ensuring the user has continuous access to the service;
- *'Change of InfoStation and user mobile device'* – most complicated scenario whereby the user may change the device simultaneously with the change of the InfoStation.

To support the third aspect of the context change (different communication type), the development of an intelligent component (agent) working within the communication layer (c.f. Fig. 2) is envisaged. This component operates with the capability to define and choose the optimal mode of communication, depending on the current prevailing access network conditions (e.g. congestion level, number of active users, average data rate available to each active user, etc.). The user identification and corresponding service personalisation is subject to a middleware adaptation for use in the particular application area. In the case of eLearning, the architecture is extended to support the three fundamental eLearning models – the educational domain model, the user/learner model, and the pedagogical model [25, 26]. The rest of the chapter focuses on the support of the first two aspects of the context-awareness and relevant adaptability.

5 Layered System Architecture

The developed layered software system architecture as depicted in Fig. 2 is a distributed architecture, meaning that its functional entities are deployed across the different tiers / nodes, i.e. on mobile devices, InfoStations, and InfoStation Centre. In this architecture the role of the InfoStations is expanded, enabling them to act (besides the mediation role) as hosts for the local mLearning services (LmS) and for preparation, adaptation, and conclusive delivery of global mLearning services (GmS). This way the service provision is efficiently distributed across the whole architecture. Each of the system network nodes have a different structure depending on their functioning within the system. However, each node is built upon a Communication Layer whose main task is to initialize, control and maintain communications between different nodes. This layer is also concerned with choosing the most appropriate mode of communication between a mobile device and an InfoStation – whether that be Bluetooth or WiFi, or indeed as the platform evolves perhaps WiMAX in the future. The software architecture of the InfoStations and InfoStation Centre includes a Service Layer on the top. The main task of this layer is to prepare the execution of the users' service requests, to activate and receive the results of the execution of different services (local and global).

The InfoStations' middle layer is responsible for the execution of scenarios and control of user sessions. It is at this layer where the user service requests are mainly processed by taking into account all contex-aware aspects and applying corresponding adaptive actions. The middle layer of the InfoStation Centre ensures the needed synchronisation during particular scenarios (c.f. Section 8). In addition, different business supporting components, e.g. for user accounting, charging and billing, may operate here.

The software architecture of the user mobile devices contains two other layers:

- Personal Assistant – its task is to help the user in specifying the service requests sent to the system, accomplish the communication with the InfoStations' software, receive and visualise the service requests' results to the user, etc. Moreover the assistant can provide information needed for the personalisation of services (based on information stored in the user profile) and/or for the synchronisation of scenario execution;
- Graphical User Interface (GUI) – its task is to prepare and present the forms for setting up the service requests, and visualise the corresponding results received back from the system.

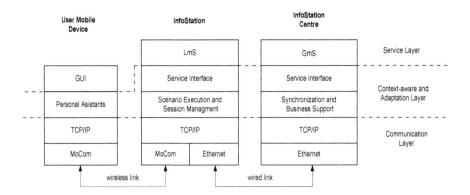

Fig. 2. The layered system architecture

6 Agent-Oriented Middleware Architecture

The main implementation challenges within this system are related to the support of distributed control, as the system should be capable of detecting all relevant changes in the environment (context-awareness) and according to these changes, facilitate the service offerings in the most flexible and efficient manner (adaptability). The system architecture presented in the previous section is implemented as a set of cooperating intelligent agents. An agent oriented approach has been adopted in the development of this architecture in order to:

- Model adequately the real distributed infrastructure;
- Allow for realisation of distributed models of control;
- Ensure pro-active middleware behaviour which is quite beneficial in many situations;
- Use more efficiently the information resources spread over different InfoStations.

Moreover, the agent-oriented architecture can easily be extended with new agents (where required) that cooperate with the existing ones and communicate by means of a standardized protocol (in this case the FIPA -Agent Communication Language (ACL) [27, 28]). Indeed the InfoStations and InfoStation Centre exist as networks of interoperating agents and services, with the agents fulfilling various essential roles necessary for system management. Within each of these platforms, agents take responsibility for selecting (optimal mode) and establishing a client-server cross-platform connection, conveyance of context information and the delivery of adapted and personalised service content. This multi-agent approach differs from the classic multi-tier architectures in which the relationships between the components at a particular tier are much stronger. Conceptually we define different layers in the system architecture in order to present the functionality of the middleware that is being developed in a more systematic fashion. Implementation-wise, the middleware architecture is considered as a set of interacting intelligent agents. Communication between the user mobile devices and the serving InfoStations could be realized in two ways: (i) an agent operating within the InfoStation discovers all new devices entering the range and subsequently initiates communication with them; or (ii) Personal Assistant agents on the user mobile devices are the active part in communication, and initiate the connection with the InfoStation. In the current implementation of the prototype architecture, the former approach is used for Bluetooth communication, whereas the latter applies for WiFi communication.

Fig. 3 highlights the main components necessary to ensure continuity to the service provision, i.e. support for the continuous provision of services and user sessions in the case of scenario change [22] or resource deficiency. The agents which handle the connection and session establishment perform different actions, such as searching for and finding mobile devices within the range of an InfoStation, creating a list of services required by mobile devices, initiation of a wireless connection with mobile devices, data transfer to- and from mobile devices. This is described in more detail in the following section. Also illustrated within Fig. 3 are the components which serve to facilitate a level of context sensitivity and personalisation to the presented services. The mechanisms by which the services achieve this context sensitivity are dealt with in more detail later in the chapter. A short description of the various agents (for Bluetooth communication) within the architecture is presented below.

The first step in the delivery of the services involves the Scanner agent, which continuously searches for mobile devices / Personal Assistant agents within the service area of the InfoStation. In addition, this agent retrieves a list of services required by users (registered on their mobile devices upon installation of the client part of the application), as well as the profile information, detailing the context (i.e. device capability and user preference information. The Scanner agent receives this information in the form of an XML file (c.f. Section 8), which itself is extracted from the content of an ACL message. The contents of this XML file are then passed on via the Connection Advisor agent, to the Profile Processor agent, which parses the received profile and extracts meaningful information. This information can in turn be utilized to perform the requisite alterations to services and service content. This information is also very important in relation to the tasks undertaken by the Scenario Manager agent. The role of this agent is to monitor and respond to changes in the operating environment, within which the services are operating (i.e. change of mobile device). In the event of a significant change of service environment, this agent gathers the new capability and preference information (CPI) via the Scanner agent. Then, in conjunction with the Query Manager agent and the Content Adaptation agent, facilitates the dynamic adaptation of the service content to meet the new service context. The main duty of the Connection Adviser agent is to filter the list (received from the Scanner agent) of mobile devices as well as requested services. The filtration is carried out with respect to a given (usually heuristic) criterion. Information needed for the filtration is stored in a local database. The Connection Adviser agent sends the filtered list to the Connection Initiator agent, who takes on the task of initiating a connection with the Personal Assistant onboard the mobile device. This agent generates the so-called Connection Object, through which a communication with the mobile device is established via Bluetooth connection. Once this connection has been established, the Connection Initiator generates an agent to which it hands over the control of the connection, called a Connection agent. From this point on, all communications between the InfoStation and the Personal Assistant are directed by the Connection agent. The internal architecture of the Connection agent contains three threads: an agent thread used for communication with the Query Manager agent, and a Send thread and Receive thread, which look after each direction of the wireless communication with the mobile device.

The Query Manager performs one of the most crucial tasks within the InfoStation architecture. It determines where information received from the mobile device is to be directed, e.g. directly to simple services, or via Interface agents to sophisticated services. It also transforms messages coming from the Connection agent into messages of the correct protocols to be understood by the relevant services, i.e. for simple services - UDDI or SOAP, or for increasingly sophisticated services by using more complicated, semantic-oriented protocols (e.g. OWL-S [16], c.f. next section). The Query Manager agent also interacts with the Content Adaptation agent in order to facilitate the Personal Assistant with increasingly contextualised service content. This Content Adaptation agent, operating under the remit of the Query Manager agent, essentially performs the role of an adaptation engine, which takes in the profile information provided by the Profile Processor agent, and executes the requisite adaptation operations on the service content (e.g. file compression, image resizing etc.)

The Query Manager agent receives user service requests via the Connection agent, and may communicate with various services. Once it has passed the request on to the services, all service content is passed back to the Query Manager via the Content Adaptation agent. The Profile Processor agent parses and validates received profiles (XML files) and creates a Document Object Model (DOM) tree [29]. Using this DOM tree the XML information may be operated on, to discern the information most pertinent to the adaptation of service content. The Content Adaptation agent receives requests-responses from the services, queries the Profile Processor agent regarding the required context, and then either selects a pre-packaged service content package which closely meets the requirements of the mobile device, or applies a full transformation to the service content to meet the constraints of the operating environment of the device.

The tasks undertaken by the Content Adaptation agent, the Scenario Manager agent and the Profile Processor agent, enable the system to dynamically adapt to changing service environments, even during a particular service session. Once the connection to a particular service has been initialized and the service content adapted to the requisite format, the Connection agent facilitates the transfer of the information to the user mobile device.

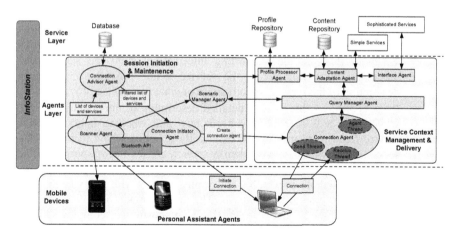

Fig. 3. The agent-oriented middleware architecture

7 Using the Ontology Web Language for Services (OWL-S)

In our architecture, communication with more sophisticated services is performed by using the standard Ontology Web Language for Services (OWL-S) protocol [30]. The Resource Description Framework (RDF) was the first step in the web-based ontology language. It facilitated the description of resources on the Web. However, RDF is only suited to the expression of lightweight ontologies. OWL-S enhances the expressivity of RDF, providing a more in-depth means of describing the various classes, property restrictions and characteristics, etc. OWL-S provides a set of constructs with which to create ontologies (i.e. specifications of concepts and relationships with agents), which are machine understandable descriptions of the service. These ontologies enable agents to discover, invoke and interoperate with the services.

Using OWL-S, the ontology structure is divided into four separate sections, each dealing with a different aspect of the service:

- *Service Profile* - this advertises the abilities of the service (i.e. what it can do), in such a way as to enable a service-seeking agent to determine if the service meets its requirements. This provides a concise high-level description of the service to a registry, which is maintained by the InfoStation Centre, and periodically disseminated copies of all service profiles to the various local InfoStations throughout the campus. The service profile outlines the limitations of the service, the quality of the service and of course the requirements placed upon the requesting agent in order to utilize the service successfully. The profile serves only to provide a description of the service to a registry. In this system, the InfoStation provides a user's Personal Assistant with access to a registry of services. Here the user can examine the available service profile information to locate a particular service. Once the user has selected a particular service, the profile has fulfilled its purpose and performs no other function. The process model specifies all interactions between the requesting agent and the service. Service profiles are comprised to three different types of information: a *description* of the service and the service provider, *functional attributes* that provide supporting information about the service and the *functional behaviour* of the service which provides a capability description of the service, specification of what the service provides.

- *Process Model* - this gives a detailed description of the operation of the service and tells a requesting agent how and when to interact with a service (read/write messages). Essentially it outlines how to initiate the service, and what happens when a service carries out its purpose. The process model is not a programme to be executed, but rather a specification of the methods in which a client must interact with a service in order for some outcome to be achieved. The process model is utilized for the invocation of the service, for planning a composition of complex actions, and for interoperation and monitoring of service functionality. The primary entity with the process model is the "Process" itself. The purpose of this process is to generate some new information, based on information provided to it and the conditions it's under. Essentially this can be thought of as generating a certain 'output', given a particular 'input' under certain conditions. For example given a user service request for a certain piece of service content, this describes the process by which that particular piece of service content is provided to the user. A process can also effect a change within its immediate environment. This can be thought of in terms of preconditions and effects of the process. For example a precondition for the process of a user utilizing a service could be that the users have successfully completed an authentication procedure. An effect of this service access process would be the requisite updating of the user profile. A process can have any number of these inputs (necessary for a process to be performed) and outputs (resultant of the execution of the process on a particular input). There can also be any number of preconditions which must be satisfied before a process can be successfully initialised. Finally, depending on the conditions in the environment of the process, any number of effects can be generated by the execution of that process.

- *Grounding* - this provides details of how an agent can access and interoperate with a service (i.e. interact with the service). Specified within the grounding, are details pertaining to message formatting, transport mechanisms, protocols, addressing, etc. When combined with the details outlined within the process model, all the information necessary for a client to utilize a particular service is presented. While both the Service Profile and the Process Model provide abstract representations of the service and its processes, the Grounding provides a concrete specification of the various elements required for interaction with the service. In essence the central function of an OWL-S grounding is to show how the various inputs and outputs of an atomic process are realized as messages which can carry those inputs and outputs in some specific transmittable format. The Web Services Description Language (WSDL) [31] is utilized within OWL-S to facilitate the initial grounding mechanism. A WSDL document provides an XML grammar for portraying services as collections of communication endpoints with the capability of exchanging messages. Essentially it specifies the location of the service and the operations the service exposes. The abstract definitions of the communication endpoints and messages are separated from the concrete specifications, and as such, the abstract definitions may be reused. The specification of the OWL-S grounding involved the cooperative utilisation of both OWL-S and WSDL. As both of these languages operate within a different conceptual space, they are each required for the specification of services. Essentially OWL-S and WSDL overlap within the grounding. Fig. 4 illustrates the convergence of these two specification languages.
- *Service* - this simply binds the other elements together into a single entity which can be published and invoked.

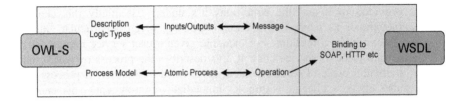

Fig. 4. The mapping between OWL-S and WSDL

Agents utilize the information contained within the Service Profile to ascertain whether or not a service meets its requirements, and adheres to certain constraints such as security, quality of service etc. While the Service Profile provides all the information needed for an agent to discover a service, the Process Model provides the information necessary for the agent to use the service. The Process Model allows the agent to perform a more in-depth analysis of the service and its capabilities, and determine if it can be utilized. As well as this, it enables the agent to compose new service descriptions through the composition and interoperation of previous existing services, to perform specific tasks. The Process Model also allows agents to monitor the execution of tasks performed by a service (or a set of services), and to coordinate the entities involved in the service execution. The Service Grounding details how agents can communicate with, interoperate with, and invoke a service. The relationships between the various service components are modelled using properties such as presents (Service to Profile), describedBy (Service to Process Model) and supports (Service to Grounding), Fig. 5.

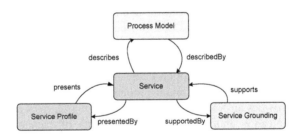

Fig. 5. The OWL-S service ontology

When all these separate parts are combined, they form an ontology/description that allows agents to discover, invoke, compose and monitor services. Utilizing the OWL-S protocol offers a good opportunity for the realisation this flexible software architecture, offering a suitable environment for the support of the mLearning services in this system. The following Fig. 6 is a sample OWL-S service description of the mLecture service. From this mLecture service instance, one can see the specification of the various relationships which exist between the elements which comprise the OWL-S. This essentially serves to bind the various elements together into a single invokable entity.

8 Context-Aware Management of Service Sessions

The hybrid InfoStation's middleware architecture, presented in previous sections, consists of two types of software components – agents and services. Services are used for the realisation of business functionality; however, they are often static and thus unsuitable for presentation within a dynamic system. On the contrary, agents are suitable for examining and responding to dynamic changes in the system environment, but unsuitable for the delivery of business functionality. As the InfoStation network provides a dynamic environment for mLearning services

```
<owl:Ontology rdf:about="">
  <owl:versionInfo>
  $Id: mLectureService.owl,v 1 17/11/2008$
  </owl:versionInfo>
  <rdfs:comment>
  This ontology represents the OWL-S service description mLecture Service.
  </rdfs:comment>

  <owl:imports rdf:resource="http://www.ece.ul.ie/trc/services/Service.owl"/>
  <owl:imports rdf:resource="http://www.ece.ul.ie/trc/services/Profile.owl"/>
  <owl:imports rdf:resource="http://www.ece.ul.ie/trc/services/Process.owl"/>
  <owl:imports rdf:resource="http://www.ece.ul.ie/trc/services/Grounding.owl"/>
  <owl:imports rdf:resource="http://www.ece.ul.ie/trc/services/mLectureProfile.owl"/>
  <owl:imports rdf:resource="http://www.ece.ul.ie/trc/services/mLectureProcess.owl"/>
  <owl:imports rdf:resource="http://www.ece.ul.ie/trc/services/mLectureGrounding.owl"/>
</owl:Ontology>

<!--###############################################
    mLectureService
    ###############################################-->

<service:Service rdf:ID="mLectureService">
  <!-- Reference to the Profile -->
  <service:presents rdf:resource="http://www.ece.ul.ie/trc/services/
mLectureProfile.owl#mLecture_Service_Profile"/>

  <!-- Reference to the Process Model -->
  <service:describedBy rdf:resource="http://www.ece.ul.ie/trc/services/
mLectureProcess.owl#mLectureServiceProcess"/>

  <!-- Reference to the Grounding -->
  <service:supports rdf:resource="http://www.ece.ul.ie/trc/services/
mLectureGrounding.owl.owl#mLectureGrounding"/>
</service:Service>

<!-- Inverse links -->

<profile:Profile rdf:about="http://www.ece.ul.ie/trc/services/
mLectureProfile.owl#mLecture_Service_Profile">
  <service:presentedBy rdf:resource="#mLectureService"/>
</profile:Profile>

<process:CompositeProcess rdf:about="http://www.ece.ul.ie/trc/services/
mLectureProcess.owl#mLectureServiceProcess">
  <service:describes rdf:resource="#mLectureService"/>
</process:CompositeProcess>

<grounding:WsdlGrounding rdf:about="http://www.ece.ul.ie/trc/services/
mLectureGrounding.owl.owl#mLectureGrounding">
  <service:supportedBy rdf:resource="#mLectureService"/>
</grounding:WsdlGrounding>

</rdf:RDF>
```

Fig. 6. An OWL-S mLecture service instance example

provision, the task of the agent-oriented middleware is to react to the environmental changes and to ensure correct execution of user requests for services. A substantial problem in this task is the management of service sessions in case of scenario change. With efficient management of service sessions, users are provided with seamless access to services from any device, using the same user profile information while moving. In order to ensure the transparency of this process to the user, a specialized Scenario Manager agent is dedicated to tackle this problem within this architecture (Fig. 3).

Due to the inherent unpredictability in the user behaviour (e.g. change of InfoStation and/or device) and the distributed nature of the system, it is impossible to support apriori a global model for scenario management. For instance, the condition for initiating the *'Change of InfoStation'* scenario could be presented by the rule depicted in Fig. 7. The difficulty here is related to the fact that particular preconditions must be checked in different InfoStations, but the validity of the rule can be determined by means of synchronisation that includes communication between relevant Scenario Manager agents on different InfoStations. The fact that a mobile device md_i is leaving the area of InfoStation IS_1 and entering the area of InfoStation IS_2 can be identified by the Scanner Agents working on these InfoStations. However, what both Scenario Manager agents have to establish by means of mutual communication, is whether this is the same device in question. Additional synchronisation is also required between these two Scenario Manager agents as to check whether an unaccomplished (open) service request in the entering mobile device awaits some results from the system. This latter synchronisation could be done in different ways:

- *InfoStation Centre Synchronisation* – in this centralised approach, the InfoStation Centre stores a central registry of all unaccomplished (open) service requests. When an execution of a new service request is started, the serving InfoStation sends relevant information on to the InfoStation Centre for storage. When a new device enters within range of an InfoStation, the latter asks the InfoStation Centre for relevant information regarding any unaccomplished service requests initiated by this device in order to continue their execution from the point reached by the previous InfoStation;

- *Mobile Device Synchronisation* – in this distributed approach, the Personal Assistant agent operating on the mobile device keeps track on all unaccomplished (open) service requests initiated by the user/device. When the user/device enters the area of another InfoStation, the Personal Assistant agent notifies the new InfoStation about any unaccomplished requests (along with relevant information such as last point reached, state, parameters values, etc) that the InfoStation should take care of. This approach could be further facilitated by installing an IEEE 802.21[1] agent on the user mobile device which may speed up the handover process from one InfoStation to another;

---

[1] **IEEE 802.21** – an emerging Media Independent Handover Services standard that provides lower layer support for device mobility across heterogeneous wireless access networks.

- *Broadcasting* – in this approach, when a mobile device with unaccomplished (open) request is leaving the area of a particular InfoStation, the latter notifies all other InfoStations by broadcasting a message including details of this request in order to prepare the fast handover of the service session to the next InfoStation;
- *Multicasting* – this is a special case of the previous approach, whereby the current InfoStation multicasts a notification message only to the group of neighbouring InfoStations, as it is more likely the mobile device that left it will re-appear within the area of some of them. Comparing to the previous one, this approach allows saving the communications resources and reducing the overhead. Disadvantage is that it could be only applied if device mobility patterns statistic is already available and used, or in cases of indoor application when users are moving inside a building along the corridors so the system may exactly predict which the next serving InfoStation will be.

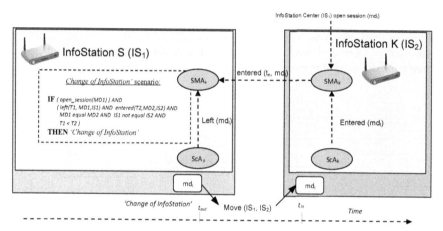

Fig. 7. The interaction between Scenario Manager Agents (SMAs) for the *'Change of InfoStation'* scenario

Whatever the chosen approach, the efficient management of the service session requires the conveyance of session information between the various architectural entities. In order to ensure the seamless session transition from one InfoStation to another, the InfoStations require information pertaining to the user, the session and the communication involved. When a user enters within wireless range of an InfoStation, the Connection Initiator agent onboard the InfoStation, initiates communication with the Personal Assistant onboard the user device. Within these initial interactions, the Personal Assistant passes on the information about that particular user. Whether the InfoStation receives the session information from the Personal Assistant or from another InfoStation, this information, in conjunction with the user details, enables the Scenario Manager agent to associate the user to a particular service session. In doing so, the new InfoStation takes on the role of facilitating the session begun elsewhere, and facilitates the user with seamless service access. Essentially we can consider any given service session consisting of the following information tuple:

$$Sn = (Uk, Dm, Cn)$$

where Sn represents the nth session comprised of Uk, Dm, and Cn, which respectively represent the kth user, the mth device (operating environment), and the service content. The session itself is managed through the utilisation of information based around this tuple. Each of these three elements serves to uniquely identify the essential components of service delivery. In order for a session to be maintained between InfoStation, the InfoStation must be aware of the user's credentials, as well as the constraints of the user's current operating environment (Scenarios Changes). However, by providing information relating to the session currently in operation by the user, the InfoStation seeking to continue the session can reinitialise, and seamlessly maintain the service session continuation accordingly. The following illustration (Fig. 8), based on the informational model put forward in [32], seeks to illustrate the informational architecture involved in the continuation of service sessions across a number of InfoStations. This illustration highlights the utilisation of user- and device information in facilitating a more comprehensive scheme for the session management.

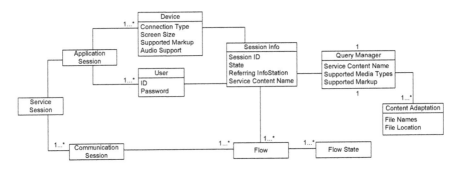

Fig. 8. The information model for service session management

9   User-Based Service Contextualisation and Adaptation

One of the fundamental elements of this system is the facility for service context-sensitivity and personalisation. In order to support this functionality we turn to the creation of device-, user- and service profiles. For the implementation of the former two profiles, which are integral components in the service adaptation procedure, we have opted to use the uniform platform-independent Composite Capability/Preference Profile (CC/PP) format [33]. This format is based on the Resource Description Framework (RDF), one of the key specifications of the Semantic Web, and is recommended by the World Wide Web Consortium (W3C) [34]. When adapting service content for a specific user, the information required for the adaptation can come from different sources - the network, the accessing device or indeed the user's own preferences/context. The InfoStation can receive these different pieces of information separately, but needs to merge the information into one model before doing content adaptation. Based on the Semantic Web and RDF, CC/PP simplifies this data integration through the use of extensible and non-centralised vocabularies.

The CC/PP specification defines a structured framework through which devices can define critical criteria relating to their capabilities. A CC/PP profile contains a description of device's capabilities as well as a specific user's personal preferences/context, which can be utilized to guide the adaptation of service content delivered to that device. This adapted and personalised mLearning allows us to offer multimedia content and activities adapted to learners' specific needs and influenced by their specific preferences and context. So when a specific user (mobile device-based Personal Assistant) submits a request to use a certain service, the source of that service (i.e. the InfoStation) customises and tailors the service content to meet the user preferences and the capabilities of his/her current mobile device. In essence, content is adapted to 'best' suit the individual user and the specific device at that particular time. Through the customisation and tailoring of the services (and their content), they can be offered to users, independently of the type of mobile devices. This is an essential factor in this type of mobile network environment, as user devices and preferences will be as varied as the users themselves.

A CC/PP profile contains a number of attributes and associated values, which are used by the InfoStations to determine the most appropriate format of the resource to be delivered to the user's Personal Assistant. The User Agent Profile (UAProf) [35] specification is a concrete implementation of the CC/PP developed by the Open Mobile Alliance [36]. This specification builds upon WAP 2.0 and facilitates the flow of capability and preference information (CPI) between the Personal Assistant, the InfoStation and the InfoStation Centre. The specification defines CPI through a structured set of components and attributes. The following are the most useful components defined within the UAProf specification:

- *Hardware Platform*: contains attributes that describe the hardware characteristics of the current user device, e.g. device type, input/output methods, screen size, colour capabilities, image capabilities, device CPU etc.
- *Software Platform:* contains attributes relating to the operating environment of the device, e.g. operating system name-vendor-version, JVM version, audio/video codecs, Java enabled etc.
- *Network Characteristics:* attributes relating to the network capabilities of the device, e.g. specification of Bluetooth or WiFi support etc.

The following Fig. 9 illustrates a sample of a device's hardware component with attributes which each have a bearing on the appearance of the final delivered contextualised service content. Utilizing the CPI contained within these CC/PP-UAProf-based profiles, the service providers (i.e. InfoStations) can tailor local services to meet the demands of a requested delivery context. The UAProf represents this CPI in a two-level hierarchy consisting of various *components* (dealing with a different characteristic of the device) and associated *attributes*.

The details specified within an instance of the UAProf, and its components, enables an InfoStation to dynamically adapt and customise the mobile services according to the specifications of that device. The hardware component provides information about the screen size, imaging capabilities, processing power, input/output methods, manufacturer etc. An InfoStation uses this information to efficiently and adequately adapt a particular service and content to the hardware environment on that device. An example of how a service may be adapted according to hardware criteria would be to examine the *ScreenSize* attribute. This attribute places constraints on the amount of information which can be shown on the screen of the device, and it would be prudent to reduce font sizes to suit the smaller screen.

```
<prf:component>
 <rdf:Description rdf:ID="HardwarePlatform">
 <rdf:type rdf:resource="http://www.openmobilealliance.org/tech/profiles/UAPROF/
 ccppschema-20021212#HardwarePlatform" />

<prf:ScreenSize>176x220</prf:ScreenSize>
 <prf:Model>V3i</prf:Model>
 <prf:ScreenSizeChar>17x11</prf:ScreenSizeChar>
 <prf:BitsPerPixel>16</prf:BitsPerPixel>
 <prf:ColorCapable>Yes</prf:ColorCapable>
 <prf:TextInputCapable>Yes</prf:TextInputCapable>
 <prf:ImageCapable>Yes</prf:ImageCapable>
 <prf:Keyboard>PhoneKeypad</prf:Keyboard>
 <prf:NumberOfSoftKeys>2</prf:NumberOfSoftKeys>
 <prf:PointingResolution>None</prf:PointingResolution>
 <prf:CPU>Motorola LTE-ARM7TDMI-S</prf:CPU>
 <prf:Vendor>Motorola</prf:Vendor>
 <prf:PixelAspectRatio>1x1</prf:PixelAspectRatio>
 <prf:SoundOutputCapable>Yes</prf:SoundOutputCapable>
 <prf:StandardFontProportional>No</prf:StandardFontProportional>
 <prf:VoiceInputCapable>Yes</prf:VoiceInputCapable>
 </rdf:Description>
</prf:component>
```

Fig. 9. An example of a CC/PP hardware component

In addition to these predefined components, we add our own components and attributes to better convey environmentally-based CPI. These additional components can provide new aspects of the personalised mLearning service offerings such as the targeted pedagogical goal and the learner's background in the current learning domain. The different entities within the system use this CPI to ensure that the user receives service/content that is tailored to the context of the environment in which it will be accessed. However, it is possible to even further customise the service to suit the preferences of the user. This is achieved through the extension of the CC/PP vocabulary. A CC/PP vocabulary defines the format or structure of the profile information, which is exchanged between a Personal Assistant and an InfoStation.

While CC/PP and UAProf define a number of components to describe the many different capabilities of the user device, we define a component containing CPI based on the users themselves, which is used to further customize and enhance the service for that individual user. The user preference component can specify anything from the user's name, the languages s/he speaks, user's age, and the format in which the user would prefer to receive information. The selection of the most appropriate format is an important option to take into account, as this defines the type of content which is to be delivered to the user, whether it is a simple text, or inclusive of various multimedia elements (audio/video). Constraints such as the capabilities of the device and the connection mechanism have a major bearing on the content delivery, but it is also important to take into consideration the user's own personal preferences. Another important attribute within the user profile is to specify the role or job title of the user, i.e. whether the individual is an educator or a learner etc. Specific groups may be allocated varying degrees of access to different resources related to the service, depending on the role they perform. This is especially the case within a University environment, where learners from different faculties may require access to the same services, but with vastly differing service content. Of course the differences between the roles of educator and the learner also imply an inherent difference in the user's utilisation of a service. For example, the mTest service would provide a learner with a means to assess their assimilation of presented content. However, the individual in the educator role would utilize this service to measure the capabilities of an entire cohort.

```
<prf:component>
<rdf:Description rdf:ID="UserPlatform">
  <rdf:type rdf:resource="http:// www.ece.ul.ie/trc/profiles/UAPROF/ccppschema-1#UserPlatform" />
<prf:Name>John Doe</prf:Name>
<prf:StudentID>0123456</prf: StudentID >
<prf:Faculty>ECE</prf:Faculty>
<prf:Course>Electronic Engineering</prf:Course>
<prf:Year>4</prf:Year>
<prf:Classes>
  <rdf:Bag>
          <rdf:li>CE4517</rdf:li>
          <rdf:li>CE4607</rdf:li>
          <rdf:li>CE4717</rdf:li>
          <rdf:li>CE4817</rdf:li>
          <rdf:li>CE4907</rdf:li>
          <rdf:li>EE4607</rdf:li>
  </rdf:Bag>
</prf:Classes>
<prf:Advisor>Dr. Ivan Ganchev</prf:Advisor>
<prf:email>0123456@STUDENT.ul.ie</prf:email>
<prf:QCA>3.47</prf:QCA>
<prf:FYP>JD09</prf:FYP>
<prf:FYPSupervisor>Dr. Ivan Ganchev</prf:FYPSupervisor>
<prf:FYPTitle>Design and Implementation of an Animated Interactive Tutorial</prf:FYPTitle>
</rdf:Description>
</prf:component>
```

Fig. 10. An example of a CC/PP user-specific component

Figure 10 illustrates how a component and a group of attributes relating to a particular user may specify vital information about that individual, which will have a bearing on how a service may be presented to a user.

This CPI is passed from the Personal Assistant to the InfoStation's Scanner agent, which in turn conveys this information to the agent within the InfoStation environment, tasked with handling such information, the Profile Processor agent.

On receipt of the ACL message containing the profile information, the Profile Processor removes the content from the message and passes the information back into its original XML form. The profile processor is charged with maintaining a repository of such profiles, and so when caching this XML file, it names the file according to the agent identifier (AID) of the original Personal Assistant. Before this XML file can be utilised for the adaptation of services it must be interpreted, as without an interpreter, this XML document is meaningless. This document must also be operated on to ensure that it adheres to various validation rules. Therefore the role of the Profile Processor agent must also include the parsing and validation of the XML document. There are a number of APIs available for working with XML in the Java language, the two most popular being the Document Object Model (DOM) and the Simple API for XML (SAX) [37]. In order to exploit the benefits of these APIs, we utilise the Java API for XML Processing (JAXP) [38],which enables the parsing, transformation and validation of XML documents independent of any particular XML processor implementation. The utilization of DOM enables the examination of the entire document and the caching of a tree structure of components and attributes in memory. The profile/DOM tree contains a root element which contains one or more component elements (HardwarePlatform, SoftwarePlatform etc.), which can themselves have one or more attributes (ScreenSize, ImageCapable etc.). With this DOM object cached in the profile repository of the InfoStation, the service request response time can be reduced in the instance of any subsequent requests. Facilitating this harnessing of context information enables the Content Adaptation agent to successfully and efficiently adapt a service and the associated content to meet the environmental constraints within the user device. The Content Adaptation agent can generate a XSLT (Extensible Style Sheet Language, Transformation part) document, which performs just like Cascading Style Sheets (CSS), in that it provides a specification for converting an XML document from one form to another. The Content Adaptation agent can utilise this XSLT document to alter the markup of the content of delivered services. This ensures that no matter the capabilities of the target device, the system can adapt and facilitate the user with dynamic adapted service content or indeed generate a package containing all the requisite content.

10 Sample/mTest Service Provision

Within the modern educational institutions, the majority of learners are accustomed to utilising a multitude of technologies to access information during the course of their studies. However, as regards forms of assessment, the utilisation of these same technologies has been limited. One of the corner-stone services within this mLearning environment is the *mTest* service, which provides a means for educators to rapidly evaluate learners assimilated knowledge and provide valuable feedback to learners concerning their progress.

The mTest facility enables the educator to more effectively shape the learning experience of the learners, ensuring the learner remains engaged in the correct material. Indeed the main benefit of using quizzes is to provide a motivation for the learners to more actively engage in the material, without the stress associated with traditional exams. Also by providing progression feedback, learners can be made aware of their progress in the assimilation of the presented course content. Of course educators too may benefit from such information. By monitoring the progression of a group of learners, the educator can dynamically modify their instruction style, should a group of learners encounter difficulties and require remedial action. This enables the educator to dramatically optimize the performance of the group, and enhance the overall learning experience. The mTest service must be capable of utilizing the full capabilities of the device on which it's being accessed. Of course, more advanced capabilities afford content developers the opportunity to be more creative in the design of multimedia mTests. On low-capability devices with limited resources, a simple text format can be adopted for the creation of the assessments. However with devices capable of supporting multi-media, assessments may incorporate elements of text, images, sounds and even videos, all of which serves to actively engage learners in the material being assessed, especially when utilised in conjunction with an *mLecture* service [39].

The sequence diagram Fig. 11, depicts sample interactions between entities involved within the provision of the mTest service through this enhanced architecture. As has been mentioned previously, the Scanner agent continuously searches for mobile devices/Personal Assistant agents within the service area of the InfoStation. The Personal Assistant, having already gathered the requisite information from the user to complete the user profile, passes the capability and preference information (CPI) on to the Scanner agent within the contents of an ACL message. The Scanner agent takes the information relating to the Personal Assistant, directs the profile information to the Profile Processor agent, and directs a list of mobile devices and applicable services to the Connection Advisor agent. The latter in turn filters this information providing the information necessary for the Connection Initiator agent to establish a communication session with the Personal Assistant onboard the mobile device. The Connection Initiator agent, having already established communication with the Personal Assistant, generates a Connection agent, which is tasked with maintaining and managing all communications between the InfoStation and the particular Personal Assistant. The Profile Processor agent, having received the profile information from the Scanner agent, gathers the relevant capability and preference information from the received XML content. This information is cached locally within the InfoStations profile repository. This agent works in conjunction with the Scenario Manager agent in order to monitor for any changes to the operating environment of the services (i.e. change in capabilities caused by changing one mobile device with another). At this stage, a communication session has been established between the InfoStation and the user mobile device, and the InfoStation is aware of the context/constraints any presented services must be adapted to.

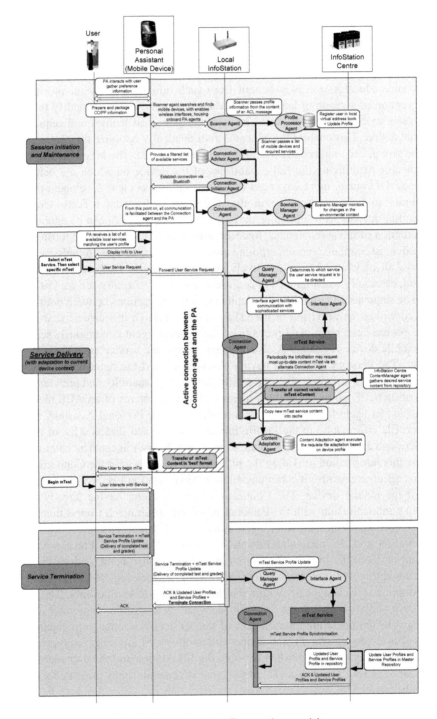

Fig. 11. The enhanced mTest service provision

Having successfully established communication between the InfoStation and the Personal Assistant, the user is presented with a list of all available services matching his/her profile. From this list the user selects a particular service, in this case the mTest service, and may then choose a particular assessment. The Personal Assistant on the user mobile device forwards the users service request on to the InfoStation, specifying the user choices. The Connection agent passes this message on to the Query Manager agent. It determines to which service a particular service request must be directed. The mTest service receives the user service request and compiles a list of applicable assessments, taking into account the personal context of the user (based on profile information). This service content is passed on to the Content Adaptation agent, which performs the role of an adaptation engine. It performs the requisite personalisation and adaptation of the service content, based on the capabilities of the access device and the preferences/context of the user. This contextualisation ensures the service is presented in its most efficient format by identifying the most relevant context information such as the screen size, audio and video capabilities etc. Once the service content has been adapted to suit the operating environment on board the mobile device, the content is packaged (MIDlet packed within a JAR file) for transfer to the user's mobile device. Depending on network constraints such as the traffic volume at a particular time and of course the communication capabilities of the target device, the Connection agent may initiate the file transfer via Bluetooth (or WiFi). The Personal Assistant presents a basic GUI to the user, enquiring as to whether or not to save the incoming file. The incoming MIDlet (JAR file) is saved to onboard memory within the mobile device, which in turn permits the user to access the service content any number of times, even whilst outside the range of the InfoStation.

Within this sequence diagram, the facility to update service content is also highlighted. Periodically, a Connection agent (one specifically generated to maintain communications between the InfoStation and the InfoStation Centre) requests the most up-to-date versions of particular service content. This facility is essential to the success of the mLearning architecture. Any updates can be propagated from the InfoStation Centre, ensuring that each InfoStation throughout the network has access to the latest versions of service content. As the learner progresses through the test, his/her user profile is maintained to reflect this progress. Furthermore we consider the possibility for the learner to do the test whilst on the move and out of range of an InfoStation. As the content is stored locally on the mobile device, the Personal Assistant can facilitate the user with continued utilisation of the service while at the same time maintaining the user profile. Thus the learner may complete the test whilst outside the radio range of any InfoStation, with the user profile reflecting the learner's progression through the material. Once the Personal Assistant eventually does enter back within range of an InfoStation, the Personal Assistant will forward on a user profile update which reflects the progress of the learner through the test content, whilst out of range of the InfoStation network. These updates are disseminated through to the InfoStation Centre so as to ensure all information across the system is up-to-date. Once the user has completed the mTest, the Personal Assistant displays the results of the assessment to the user, providing valuable feedback on their own progress and performance.

11 Implementation Issues

The proposed agent-oriented middleware architecture is implemented based on the Java Agent DEvelopment (JADE) framework [40, 41], with inter-agent interactions facilitated through Bluetooth communication (in conformance with JSR-82 [42]). The JADE framework simplifies multi-agent systems implementation through the provision of a predefined set of services and management tools in addition to the runtime library and agent programming library. Using this enabling technology, we have developed agents based within a mobile device as well as within a desktop connected to the wired network.

In order to examine the prototype, a proof-of-concept testbed environment was created, in which to investigate the provision of service applications in various different scenarios. The InfoStation Centre Synchronisation approach was simulated in this testbed environment. As an example here we consider the developed Mail-Checker service used to notify users for the presence of new messages in their mailboxes. The notification is sent to the user mobile phones over Bluetooth.

The MailChecker application consists of two parts: a MailChecker server and a MailChecker MIDlet. The server works on a Bluetooth-type InfoStation and is configured to scan periodically users' mailboxes (on the InfoStation Centre) to check for new messages. If new messages are detected, a notification is sent to the corresponding user mobile device(s). Users may register more than one device to receive these notifications. The MIDlet works on the user mobile device. It is registered with the MailChecker server and with the AMS (Application Management System) on the device as a handler of a specific Bluetooth service request (AMS Service Registry). The MIDlet itself is not active, allowing the user to operate normally with his/her mobile device.

When the MailChecker server attempts to connect to the device, the AMS on the device recognizes the request and automatically launches the registered MIDlet. When the MIDlet is activated, it automatically completes the connection with the server. Then it receives information about the new messages and presents it to the user as a list consisting of 'sender-subject' pairs. The MIDlet also generates an audio signal and vibration to alert the user.

Some of the issues were encountered during the MailChecker development and testing are listed below:

- *Slow discovery of Bluetooth devices and services* – thus this seems suitable for only one-time initialisation and registering of a single device or a few devices;
- *Unreliable Bluetooth discovery* – no discovery of nearby devices or discovery of devices cached by the operating system that are actually not present;
- *Long time needed to connect to Bluetooth devices* even when the device name, address, and service are known in advance;

- *Not all Bluetooth-enabled mobile phones support automatic start of the MIDlet*, when Bluetooth connection is attempted (Bluetooth Service Registry). This means that in order to be usable, the MIDlet has to be manually activated by the user, which (in most cases) prevents the normal user operation with his/her device. Another drawback is the increased power consumption;
- Even when the mobile device supports the Bluetooth Service Registry, when connection is attempted, the AMS on the device first asks the user for permission to actually start the MIDlet. The alert for starting the application is usually silent and unless the user is using the device at the moment, s/he may not notice it. In case when the user notices the alert and allows AMS to start the MIDlet, the connection attempt may have been expired due to time-out. In these cases, the availability of Bluetooth Service Registry gives no practical benefits, because a *user interaction is required to actually receive notifications about new messages*, which diminishes the original idea of automated notifications. Just a limited number of high-end smart mobile phones have options to bypass user confirmation and to start MIDlets really automatically;
- *The Bluetooth connection is slow and limited in range and easily affected by obstacles* such as walls. This implies that it is preferable that the user is not moving while data transfer is in progress, and that the transferred data should be relatively small;
- Bluetooth supports up to 7 active connections. For the MailChecker service this is not an issue, but other services with multiple Bluetooth peers must *implement sophisticated connection pooling*;
- Various *technical problems with Java-enabled mobile phones*. For instance, Nokia N73 dims the screen (turns off the backlight) while working with the MIDlet and turns it on again only when the main menu is activated;
- *Different Java ME compatibility of mobile phones* – some phones may not support the Java Media API, which is required for alerting the user with sound which essential for this service.

The listed implementation findings provide valuable insight for the successful development of other Bluetooth-based service applications, e.g. the mTest service.

12 Conclusion

This chapter has presented a mLearning InfoStation-based system and the agent-oriented prototype implementation of a context-aware and adaptable middleware integrated into it. The process control in this middleware is based on four basic scenarios. The scenarios realisation is controlled and managed by specialised agents - Scenario Management Agents – operating on the InfoStations.

Important in supporting the scenarios management are the issues of time-criticality. Next step in our research and development is focused on extending the Scenario Manager Agent's implementation with more real-time possibilities for scenario management. For this, we have adopted the compositional approach, presented in [43] and used by the AnaTempura tool - an interpreter for executable

Interval Temporal Logic (ITL) [44]. The specifications, written in Tempura, are a subset of ITL. AnaTempura can generate a state-by-state analysis of the system behaviour during the runtime.

In addition, we are currently developing a new version of a WiFi-based system prototype where communication is initiated differently comparing to the Bluetooth-based prototype. The approach adopted for the WiFi implementation is interesting from the viewpoint of the agent-oriented applications because the initiative for establishing the communication between a mobile device and an InfoStation is pro-actively taken by the Personal Assistant agent operating on the device (by comparison, in the Bluetooth prototype the Scanner agent on the InfoStations is the pro-active part in communication.) We plan a series of experiments in order to compare the two approaches in order to establish their feasibility and appropriateness for usage in each particular case/scenario etc.

Acknowledgments. This work was supported by the Irish Research Council for Science, Engineering and Technology (IRCSET), the Bulgarian National Fund under Grant No. Д002-149/16.12.2008, and the Telecommunications Research Centre, University of Limerick (http://www.ece.ul.ie/trc/).

References

1. Barker, P.: Designing Teaching Webs: Advantages, Problems and Pitfalls. In: Proc. of ED-MEDIA 2001 World Conference on Educational Multimedia, Hypermedia & Telecommunication, Association for the Advancement of Computing in Education, Charlottesville, VA, pp. 54–59 (2000)
2. Maurer, H., Sapper, M.: E-Learning Has to be Seen as Part of General Knowledge Management. In: Proc. of ED-MEDIA 2001 World Conference on Educational Multimedia, Hypermedia & Telecommunications, Tampere, pp. 1249–1253. AACE, Chalottesville (2001)
3. O'Droma, M., Ganchev, I.: Toward a Ubiquitous Consumer Wireless World. IEEE Wireless Communications 14, 52–63 (2007)
4. Passas, N., et al.: Enabling technologies for the 'always best connected' concept: Research Articles. Wirel. Commun. Mob. Comput. 6, 523–540 (2006)
5. Frenkiel, R., Imielinski, T.: Infostations: The joy of 'many-time, many-where' communications. WINLAB Technical Report (1996)
6. Ganchev, I., et al.: An InfoStation-Based Multi-Agent System Supporting Intelligent Mobile Services Across a University Campus. Journal of Computers 2, 21–33 (2007)
7. Huhns, M.: Software development with objects, agents, and services. In: Third International Workshop on Agent-Oriented Methodologies, Vancouver, Canada (2004)
8. Chang, J.-W., Lee, H.-J.: Context-Aware Architecture for Intelligent Application Services in Ubiquitous Computing. Presented at the Proceedings of the International Conference on Semantic Computing (2007)
9. Goh, E., et al.: A Context-Aware Architecture for Smart Space Environment. Presented at the Proceedings of the 2007 International Conference on Multimedia and Ubiquitous Engineering (2007)

10. Chen, H.L.: An Intelligent Broker Architecture for Pervasive Context-Aware Systems. Phd, Department of Computer Science and Electrical Engineering, University of Maryland, Baltimore (2004)
11. Qingsheng, Z., et al.: Research on context-aware architecture for personal information privacy protection. In: Proceedings of the IEEE International Conference on Systems, Man and Cybernetics, Montréal, Canada, pp. 3912–3916 (2007)
12. Capilla, R.: Context-aware Architectures for Building Service-Oriented Systems. In: Proceedings of the Conference on Software Maintenance and Reengineering, pp. 300–303 (2006)
13. Schmohl, R., Baumgarten, U.: A Generalized Context-aware Architecture in Heterogeneous Mobile Computing Environments. In: Proceedings of the 2008 The Fourth International Conference on Wireless and Mobile Communications, pp. 118–124 (2008)
14. Costantini, S., et al.: DALICA: Agent-Based Ambient Intelligence for Cultural-Heritage Scenarios. IEEE Intelligent Systems, 34–41 (2008)
15. Tumer, J., Agogino, A.: Improving Air Traffic Management with a Learning Multiagent System. IEEE Intelligent Systems, 18–21 (2009)
16. Rehak, M., et al.: Adaptive Multiagent System for Network Traffic Monitoring. IEEE Intelligent Systems, 16–25 (2009)
17. Zhang, Z., et al.: An Agent-Based Hybrid System for Microarray Data Analysis. IEEE Intelligent Systems, 53–63 (2009)
18. Zhou, L., et al.: Context-Aware Middleware for Multimedia Services in Heterogeneous Networks. IEEE Intelligent Systems, 40–47 (2010)
19. Sheu, R.-Y., et al.: Multiagent-based adaptive pervasive service architecture (MAPS). In: 3rd Workshop on Agent-Oriented Software Engineering Challenges for Ubiquitous and Pervasive Computing, London, United Kingdom, pp. 3–8 (2009)
20. Qureshi, N., Perini, A.: An Agent-Based Middleware for Adaptive Systems. In: International Conference on Quality Software, Oxford, UK, pp. 423–428 (2008)
21. Ganchev, I., et al.: On InfoStation-Based Mobile Services Support for Library Information Systems. In: 8th IEEE International Conference on Advanced Learning Technologies (IEEE ICALT 2008), Santander, Cantabria, Spain, pp. 679–681 (2008)
22. Ganchev, I., et al.: InfoStation-Based Adaptable Provision of m-Learning Services: Main Scenarios. International Journal "Information Technologies and Knowledge" (IJ ITK) 2, 475–482 (2008)
23. Stoyanov, S., et al.: An Approach for the Development of InfoStation-Based eLearning Architectures. Compt. Rend. Acad. Bulg. Sci. 61, 1189–1198 (2008)
24. Dey, A.K., Abowd, G.D.: Towards a better understanding of context and context-awareness. In: Workshop on the What, Who, Where, When and How of Context-Awareness, New York (2000)
25. Stoyanov, S., et al.: From CBT to e-Learning. Journal "Information Technologies and Control" 4, 2–10 (2005)
26. Ganchev, I., et al.: InfoStation-based mLearning System Architectures: Some Development Aspects. In: 8th IEEE International Conference on Advanced Learning Technologies (ICALT 2008), Santander, Spain, pp. 504–505 (2008)
27. Foundation for Intelligent Physical Agents (FIPA) - [Online], http://www.fipa.org (accessed: January 10, 2010)
28. FIPA, ACL Message Structure Specification. Foundation for Intelligent Physical Agents, Geneva, Switzerland SC00061G (December 3, 2002)
29. W3C, Document Object Model (DOM) [online], http://www.w3.org/DOM/ (accessed: March 03, 2010)

30. OWL-S: Semantic Markup for Web Services [Online],
    http://www.w3.org/Submission/OWL-S/ (accessed: March 3, 2010)
31. Christensen, E., et al.: Web Services Description Language (WSDL) 1.1 [Online],
    http://www.w3.org/TR/2001/NOTE-wsdl-20010315
    (accessed: March 10, 2010) W3C2001
32. Kim, G.-H., Lee, B.-H.: Seamless streaming service session migration support architecture for heterogeneous devices. In: Balandin, S., et al. (eds.) ruSMART 2010. LNCS, vol. 6294, pp. 473–484. Springer, Heidelberg (2010)
33. W3C, Composite Capability/Preference Profiles (CC/PP): Structure and Vocabularies 2.0. World WIde Web Consortium (W3C) (December 8, 2006)
34. World Wide Web Consortium (W3C) [Online], http://www.w3.org/ (accessed: January 15, 2010)
35. Wireless Application Group User Agent Profile Specification (WAG UAPROF). Wireless Application Protocol Forum, Ltd. (November 10, 1999)
36. Open Mobile Alliance (OMA) [Online],
    http://www.openmobilealliance.org/ (accessed: December 15, 2009)
37. SAX, Simple API for XML (SAX) [online], http://www.saxproject.org/ (accessed: March 03, 2010)
38. Sun Microsystems, Java API for XML Processing (JAXP) [online],
    https://jaxp.dev.java.net/ (accessed: March 03, 2010)
39. Meere, D., et al.: Adaptation for Assimilation: The Role of Adaptable M-Learning Services in the Modern Educational Paradigm. International Journal "Information Technologies and Knowledge" (IJ ITK) 3, 101–110 (2009)
40. Bellifemine, F.L., et al.: Developing Multi-Agent Systems with JADE. Wiley Series in Agent Technology. John Wiley & Sons, Chichester (2007)
41. JADE. Java Agent Development Framework Project [Online],
    http://jade.cselt.it (accessed: January 10, 2010)
42. JSR-82: Java Bluetooth [Online], http://www.jsr82.com (accessed: March 10, 2010)
43. Solanki, M., et al.: Augmenting semantic web service descriptions with compositional specification. Presented at the Proceedings of the 13th International Conference on World Wide Web, New York, NY, USA (2004)
44. Moszkowski, B.: Executing temporal logic programs. Cambridge University Press, Cambridge (1986)

Chapter 12
Identifying Novel Topics Based on User Interests

Makoto Nakatsuji

NTT Cyber Solutions Laboratories, NTT Corporation, Hikarinooka, Yokosuka-Shi,
Kanagawa 239-0847, Japan
nakatsuji.makoto@lab.ntt.co.jp

Abstract. In this chapter, we introduce an agent that builds user interests as a hierarchy of classes where a rating value of the user is assigned to each class and item. The agent measures the similarity of users using user ratings against items as well as those against classes and then generates a user group that has high similarity to the user. Finally, the agent identifies novel topics, those that include new classes that are likely be interesting to the user even though those classes are not present in the user profile. The novel topics for the user are identified by determining a suitable size of the user group and analyzing the items possessed by the users in the user group. Thus, highly accurate recommendation results are guaranteed. Furthermore, our agent presents recommendations with a new measure, score of novelty, so that the user may better understand how novel the recommended items are. By letting the user browse topics against novel items with scores of novelty, we try to expand user interests significantly.

1 Introduction

Social media such as blogs and microblogging are becoming more popular for publishing and discussing shared interests among users. Information sharing systems for social media could enable users to expand their interests by browsing the collections of blog entries published by other users. However, to retrieve information from blog entries, current services simply employ keyword searches of blogs using Google or simple metadata attached to blog-entries, i.e. RSS metadata such as titles, creators, dates and so on. Unfortunately, neither approach offers detailed semantics about the description content in blog entries. Moreover, there is no function to generate personalized searches easily, users are restricted by their own knowledge or imagination when entering search keywords. Such keyword searches are time consuming and troublesome. For example, users cannot perform a keyword search if they do not understand what they want to search for to some degree beforehand. Thus, when keywords cannot be specified, information retrieval from blog entries often cannot be performed even if the database contains topics that the user might become interested in.

A. Elçi, M.T. Koné, and M.A. Orgun (Eds.): Semantic Agent Systems, SCI 344, pp. 273–292.
springerlink.com   © Springer-Verlag Berlin Heidelberg 2011

To counteract the above problems, the study on Adaptive Information Filtering (AIF) [4] cooperates with the user in constructing a user profile; recommendations are offered based on the profile. Making a user profile interactively beforehand is good for offering recommendations to users, as indicated by the high-accuracy of AIF. Unfortunately, a common complaint about AIF is the user's need to make his/her own profiles, and often known information is encountered many times. This is because recommendation systems with conventional AIF only check the possibility of the user being interested the document and fail to identify if the information has already been presented to the user or not.

For filtering these redundant documents, novelty-detection researchers [21] define a novel document as a document that includes new information that is relevant according to the user profile. They extract relevant documents from a document stream and then classify the documents as novel or not; novel documents are provided to the user. Novelty detection can, however, provide documents that offer new information about classes that have already present in the user profile.

In our study [13, 14], we define novel topics as topics that include new classes that are likely be interesting to the user even though those classes are not present in the user profile. The goal is to expand the user's interests significantly by identifying novel topics and recommending those to the user. In particular, we first focus on building user agents that automatically understand users' interests from blogs according to the taxonomy of items and the identification of novel topics from blog entries. This is because blogs have become a popular method of publishing and searching for information that can appeal to the user.

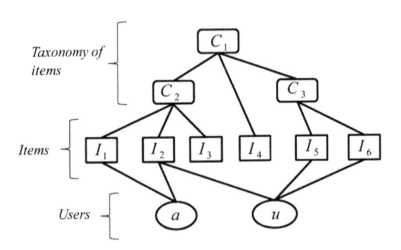

Fig. 1. An explanation of score of novelty.

For achieving the above-mentioned goal, we use the following approaches.

- We start with a proposal to introduce user agents that understand the interests of a user according to a taxonomy of items. We consider that users who like items, may like the classes that include those items. Our agent thus reflects the rating values of the user on items to those of classes that include such items (see Section 4). In the taxonomy, items are objects that the user is interested in, such as music artists, music songs, movie titles and so on. On the other hand, the classes are defined using a taxonomy of items in a service domain. For example, we can set classes as genres which are defined by item sets. By classifying blog entries into each class and item in the taxonomy, the agent can automatically generate user interests according to the taxonomy. In classifying user entries according to the taxonomy of items, the agent removes classification mistakes automatically by using the taxonomy of items and continuity of descriptions about user interests as explained in our previous paper [14, 13]. Of course, the agent can also build user interests according to the taxonomy by using buying histories and listening histories of users.

- Next, the agent measures the similarity of users by considering the degree of interest agreement between each class and item. Most previous techniques for measuring similarity of users use Pearson correlation coefficient or cosine-based similarity against items rated by both users as we will explain in Section 3. In this chapter, the agent measures similarity of users by using not only co-rating behaviors against items but also those against classes in the taxonomy (see Section 5). By considering the degree of interest agreement between each class and item, it can measure the similarity of users through the width and depth of a user's interests as discerned by using the taxonomy of items. As a result, the agent can identify *many items accurately* for the user by analyzing the items of users who share the same items and/or same classes with the user (please refer to the evaluation result in Section 7.1.4). We also establish a new evaluation method that determines a suitable size of user group Gu, whose users are similar with the active user, a, who receives recommendations, by observing the difference between the interests of user a and the interests of users among Gu while changing the size of Gu (see section 7.2). Finally, novel topics for active user a are identified by analyzing the classes, C, that are of interest to users in user group Gu even though a did not explicitly show any interest in C.

- We introduce a measure, *the score of novelty*, to understand *how novel* the recommended items are for the user and try to identify items of high novelty for the user, while also guaranteeing highly accurate recommendations [13, 14]. Accuracy is also important because users trust accurate recommendation results and tend to use such services [12]. We define the score of novelty as the smallest number of hops from the class user accessed before to the class that includes a possible item over the taxonomy. In Fig. 1, if active user a accessed items I_1 and I_2 in class C_2, items I_5 and I_6 have, for a a novelty score two because there are two classes, C_1 and C_3, on the hop from C_2 to those items. By accurately identifying items that are highly novel to the user, and recommending those to him, he may accept those items and widen his interests. We show two evaluation steps based on users' implicit ratings against music items extracted from a large number of blog entries as collected by the blog portal Doblog[1]. The taxonomy of music artists is provided by ListenJapan[2].

- The first step is an offline experiment that evaluates the accuracy in predicting users' hidden interests using our implicit rating dataset and investigates the distribution of user interests extracted from blogs according to the score of novelty. The results show that our method can identify items with higher accuracy than the previous methods including a previous taxonomy-based method [22]. They also show that our method can identify items with higher novelty than the recommendations manually created by the designers working for the service provider.
- The second step is an online experiment that analyzes user reactions to topics recommended based on our assessments of an online experimental service[3]. Most prior works used only offline synthetic data to evaluate their recommendation techniques. However, analyzing the reactions of actual users to recommendations is very important for confirming whether the recommended novel topics are actually effective. By analyzing the frequency of user access to items output by our recommendation scheme over time, we confirmed the effectiveness of our novel topic recommendation. We found that the novel topics recommended by our technique were used for creating new communication links between users; this was confirmed by evaluating the frequency of comments between users who came to know each other through our online recommendations.

We now describe the impact of our method on applications.

Most recommendation schemes fail to consider the semantic relationships between a user and items that are recommended to the user. Thus, the user can't easily understand why particular items were recommended. The semantic relationships described above are necessary if the user is to accept the recommended items, especially when the user has not thought of those items before.

Our agent assesses the score of novelty for each recommended item, which indicates how novel the item is to the user. It indicates the relationships between the present interests of a user and the items that are recommended to the user, by using a taxonomy of items defined by service designers. That is, our method can recommend to user a content items that belong to the class that user a does not know of, together with their score of novelty. Here, we consider class as defined in the taxonomy created in each service domain. By presenting items with supporting information such as the novelty of those items, the user can more readily become interested in items not stored in his/her profile and so acquire new interests.

Some examples might help the reader's understanding. Consider user a who has items I_1 and I_2 under the class "Rock" in her interests, and the agent of user a extracts users X whose interests are similar to those of a according to the results of

---

[1] http://www.doblog.com was one of the biggest blog portals in Japan. Unfortunately, Doblog terminated services on May 2009.

[2] http://listen.jp/

[3] We provided an experimental service DoblogMusic at http://music.doblog.com/exp/ for Doblog users from August to December 2006.

similarity measurements between user a and other users. If there are many users in X who are interested in item $I5$ under the class "Classic", we can recommend item $I5$ of class "Classic" to user a together with information indicating its score of novelty, even though Classic and Rock are semantically dissimilar given the definition in taxonomy of items in music domain. Thus, the agent can recommend items to user a with the phrase "you may not have heard about item $I5$ in Classic genre, but users whose interests are similar to yours, are interested in item $I5$". By presenting some unknown items to the user together with the score of novelty, or using phrases like the one described above, user a may develop an interest in $I5$ even though its class may not be known to user a, i.e. not stored in a profile of a. However, user a has a chance to expand his/her interests significantly, if he/she accesses novel item $I5$.

The chapter is organized as follows. Section 2 introduces related works and Section 3 explains the technical background of the study. Section 4 describes our model of user interests according to the taxonomy of items. Section 5 describes our similarity measurement of users using the taxonomy of items and Section 6 explains identifying novel topics based on similarity measurement results. Sections 7 and 8 describe our offline and online experimental studies, respectively. Section 9 concludes this chapter.

2 Related Works

Many online content providers such as Amazon[4], offer recommendations based on Collaborative Filtering (CF) [17, 1, 16], which is a broad term for the process of recommending items to users based on the intuition that users within a particular group tend to behave similarly under similar circumstances. One advantage of CF techniques is that they can recommend relevant items that are different from those in a user's profile. However, the existing CF techniques don't consider the semantic relationships between user a and content items that are recommended to a by using the taxonomies attached to content items. As a result, the user cannot understand semantic reasons why those items are recommended and how novel the recommended items are, and so is less likely to access the recommendation.

Some CF researchers use a taxonomy of items to raise the accuracy of prediction results [22]. Their method was shown to be useful when the transaction data of users was sparse. However, in measuring user similarity, their method focuses only on classes that include items rated by both users and their super classes. As a result, this method naively assumes that users who share many items are highly similar with the user; those users may have many good as well as many not so good items for the user. Ziegler and McNee also tried to improve the recommendation list for a use by increasing the diversity of the items in the list [23]. The list includes items in several classes defined by a taxonomy. Their online evaluation indicates that users were satisfied with the diverse items, though the accuracy of item prediction was degraded. Their method, however, does not aim to identify, for the user, novel topics.

---

[4] http://www.amazon.com

Herlocker and his co-workers [6] described that *novel* items and *serendipitous* items are different though both are defined as items that are not known to the user but interesting for the user. The difference is that the former is more easily found by the user than the latter. Our method does not classify novel items and serendipitous items. It just makes users aware of how far the recommended items are from their present interests through our proposed measure, *the score of novelty*. However, as the reader can naturally imagine, items with high novelty for the user can not be easily discovered by the user. For example, the user who, up to now, has demonstrated an interest only in music items in "Classic", is unable to easily discover interesting items in "Jazz" by himself. Our evaluation, described later, also shows that existing CF methods have difficulty in accurately identifying items with high novelty for the user. Indeed, our offline evaluation did not explicitly treat *serendipitous* items because the evaluation data set were taken from user access histories. However, the online evaluation presented novel (or serendipitous) items to users that were not included in the users' access histories, and confirmed that the actual users were excited in those items.

In a research study of novel item identification, Hijikata et al. [7] prepared a dataset that had items with two types of labels, known or unknown, attached by users. They used that dataset as a training dataset by employing the CF approach to classify the predicted items are known or not known to the active user. Their evaluation showed that their method can identify novel items (here, item novelty follows the definition of Herlocker [6]), unfortunately the prediction results are rather inaccurate. Furthermore, assigning such labels is time-consuming for the users. Their approach also cannot analyze the novelty of predicted items in detail an omission rectified by our study.

In research studies of ontology mapping [15, 2, 10], similarity measurements considering approximation of classes and class topologies are proposed in [10]. In addition to class topology, we consider each user's ratings assigned to each class and item. Furthermore, in analyzing conjunctions in class topologies in the taxonomy with high similarity scores, we identify novel items, those that other users have in their interests but the user does not.

3 Collaborative Filtering

Our method extends CF to identify novel topics. Thus, we explain CF in this section.

CF methods can be classified into two approaches: memory-based CF and model-based CF. Memory-based CF is based on the assumption that each user belongs to a larger group of similarly behaving users. Indeed this method is referred to as user-oriented memory-based CF [5]; an analogous method which builds item similarity groups using co-purchase history is known as item-oriented [17]. On the other hand, model-based CF generates predictions by using a model that is optimized by training data. Clustering [9, 19], Bayesian network models [8, 20] are examples of the model-based approach.

In computing similarity of users, basic user-oriented memory-based CF methods often use the Pearson correlation approach [18, 16] or the cosine-based approach [1]. If we define M as number of items rated by user a and u, r_{a,I_i} is the rating value of user a for item I_i, and $\bar{r}_a$ is the average value of item ratings given by a, the Pearson correlation coefficient measures the similarity $S(a, u)$ between a and u according to equation (1).

$$S(a,u) = \frac{\sum_i^M (r_{a,I_i} - \bar{r}_a)(r_{u,I_i} - \bar{r}_u)}{\sqrt{\sum_i^M (r_{a,I_i} - \bar{r}_a)} \sqrt{\sum_i^M (r_{u,I_i} - \bar{r}_u)}} \tag{1}$$

When we use the cosine-based approach, we set $\bar{r}_a$ and $\bar{r}_u$ as zero in equation (1). The advantage of the Pearson correlation approach is that it takes into account that different users might have different rating schemes.

If we assume that N is the set of users that are most similar to the active user a, the predicted rating of a on item I_i, P_{a,I_i} is obtained by the following equation (2).

$$P_{a,I_i} = \bar{r}_a + \frac{\sum_u^N (r_{u,I_i} - \bar{r}_u)s(a,u)}{\sum_u^N S(a,u)} \tag{2}$$

4   Modeling User Interests According to the Taxonomy

Our method starts by modeling user interests according to a taxonomy of items.

The agent builds user interests according to this model. Taxonomies are becoming more readily available; examples include the taxonomies of music, movies, and game content generated by All Media Guide[5] and ListenJapan[6]. We consider that modeling user interests according to these taxonomies is reasonable because content providers make significant efforts to optimize the granularity and branching factors of classes so that their customers can readily access their items according to the customers' interests.

Our approach is based on the observation that users who are interested in some items, are also interested in classes that include those items, and the rating values of the items are then automatically reflected to that of the class that includes those items. The rating value for an item is implicitly and automatically assigned according to the frequency of a user's access to the item, or explicitly assigned by the user.

The agent of a user assesses the user rating for the class from the user ratings of items in the class. Formally, let I be an item in class C_i, the rating value of the class, r_{u,C_i}, is computed as $\sum_{I_i \in I} r_{u,I_i}$. In the example in Fig.2, if user u assigns rating value 4.0 against I_5, and 4.5 against I_6, the rating value of class C_3 for

---

[5] http://www.allmediaguide.com/
[6] http://listen.jp/

u is 8.5. By computing the rating values of a class in this way, we can assign high rating values for the class if the user is interested in many items with high rating values under the class. The rating values of the super classes are computed in the same way; a key point is that our agent uses the rating to each class instead of the rating to each item.

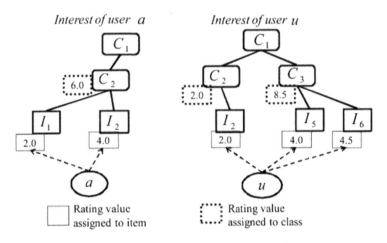

Fig. 2. Measuring similarity of user a and u.

5 Measuring Similarity of Users

Next, we explain how to determine the similarity of users a and u according to the taxonomy of items. We first explain the approach of our method, and then show our algorithm.

5.1 Approach

We explain our approaches in several steps.

- Our agent first computes $S(a, u,Ci)$, the score of interest agreement between user a and u against class Ci. This rating value takes a smaller value in ra,Ci and ru,Ci . Thus we can filter users of low-interest when measuring the score of interest agreement.
- Next, it computes the similarity of rating behaviors against all classes between a and u, denoted as $SC(a, u)$, with $S(a, u,Ci)$. We utilize the idea of the Jaccard coefficient [11] since it can effectively separate user u who assigns ratings against many classes from a who assigns ratings to fewer classes. The Jaccard coefficient considers not co-rating classes as well as co-rating classes by utilizing the union of class sets. In other words, it considers the similarity of the widths of users' interests. The Pearson correlation approach and cosine-based approach do not have this property since they only consider the classes that are assigned ratings by both a and u (see equation (1)).

- The agent then measures $SI(a, u)$, the similarity of rating behaviors against items between a and u. We use the Pearson correlation approach because it can handle the difference in the rating schemes of each user against items as explained in equation (1). But our approach can also employ the Jaccard coefficient approach.
- Finally, the agent combines the two above similarities, $Sc(a, u)$ and $SI(a, u)$, to evaluate the similarity of rating behaviors against classes and items between users.

The proposed method can effectively measure the similarity of the *widths* of users' interests while offsetting the difference in the rating schemes of users. Thus, it achieves high accuracy as demonstrated in our evaluation section 7.1.4.

5.2 Algorithm

We introduce the algorithm of our method below. In this algorithm, we use $C_i(a)$ as the subclasses of class C_i rated by user a.

1. First, it computes $S(a, u, C_i)$ as $min(r_a, C_i, r_u, C_i)$.
2. Then, it measures similarity scores $Sc(a, u)$ by the following equation.

$$S_C(a,u) = \sum_C \frac{\sum_{C_j \in \{C_i(a) \cap C_i(u)\}} S(a,u,C_j)}{|C_i(a) \cap C_i(u)|} \tag{3}$$

3. Next, it uses the Pearson correlation approach to compute similarity scores $SI(a, u)$ using equation (1).
4. Finally, it normalizes $Sc(a, u)$ and $SI(a, u)$ among users similar to a, and determines the similarity of user pair $S(a, u)$ as $Sc(a, u)+SI(a, u)$.

5.3 Example

We explain our algorithm by using the example in Fig.2. (1) $S(a, u, C_2)$ is computed as $min(6.0, 2.0) = 2.0$. $S(a, u, C_3)$ is computed as $min(0.0, 8.5) = 0.0$ (2) Then, $Sc(a, u)$ is computed as $(2.0/2) +0.0 = 1.0$. (3) Next, $SI(a, u)$ is computed as $(4-3)(2-3.5) / \sqrt{(4-3)2} \sqrt{(2-3.5)2} = -0.667$.(4) Finally, $S(a, u)$ is measured following procedure four of our algorithm.

6 Novel Topic Identification

The agent uses similarity measurement results for novel topic identification and its application can activate user-user communications.

1. The agent calculates the similarity between active user a and the other users. By using the heuristic threshold X, it derives X users who have high similarity to user a as the interest-sharing group G_U.

2. It then analyzes the difference in the interests of user a and the interests of Gu. It also analyzes the score of novelty of items interested by users in Gu for user a. In Fig. 3, the score of novelty of item "Elf Power" for user a is three, because we need three hops to go from the different item "Elf Power" accessed by user u to class "Rock" accessed by user a. By recommending items such a high score of novelty, the interests of users may be significantly expanded. Lowering the score of novelty may produce more *comfortable* new classes but these will prove to be less satisfying.

3. Finally, the agent extracts novel items Gi, which are unknown to user a, but that are well-known to users in Gu; the novel blog entries about Gi are recommended to user a together with their novelty scores. Here, determining the most suitable size of Gu is very important for identifying attractive and novel items. If the size of Gu is reduced, the difference in user interests decreases, and items in Gi may be close to the classes in interests of each user. However, there may be few novel items in Gi. On the other hand, if the size of Gu is increased, the difference in user interests increases, and items in Gi may be too novel for the user. Thus, we observe the difference between interests of active user a and those of Gu while changing the size of Gu. The most suitable size of Gu is the point at which there is a rapid increase in the number of Gi. Details of this process are given in Section 7.2.

An example of community creation is depicted in Fig. 3. User b is included in user group Gu whose interests are determined to be similar to the interest of user a. If users in Gu often have an interest in "Elf Power", user a has the potential to be interested in "Elf Power" even though the class "Elephant 6" that includes "Elf Power" is many hops from the class "Rock" that user a has a known interest in. Furthermore, by browsing blog entries concerning these novel items, users may expand their interests and share interests with each other.

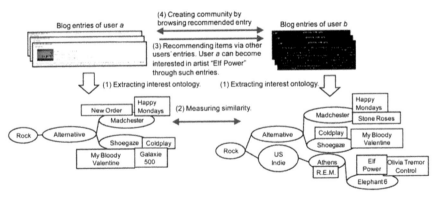

Fig. 3. Community creation service of presenting blog entries concerning the novel items to users.

7 Offline Experiments

We now present the results of offline experiments that demonstrate the accuracy of our taxonomy-based method in predicting users' hidden interests. We also evaluate the suitable size of similar users to the active user a by observing the difference of interests of users, and also investigate the distribution of user interests extracted from blogs according to the score of novelty.

7.1 Investigating Accuracy

We first explain the dataset used in evaluating the accuracy of our method.

7.1.1 Dataset

User implicit ratings using non-Japanese taxonomy:
   This dataset includes 48,695 implicit ratings of 3,545 users according to a taxonomy extracted in the experiment of Nakatsuji et al. [14] from the blog portal Doblog and the taxonomy of non-Japanese music artists provided by ListenJapan.
   The taxonomy contains 279 genres as classes and 21,214 artists as items[7]. Nakatsuji et al. created rating values for each item of a user by analyzing the description frequency of each item among the user's blog entries. The average number of ratings assigned to an item is 6.3. We linearly scaled up each rating value such that the maximum user rating corresponded to 5 and the minimum corresponded to 1 following the range of ratings in MovieLens dataset. The class hierarchy in this taxonomy is deep; it has, on average, four hierarchies, and sometimes has a fifth hierarchy under the root class "Music" with detailed end classes such as "Space rock" and "Acid jazz". This represents detailed expert knowledge that can be used to accurately measure the similarity of users.

User Implicit Ratings Using Japanese Taxonomy:

We also used 58,104 implicit ratings of 2,800 users extracted from blog entries in Doblog using a Japanese taxonomy provided by ListenJapan in the same way as Nakatsuji et al. did for the non-Japanese taxonomy. Japanese taxonomy contains 153 genres as classes and 7,421 artists as items. The class hierarchy in this taxonomy is also as deep as that of the non-Japanese taxonomy. The average number of ratings assigned to an item is 10.8.

7.1.2 Methodology

We randomly divided dataset D that includes items with user ratings into two datasets: training dataset T and predicted dataset P. Thus, we could acquire users who had items whose classes are in P though those were not included in T. Our agent then measured the similarity of users using T. We examined four ratios of T

---

[7] The music taxonomies in our evaluation can be accessed from ListenJapan home page: http://listen.jp/

to D, TD: 0.1, 0.3, 0.5, and 0.7. When TD is small, the dataset of user ratings is sparse.

Following the standard evaluation methodology for CF, we predicted the user ratings only on the withheld ratings in T and computed Mean Absolute Error (MAE), which penalizes each miss by the distance to the actual rating. This measure is written below, where n is the number of entries in P, and P_i and R_i are the predicted and actual ratings of the ith entry, respectively.

$$MAE = \frac{\sum_{i=1}^{n} |P_i - R_i|}{n} \tag{4}$$

The accuracy was assessed only for some of the ratings that have already been rated by the active user, because we usually do not have relevance judgments for all ratings by each user. Several ratings not included in dataset D might actually be relevant to the interests of the user, i.e. belong to the neighborhood of items accessed by the user and thus should be rated highly in her prediction values. However, since those items have not been accessed yet, they are considered as non-relevant. To evaluate the effect of presenting such items, we evaluated actual user reaction against novel items in our online evaluation.

7.1.3 Compared Methods

We compared our similarity measurement method to the following similarity measures.

- Pearson correlation coefficient (Pearson): similarity of users is measured by Pearson correlation coefficient.
- Cosine-based approach (*Cosine*): similarity of users is measured by cosine-based approach.
- Method by Ziegler (*Ziegler*): similarity of users is measured by the method proposed by Ziegler et al. [22]. This method measures the similarity of users without regard to the width of user interests, or the similarity score of ratings between users against items. We set parameter χ used in this method properly against each dataset to achieve the highest result accuracy. As a result, we set χ as 0.2 against those datasets.
- Taxonomy (Jaccard & Pearson) (*T(J&P)*): this method is explained in detail in the method section.
- Taxonomy (Jaccard) (*T(J)*): the similarity score of ratings between users against classes and those against items are measured using our Jaccard-based method.

We select *Pearson* and *Cosine* as baseline methods because they are the most frequently used methods in collaborative filtering studies [17, 18, 16, 1], and we also select *Ziegler* because it is the most famous taxonomy-based similarity measure though it does not consider the *width* of user interests as we do.

7.1.4 Results

We first estimated the number of users, N, similar to the user, and determined the MAE. As a result, we set N to 30 against the music dataset because MAE gradually worsens if N is large.

Next, we evaluated the accuracy of our method by changing the ratio of training data, TD Results against non-Japanese dataset and those against Japanese dataset are shown in Table 1-(a) and Table 1-(b), respectively.

Table 1. MAE against music dataset.

(a) MAE against non-Japanese music dataset

T/D	0.1	0.3	0.5	0.7
Pearson	*1.36*	1.26	1.20	1.23
Cosine	1.38	1.33	1.25	1.23
Ziegler	1.42	1.33	1.24	1.22
T(J&P)	1.38	*1.19*	*1.11*	*1.14*
T(J)	1.39	1.22	1.13	1.15

(b) MAE against Japanese music dataset

T/D	0.1	0.3	0.5	0.7
Pearson	*1.29*	1.19	1.20	1.18
Cosine	1.32	1.16	1.13	1.16
Ziegler	1.39	1.17	1.13	1.15
T(J&P)	*1.29*	*1.11*	1.12	1.15
T(J)	1.32	1.12	*1.11*	*1.13*

Most of the results of our method, $T(J\&P)$ and $T(J)$, are better than those of other methods including the previous taxonomy method. Those music datasets are rather sparse datasets, there are not so many ratings assigned to an item averagely, thus we consider that our method is suitable in measuring similarity of users when the rating dataset is not dense.

Interestingly, $T(J\&P)$ achieved the highest accuracy against the non-Japanese music dataset. However, for the Japanese music datasets, $T(J)$ tended to achieve higher accuracy. This is because the non-Japanese taxonomy is more detailed than the Japanese taxonomy, and this situation well suits our assumption that users who like an item also like the classes that include that item. Furthermore, when the dataset is not sparse, users tend to have many items in each class. In such situations, the end class of the taxonomy can be classified into more detailed classes, and $T(J)$ works better because it considers the *width* of user interests against items. For example, in Fig. 1, $T(J)$ considers that the *width* of a and that of a user who likes items, I_1, I_2 and I_3, are not the same. As a result, it can divide similar users who like items I_1 and I_2, from users who like items I_1, I_2, and I_3.

Finally, we recently confirmed that our approach is well applied to items other than music [3].

7.2 Analyzing Suitable Size of User Group to Identify Novel Topics

We next determined the suitable size of Gu, as described in Section 6, by observing the difference between the user interests of the active user a and those of Gu while changing the size of Gu. We optimize the number of users, N, similar to the active user, in Section 7.1, from the view point of the accuracy. In this section, we optimize the suitable size of Gu from the view point of identifying novel topics. The evaluation dataset is almost the same as that used in Section 7.1.1 though it uses a smaller taxonomy than the previous subsection[8]. The taxonomy we used here contains 114 classes as genres, covering a wide range of genres in the music domain, Rock, Classic, Jazz, and Soul and the items are 4,300 artists.

First, we selected user a from among all users extracted and analyzed a suitable size of Gu by changing parameter X, which represents the number of users who have high similarity to user a in interest-sharing group Gu, see Section 6. In this evaluation, we divided novel items Gi into 3 item groups in order of the appearance rate of items when we set X to 70: a very popular item group, a moderately popular item group, and item group with a small number of fans. We then calculated the number of users who were interested in the artists of each item group while changing X from 10 to 70.

Graphs of the number of users who were interested in each item group obtained while changing X are shown in Fig. 4-(a). Next, we focused on users who had interested in many items. Graphs of the number of such users obtained while changing X are shown in Fig. 4-(b). The very popular item group was recommended to users regardless of the value of X, see Fig. 4-(a). The item group with a small number of fans, on the other hand, was recommended most often when X was ten (Fig. 4-(b),); the moderately popular item group was recommended more often as X was increased. This is because users who had a lot of item interests tend to discuss items in the item group with a small number of fans, rather than discussing items in the famous item group. Furthermore, the number of users in each item group increased suddenly when X is greater than 60. This is because the difference between a user interests and those of Gu is larger when X is greater than 60, and items with low probability of being interesting come to be recommended more often. From this result, novel topics are effectively detected with respect to detailed user interests when X is smaller than 60 given the datasets used in our experiment. This result also suggests that the suitable size of Gu is given by $X = 60$ because the number of items of each group radically increased when X exceeded that point.

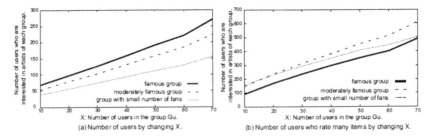

Fig. 4. (a) number of users obtained by changing X. (b) number of users that have high interest weight after changing X.

---

[8] The dataset is described in [14] in detail.

7.3 Investigating User Interest Distribution According to the Score of Novelty

We next evaluated novel item detection. In the evaluation, we compared the proportion of novel items in the manually defined recommendation lists created by "you might like these artists" in a music portal ListenJapan to the proportion of novel items in the recommendation lists created by our methods. Designers of music portal ListenJapan have manually defined artists (A_n) that are considered to relevant to artist (A_i).

We checked the 75 users, out of the total of 1503 users, who were judged to be interested in the music artists in our taxonomy. First, we identified X users who had high similarity to user a as described in Section 7.2[9]. We took from the recommendation lists created by our method the top 150 items that appeared frequently in the interests of those X users. The manually defined recommendation lists were generated by passing the user interests, extracted by our algorithm, to the portal's recommendation system. The manual recommendation lists included, on average, 23 items.

Table 2. Experimental results of distribution of topics according to item novelty.

(a) Comparing score of novelty in recommendation lists.

Score of novelty	0	1	2	3
Percentages of instances	57.6%	15.2%	23.2%	4.0%

(b) Comparing score of novelty in our detections.

Score of novelty	0	1	2	3
Percentages of instances	23.4%	23.1%	44.3%	9.2%

Table 2-(a) and (b) show the percentages of recommended items and their score of novelty for the manually generated recommendation lists and our lists, respectively. These results indicate that our technique recommends more items with a higher score of novelty than the manually created recommendation lists. Another conclusion that can be drawn is that users actually have a much wider range of interests than predicted by the music portal experts.

8 Online Experimental Results

To evaluate the effectiveness of novel topic identification, we offered an experimental service DoblogMusic to Doblog users. We used a larger taxonomy (325 classes and about 25,000 items) of the music domain than considered in the offline evaluation, since the service covered both Japanese music and non-Japanese music.

8.1 Explaining Our Online Experiment

As shown in Fig. 5, when doblog users, who were predicted to have an interest in artists in our taxonomy, logged into Doblog, DoblogMusic set "Recommendations"

---

[9] According to Section 7.2, we set X to 60.

tabs in the blog site of each user. Clicking the "Recommendations" tab yielded a page listing recommended novel artists. Neighboring users who were measured as similar to the user were also presented. If the user clicked the "Entries of Neighbors" tab, he was presented with summarized blog entries of the neighboring users on the novel artists (See procedure (2) in Fig. 5). A user could select one of these summaries to browse the complete blog entry at the Doblog portal. Furthermore, users that logged in and were analyzed to be interested in the artists in our taxonomy could also browse DoblogMusic through the "Recommendations" tab. As shown in Fig. 5, in our experiment, the score of novelty ranged from zero to three. The agent re-builts the user interests once a week for the service if users created blog entries and described music artists in our taxonomy.

The following subsections evaluate the performance of recommendation of novel topics, and generation of a virtual community through recommendations.

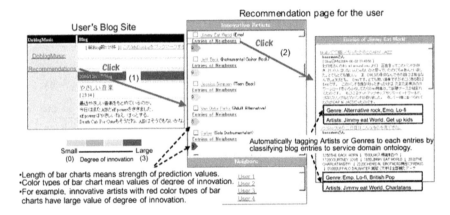

Fig. 5. Snapshot of online experimental service DoblogMusic.

8.2 Investigating Continuance of User Access to Our Recommendation List

Graphs of the number of users accessing DoblogMusic are shown in Fig. 5. There are about 55,000 Doblog users, of which about 1000 log in each day. We analyzed 3,632 Doblog users who were interested in the music of our music taxonomy in December 2006. Considering the ratio of the number of users who logged in to DoblogMusic to the number of all users in Doblog, we can expect that about 66 users among those logged in to Doblog each day had an opportunity to access DoblogMusic. From the results depicted in Fig. 6-(a), the average number of users that accessed DoblogMusic each day was about 63. Furthermore, we found that about 65% of those users were interested in the artists in our music taxonomy. From the results depicted in Fig. 6-(b), more users were presented with their own DoblogMusic recommendation list in November 2006. This is because we increased the number of artists in our music taxonomy at that time. These results indicate that recommending not only accurate but also novel items is effective for encouraging users to continue to access DoblogMusic, i.e. they do not become bored.

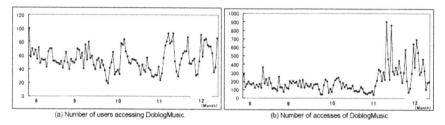

(a) Number of users accessing DoblogMusic.   (b) Number of accesses of DoblogMusic.

Fig. 6. User accesses to DoblogMusic.

8.3 Evaluating Identification of Novel Topics

We also evaluated the identification of novel topics. In the recommendation page of DoblogMusic, users could select novel artists according to their score of novelty (see procedure (2) in Fig. 5). We found that 75% of the selections made were for artists with a score of novelty of two or three. The offline evaluation, see Section 7.3 Table 2-(b), showed that content items that have the score of novelty of two or three, are presented to users most frequently, about 53.5%. It is obvious that users prefer novel recommendations.

To summarize, users were interested in recommendations even though they were comparatively far from the users' known interests from the viewpoint of the class hierarchy in our taxonomy. The results of Section 8.2 and Section 8.3 confirm that novelty identification is effective for inducing users to expand their interests by browsing the blog entries of novel artists.

8.4 Evaluating Activation of Blog Community

We found that presenting blog entries about the novel artists recommended to a user stimulated communication between users. If user a selected a summarized blog entry of user b presented as part of the recommendation service, user a could browse the complete blog entry on the blog site of user b. In this experiment, we checked the number of message exchanges made between users in two cases: (1) the number of exchanges (between users A and B) in the two month period before user a accessed the blog site of user b through the recommendation page of DoblogMusic and (2) the number of exchanges in the two month period immediately after the access. Fig. 7-(a) compares (1) and (2). It shows that there was a 30% increase in the number of exchanges due to DoblogMusic. Next, we focused on the users who exchanged comments no more than twice in case (1), see Fig. 7-(b). This comparison showed a roughly 250% increase in exchange number after Doblog-Music access. These results suggest that our novel blog-entry recommendation system is effective for activating the communications between users who love music.

Furthermore, we consider that users were interested in discovering novel items or users who described novel items presented by our method, because the users subsequently exchanged many comments. (Sending a comment to an unknown user is not an intuitive action.)

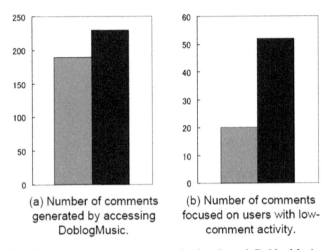

(a) Number of comments generated by accessing DoblogMusic.

(b) Number of comments focused on users with low-comment activity.

Fig. 7. Increasing user-user communication through DoblogMusic.

9 Conclusion

This chapter introduced the agent that automatically understands user interests according to the taxonomy of items and measures the similarity of users based on the taxonomy of items. The agent uses the similarity measurement results to identify novel topics. This agent can identify attractive and novel topics for a user by observing the difference between the user-interests of the user and those of the users similar to the user. We also performed offline and online experiments based on a large number of blog entries from an actual blog portal.

- In the offline experiments, we showed that our method can identify items more accurately than existing methods including the previous taxonomy-based method. We also examined the appropriate size of user group Gu to identify novel items for each user by changing the number of users in Gu. In an experiment on confirming the identification of novel topics, our technique identified more items with a higher score of novelty than were found in the manually created "you might like these artists" recommendation lists provided by the music portal site.
- In the online evaluation, our recommendations, novel items, were accessed repeatedly by users who were analyzed to be interested in items in our music taxonomy during the period of the online experiment. Furthermore, about 75% of accesses among those users were to items with high novelty. We also evaluated the number of messages exchanged between users who came to know each

other through our recommendations. As a result, users who seldom or never exchanged messages before seeing the recommendation page of novel items, exchanged significantly more messages. Thus, we conclude that our proposed algorithms can well identify novel items in the blog entries of other users and effectively create new virtually communities.

Future work is required to study how agents control the profile of user-interests by providing feedback about their collective knowledge with the goal of updating the taxonomy of items. The long-term goal is to integrate the user interests over several key domains and offer cross-domain recommendations for the user.

Acknowledgments. I appreciate the constructive discussions and the kindly advice of Professor Toru Ishida at Kyoto University. I wish to thank my colleagues at NTT Cyber Solutions Laboratories and NTT Network Service Systems Laboratories for their support in numerous ways.

References

1. Breese, J.S., Heckerman, D., Kadie, C.: Empirical Analysis of Predictive Algorithms for Collaborative Filtering. In: Proc. UAI 1998, pp. 43–52 (1998)
2. Doan, A., Madhavan, J., Domingos, P., Halevy, A.: Learning to map between ontologies on the semantic web. In: Proc. WWW 2002, pp. 662–673 (2002)
3. Nakatsuji, M., Fujiwara, Y., Tanaka, A., Uchiyama, T., Fujimura, K., Ishida, T.: Classical Music for Rock Fans?: Novel Recommendations for Expanding User Interests. In: Proc. CIKM 2010 (to appear, 2010)
4. Godoy, D., Amandi, A.: User Profiling in Personal Information Agents: A Survey, Knowledge Engineering Review. Cambridge University Press, Cambridge (2005)
5. Herlocker, J.L., Konstan, J.A., Borchers, A., Riedl, J.: An algorithmic framework for performing collaborative filtering. In: Proc. SIGIR 1999, pp. 230–237 (1999)
6. Herlocker, J.L., Konstan, J.A., Terveen, L.G., Riedl, J.T.: Evaluating collaborative filtering recommender systems. ACM Trans. Inf. Syst. 22(1), 5–53 (2004)
7. Hijikata, Y., Shimizu, T., Nishida, S.: Discovery-oriented collaborative filtering for improving user satisfaction. In: Proc. UAI 2009, pp. 67–76 (2009)
8. Konstas, I., Stathopoulos, V., Jose, J.M.: On social networks and collaborative recommendation. In: Proc. SIGIR 2009, pp. 195–202 (2009)
9. Kohrs, A., Merialdo, B.: Clustering for Collaborative Filtering Applications. In: Computational Intelligence for Modelling, Control & Automation. IOS Press, Amsterdam (1999)
10. Maedche, A., Staab, S.: Measuring similarity between ontologies. In: Gómez-Pérez, A., Benjamins, V.R. (eds.) EKAW 2002. LNCS (LNAI), vol. 2473, pp. 251–263. Springer, Heidelberg (2002)
11. Manning, C.D., Schuetze, H.: Foundations of Statistical Natural Language Processing, 1st edn. The MIT Press, Cambridge (1999)
12. McNee, S.M., Riedl, J., Konstan, J.A.: Being accurate is not enough: how accuracy metrics have hurt recommender systems. In: Proc. CHI 2006, pp. 1097–1101 (2006)

13. Nakatsuji, M., Miyoshi, Y., Otsuka, Y.: Innovation Detection Based on User-Interest Ontology of Blog Community. In: Cruz, I., Decker, S., Allemang, D., Preist, C., Schwabe, D., Mika, P., Uschold, M., Aroyo, L.M. (eds.) ISWC 2006. LNCS, vol. 4273, pp. 515–528. Springer, Heidelberg (2006)
14. Nakatsuji, M., Yoshida, M., Ishida, T.: Detecting innovative topics based on user-interest ontology. J. Web Sem. 7(2), 107–120 (2009)
15. Noy, N.F., Musen, M.A.: Anchor-PROMPT: Using Non-Local Context for Semantic Matching. In: The 17th International Joint Conference on Artificial Intelligence Workshop on Ontologies and Information Sharing, pp. 450–455 (2001)
16. Resnick, P., Iacovou, N., Suchak, M., Bergstorm, P., Riedl, J.: GroupLens: An Open Architecture for Collaborative Filtering of Netnews. In: Proceedings of ACM 1994 Conference on Computer Supported Cooperative Work, Chapel Hill, North Carolina, pp. 175–186. ACM, New York (1994)
17. Sarwar, B., Karypis, G., Konstan, J., Riedl, J.: Item-based Collaborative Filtering Recommendation Algorithms. In: Proc. WWW 2001, pp. 285–295 (2001)
18. Shardanand, U., Maes, P.: Social Information Filtering: Algorithms for Automating "Word of Mouth". In: Proc. CHI 1995, pp. 210–217 (1995)
19. Ungar, L., Foster, D.: Clustering Methods For Collaborative Filtering. In: Proceedings of the Workshop on Recommendation Systems. AAAI Press, Menlo Park (1998)
20. Yildirim, H., Krishnamoorthy, M.S.: A random walk method for alleviating the sparsity problem in collaborative filtering. In: Proc. RecSys 2008, pp. 131–138 (2008)
21. Zhang, Y., Callan, J., Minka, T.: Novelty and redundancy detection in adaptive filtering. In: Proc. SIGIR, pp. 81–88 (2002)
22. Ziegler, C.N., Lausen, G., Thieme, L.S.: Taxonomy-driven computation of product recommendations. In: Proc. CIKM 2004, pp. 406–415 (2004)
23. Ziegler, C.N., McNee, S.M.: Improving recommendation lists through topic diversification. In: Proc. WWW 2005, pp. 22–32 (2005)

Part IV
Future Outlook

Chapter 13
Semantic Agents with Understanding Abilities and Factors Affecting Misunderstanding

Tuncer Ören[1] and Levent Yılmaz[2]

[1] The McLeod Modeling and Simulation Network (M&SNet) of SCS,
School of Information Technology and Engineering, University of Ottawa, Ottawa,
Ontario, Canada
oren@site.uottawa.ca
[2] M&SNet, Auburn Modeling and Simulation Lab,
Computer Science and Software Engineering, Auburn University, Auburn, AL, USA
yilmaz@auburn.edu

Abstract. Ability to understand is a pivotal feature of intelligent systems in general and in software agents and especially in semantic agents in particular. Levels of understanding are understanding truth, understanding metaphors, and understanding pataphors. Currently, we focus on understanding truth. In this chapter, about 90 concepts and terms related with machine understanding are presented. An important aspect of machine understanding, namely misunderstanding and almost 60 concepts and terms related with it are elaborated systematically for the first time in the literature; along with rationale why discriminating and identifying types and factors influencing misunderstanding has pragmatic importance for agents with understanding abilities.

Keywords: Semantic agents, machine understanding, single understanding, multi-understanding, switchable understanding, misunderstanding.

1 Introduction

Machine understanding is also named computer understanding, computerized understanding, and computational understanding. We refer to it as understanding in this chapter. Understanding is a pivotal ability for agents in general and for semantic agents in particular. In this chapter, we provide a general framework for understanding applicable to agents in general and to semantic agents in particular. Several modes of machine understanding such as single understanding, multi-understanding, and switchable understanding are covered and for each mode, several types of understanding are explained. For example, 90 types of machine understanding are covered. The discussion of almost 60 types of misunderstanding can be useful to detect and avoid them, as a basis for failure avoidance in machine understanding; and as a basis for emulation and simulation of human behavior. Other benefits of studying misunderstanding are clarified in section 6.

A. Elçi, M.T. Koné, and M.A. Orgun (Eds.): Semantic Agent Systems, SCI 344, pp. 295–313.
springerlink.com   © Springer-Verlag Berlin Heidelberg 2011

The term "to understand" denotes associated processing of knowledge to have an "understanding" which is the result or product of this knowledge processing. However, in English, the –ing form can be used as a noun, an adjective or like a verb. Hence, when necessary, we will use the term "understanding" both as the product and the process of the verb to understand; e.g., understanding ability (instead of ability to understand), understanding system (instead of system to understand), machine understanding (instead of to understand by a machine), and understanding process (instead of process to understand).

Technology is bounded by science; however, science does not need to be bounded by technology. Hence, it is quite normal that some of the modes and types of understanding covered in this chapter are not yet available in practice; and hopefully will in the future.

1.1 Machine Understanding

As it is the case with many terms used in everyday language as well as technical terms, to understand and understanding need to be clarified. In earlier publications, we revised definitions of everyday and technical usages, as well as philosophical backgrounds [18, 22]. In our work, we take knowing and computerized understanding as synonyms. Hence, to understand an entity is to get appropriate knowledge about it. The following quote from a previous publication clarifies the qualifier appropriate: "The following is a good starting point for the specification of the scope of machine understanding: ' . . . if a system knows about X, a class of objects or relations on objects, it is able to use an (internal) representation of the class in at least the following ways: receive information about the class, generate elements in the class, recognize members of the class and discriminate them from other class members, answer questions about the class, and take into account information about changes in the class members' [33]" [22]. However, one should remark here that knowing (something, somebody, some event, etc.) refers to the result of the process of acquiring knowledge and not the knowledge processing activity required to know" [18]. From a philosophical background, understanding and having a meaning are closely related. For example, "Dewey [9] relates understanding with meaning by stating that if A cannot understand B, B does not have a meaning for A" [18].

1.2 Motivation: The Role of Understanding in Decision Support

Understanding involves perception of elements in a particular environment within time and space, the comprehension of their meaning and the projection of their status in the near future. Machine understanding requires a set of mechanisms that enable attention to cues in the environment, as well as expectancies regarding future states. In realistic settings, establishing an ongoing awareness and understanding of important situation components pose the major task of the decision maker. Therefore, situation awareness is the primary is basis of the decision making process in experience-based decision making process (i.e., Naturalistic Decision Making).

The recognition, revision, and exploration phases of situation awareness, shown in Figure 1, suggest three main functional areas that revolve around a mental model of the problem domain. More specifically, a well-defined mental model provides:

1. knowledge about the concepts, attributes, associations, and constraints that pertain to the application domain,
2. a mechanism that facilitates integration of domain elements to form an understanding of the situation, and
3. a mechanism to project to a future state of the environment given the current state, selected action, and the knowledge about the dynamics of the environment.

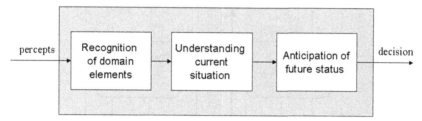

Fig. 1. Elements of situation awareness

Situation awareness is an important cognitive skill that is essential for expert performance in any field involving complexity, dynamism, uncertainty, and risk. The failure to perceive a situation correctly may lead to faulty understanding. Ultimately, this misunderstanding may degrade an individual's ability to predict future states and engage in effective decision making.

1.3   Agents, Semantic Agents, and Pivotal Role of Machine Understanding

Understanding is an essential characteristic of intelligent systems, including intelligent agents. Agents are becoming more and more important in advanced knowledge processing. For this reason, another aspect of understanding, namely misunderstanding is also systematically elaborated in this chapter to be able to take necessary precautions for failure avoidance in machine understanding [23].

"The Semantic Web relies on structured sets of metadata and inference rules that allow it to *understand* the relationship between different data resources. The technologies that form the basis of the Semantic Web by adding these metadata and inference rules are RDF (Resource Description Framework), RDFS (RDF Schema) and OWL (Web Ontology Language)" [27]. A general and flexible framework is needed to satisfy the understanding requirements of semantic agents. As stated by Allemang and Hendler "A consequence of the AAA slogan is that there could always be something new that someone will say; this means that we must assume that there is always more information that could be known" [1]. They define "The AAA slogan" as "Anyone can say Anything about Any topic" and qualify it as "One of the basic tenets of the Web in general and the Semantic Web in particular." "The right way to understand what a statement or a set of statements means in RDFS is to understand what inferences can be drawn from them." as stated by [1, p. 117] is yet another practical implication of understanding in Semantic Web agents.

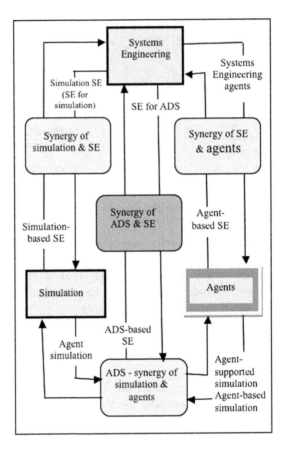

Fig. 2. First and second order synergies of agents, systems engineering, and simulation [from 19]

1.4 Synergies of Agents and Semantic Agents with Simulation and Systems Engineering

In some application areas of knowledge processing, synergies of agents, simulation, and systems engineering appear to be very desirable [24, 32]. Figure 2 – from Ören [19] – highlights such synergies.

Some of the synergies with direct relevance to semantic agents can be systems engineering agents, agent-based systems engineering, as well as synergies of agents and simulation to provide experimentation abilities to agents. Agents with understanding abilities would have important implications in these cases. With increasing complexity of the applications of Semantic Web agents, one may expect to consider the benefits of the synergy of systems engineering and agents.

2 Machine Understanding Systems and Agents with Understanding Abilities

Machine understanding of an entity is getting appropriate knowledge about this entity by a system. For a system **A** to understand an entity **B**, three conditions should be met:

1. System **A** should have access to a meta-model **C** of **B**s.
   (i.e., system **A** should have knowledge about a class of entities like **B**.)
2. System **A** should be able to form **D**, a perception of **B** by analyzing **B**.
3. System **A** should be able to interpret its perception **D** with respect to the meta-model **C**.
   (i.e., should be able to compare and evaluate its perception **D** with respect to the meta-model **C**. The evaluation of features in **C** can be done with respect to existing and non-existing attributes in **D**.)

The elements of a machine understanding system are given in Figure 3.

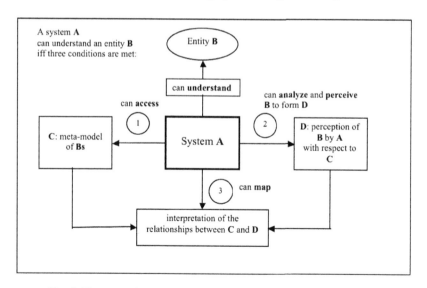

Fig. 3. Elements of an understanding system [adopted from 18 and 25]

Several modes of machine understanding are possible; they are single understanding, multi-understanding, and switchable understanding.

3 Types of Single Understanding

There are three levels of understanding: (1) understanding reality, (2) understanding metaphors, and (3) understanding pataphors. Currently, most understanding systems are at the level of understanding reality. Metaphors are used to indicate that an entity **A** has some well known characteristics of (or is similar to) an entity **B**. "The *pataphor* is an extreme form of metaphor, taking the principle to its limit, where the basic metaphor is typically not mentioned but extensions to it are used without reference. . . . Metaphor is used to bring novelty, interest and elucidation to writing and speech. In the extreme form, pataphors may bring greater novelty and perhaps interest, though potentially at the price of clarity and broad understanding" [5]. Especially, in natural language understanding, understanding metaphors and pataphors would be very important and challenging for machine understanding. For powerful Semantic Web applications involving text processing, these two levels of understanding may come into play.

In this and many other publications we concentrate on the fundamental issue of understanding reality. A previous reference covers essence of machine understanding as well as different types of single understanding [18] and especially ontology-based (relational) dictionary of machine understanding [22]. For this reason, to avoid repetition, only the highlights will be given in this chapter. A list of 90 concepts and terms related with machine understanding is given as Appendix A. Machine understanding can be considered from different points of view such as: (1) product of understanding, (2) process of understanding, (3) meta-model used in understanding, and (4) characteristics of understanding system.

3.1 Machine Understanding from the Point of View of Product of Understanding

Product of understanding is the result of the process to understand. The following criteria can be used to over 20 types of machine understanding: domain, nature, scope, depth, granularity, reliability, and processing of product of understanding.

- Based on the *domain* of understanding, internal and external understandings are possible.
- Based on the *nature* of understanding, the following types are possible: lexical, syntactic, morphological, semantic, and pragmatic understandings.
- Based on the **scope**, focused, broad, and multi-aspect understandings are possible.
- Based on *depth* of understanding, shallow and deep understandings can be identified.
- Based on *granularity*, two types of understanding are possible: coarse and in-depth understanding (i.e., detailed understanding).
- Based on *reliability*, there are two types of understanding: reliable understanding, such as valid, verified, and incorrupt understandings; and unreliable understanding such as invalid, unverified, and corrupt understanding.
- Additional *processing* identifies associative, generalized, and instantiated understanding.

3.2 Machine Understanding from the Point of View of Process to Understand

Process of understanding (process to understand) is the necessary knowledge processing activities to obtain a valid understanding of an entity. Different types of understanding processes can be identified based on the following criteria: directness, direction, precedence, modality, and dependability of the understanding process, and the accumulation of knowledge.

- Based on **directness** of understanding process, we can identify apprehension (or direct understanding) and comprehension (or indirect or mediated understanding) such as logical understanding.
- **Direction** of understanding process defines top-down and bottom up understandings.
- Based on **precedence** of understanding process, we have sequential and parallel understandings.
- Based on **modality** of understanding process, there are unimodal (i.e., understanding one modality at a time) and multimodal understandings.
- Based on **dependability** of understanding process, there are robust and brittle understandings.
- Based on **accumulation of knowledge** during the understanding process, there are two possibilities: tabula rasa understanding (i.e., re-initialized understanding) and cumulative understanding.

3.3 Machine Understanding from the Point of View of Meta-Model of Understanding

To be able to understand an entity, a system needs some background knowledge about it and similar ones. This fundamental knowledge is called a meta-model of the entities. Two characteristics of meta-model can be used to discern several types of understanding; these characteristics are its variability and nature. According to its variability, a meta-model can be: (1) fixed, (2) replaceable, and (3) evolvable. According its nature, a meta-model can be (1) analogical and (2) tailored.

- **Fixed meta-models** limit understandings to single-vision understanding; which can be dogmatic understanding.
- **Replaceable meta-models** allow multi-vision understanding which can be the basis of switchable understanding.
- **Evolvable meta-models** allow evolving (or learning) understanding.
- Use of **analogical (equivalent) meta-model(s)** would allow analogical (associative) understanding.
- **Meta-models tailored** to specific function(s) result in tailored understanding.

3.4 Machine Understanding from the Point of View of Characteristics of Understanding System

The characteristics of an understanding system are its initiative in understanding, its locality, its number, role of prejudice (emotions) in understanding, its knowledge sharing features, dissemination of the results of understanding.

- Based on the *initiative* of the understanding system, we identify autonomous and delegated understandings.
- Based on the *locality* of the understanding system, there are local and remote understands.
- Based on the *number* of the understanding system, we identify individual, group, and distributed understandings.
- Based on the **prejudice** (emotions) of the understanding system [13], there are objective and subjective understandings.
- Based on the *knowledge sharing feature* of the understanding system, there are repetitive, partially repetitive, and cooperative understandings.
- Dissemination of the results by an understanding system, delimits understanding per command, understanding for subscribers, broadcast understanding, blackboard understanding, and legacy understanding.

4 Multi-understanding

Multiple understanding (or multi-understanding) *ability* denotes a special type of knowledge processing ability to have more than on understanding of an entity. As a *process,* multi-understanding is a sequential or parallel activity to generate more than one understanding of an entity. Minsky posited the following statement about multi-understanding "If you 'understand' something in only one way, then you scarcely understand it at all –because when you get stuck, you'll have nowhere to go. But if you represent something in several ways, then when you get frustrated enough, you can switch among different points of view, until you find one that works for you!" [17].

Multi-understanding ability necessitates one or more of the following:

- Existence of and access to multiple meta-models
- Multi-perception ability
- Multiple interpretation ability

4.1 Role of Meta-Models in Multi-understanding

As clarified in section 3.3 (Machine understanding from the point of view of meta-model of understanding), two types of meta-models can allow multi-understanding. They are: (1) replaceable (or multi-)meta-models, and (2) evolvable meta-models (or learning) meta-models. Accordingly, multi-vision understanding, and evolving understanding (learning understanding) are possible.

4.2 Role of Perception in Multi-understanding

Perception is very important in understanding; it is influenced by background knowledge (a system cannot understanding entities for which it has no knowledge). Perception depends on the context and is also influenced by several biases (not always in the negative way). A good overview and list of references can be found at Stanford Encyclopedia of Philosophy [6].

4.3 Role of Interpretation in Understanding

Interpretation is one of the three pillars of understanding and is open to many factors that are clarified in section on misunderstanding.

5 Switchable Understanding

Switchable understanding is a special case of multi-understanding. An advanced understanding system can explore to have several understandings and select the most appropriate one fit for the context. A basis for switchable understanding for agents was covered elsewhere [25] and is not repeated in this chapter.

6 Misunderstanding

While ability to understand has a pivotal role in any intelligent systems in general and in advanced software agents and semantic agents in particular, another topic is equally important: How to assure reliability of understanding systems. This question is treated from a general failure avoidance point of view [23]. Misunderstanding, in machine understanding means (1) a failure to understand an entity correctly or (2) failure to understand. Identifying the factors and mechanisms of misunderstanding, especially by software agents, is important from several points of view:

(1) in computerized understanding, one can avoid errors and
(2) in simulating or emulating human behavior, one can have realistic human understanding, including its flaws.
(3) Another important implication of misunderstanding is the crucial role in conflict management. To be able to reconcile differences of opinions, understanding them and their root causes are essential even before offering alternatives to find a common ground to agree on. Hence, factors affecting misunderstandings are also essential in finding alternatives as bases for persuasion. Advanced applications of Semantic Web agents for analysis of texts written in natural language may be significant for this purpose.

Categories of sources/causes of misunderstanding are outlined in the sequel; and a list of concepts (and terms) related with machine misunderstanding is given as Appendix B. *Biases* of the understanding system may exist in all three types of abilities and corresponding processes. *Errors* in all three major elements of

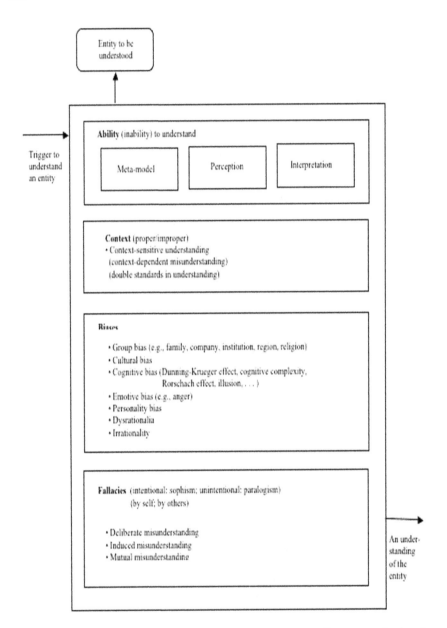

Fig. 4. Filters that can induce misunderstanding and corresponding types of misunderstanding

understanding may lead to additional types of understanding which may lead to erroneous understandings and misunderstanding [25]. Misunderstanding can be caused by several factors such as features of the available meta-model(s) and processes for perception and interpretation. These three elements, in turn, can be affected by cultural, personal, cognitive, and emotive biases as well as by irrationality and dysrationalia. Furthermore, both intentional and unintentional types of logical errors (or fallacies) in knowledge processing needed in perception and interpretation processes may also result in misunderstanding. Figure 4 depicts filters that can induce misunderstanding and corresponding types of misunderstandings.

6.1 Ability/Inability to Understand

6.1.1 Role of *Meta-Model* in Misunderstanding

Meta-model can have knowledge unfit for the goal of understanding (e.g., erroneous, incomplete, inconsistent, irrelevant, or corrupt meta-model). Furthermore, it can have cultural and cognitive biases (sometimes implicitly; due to the corresponding biases of its developer). Hence, the following types of misunderstandings are possible: erroneous understanding, incomplete understanding, inconsistent understanding, irrelevant understanding, as well as effects of deliberate use of unfit meta-model in understanding, and effect of corrupt meta-model in understanding.

6.1.2 Role of *Perception* in Misunderstanding

"An understanding system needs the ability to perceive the entity that it intends to understand; for this it needs to analyze characteristics relevant to the goal of understanding. Hence, the system needs to be able to analyze and perceive. Sources of multiple perceptions include:

1. focus on several (some irrelevant) aspect (domain, nature, scope, granularity, modality)
2. use of several (some irrelevant) meta-model
3. lack of appropriate ability to analyze
4. lack of appropriate ability to discriminate" [25]

Misperception (as well as *misinterpretation)* *of motivation* and perceptual confusion are also causes of misunderstanding.

6.1.3 Role of *Interpretation* in Misunderstanding

Misinterpretation is a source of misunderstanding. It may be caused by using a meta-model which is not appropriate for the context, by lack of pertinent knowledge processing abilities in perception and/or in interpretation, or by logical errors (fallacies) in the interpretation.

6.2 Role of Context in Misunderstanding

Context-dependent misunderstanding would be a result of an understanding process within a wrong context. It may also be caused by fallacies (sophism and paralogisms) induced by self or by others.

6.3 Role of Biases in Misunderstanding

6.3.1 *Group Bias* in Misunderstanding

Acceptance of reality is often context-sensitive and depends on the group within which it is uttered. The group can be limited by a family, company, institution, region, nation, interest or affinity, and/or religion. Hence, not only what is said is important but also who (i.e., a member or an outsider of the group) said it are often important. For example, consider the following statement from Scientific American: "America is an absurdly backward country when it comes to passenger trains." [3]. In this sentence –taken as an example– "America" can be replaced by any other country; and after "when" another statement can be put.

The important point is whether the interlocutor is within the group or an outsider. Hence, the statement can be accepted, can be denied, or even considered offensive (all alternatives would be based on different "understandings." Let α denote an individual of a group A and let β denote an individual of a group B. A same statement expressed by α about A, can be accepted as a true fact or even a treason to (some of) the members of group A; while it can be ignored or hailed by some members β of B. In a fragmented group A, an individual α can also be considered an insider or an outsider. An insider's statement can be accepted; while an outsider's statement can be accepted, tolerated, or can be interpreted as offensive.

6.3.2 *Cultural Bias* in Misunderstanding

Values and symbols differ for various cultures; hence a same entity may be interpreted differently based on the cultural background [11, 16]. Geert Hofstede^{TM} Cultural Dimensions provide comparative cultural characteristics of large number of countries [12]. A book series, called "culture smart" provides profiles of several countries [7]. Cultural bias in understanding leads to culture-induced misunderstanding.

6.3.3 *Cognitive Bias* in Misunderstanding

Cognitive bias is a "common tendency to acquire and process information by filtering it through one's own likes, dislikes, and experiences" [4]. Some of the cognitive biases that may cause misunderstanding are explained in the sequel:

A well known case is *Dunning-Kruger effect* [15]. As concluded by Kruger and Dunning in 2009 "those with limited knowledge in a domain suffer a dual burden: Not only do they reach mistaken conclusions and make regrettable errors, but their incompetence robs them of the ability to realize it."

As another example, *high cognitive complexity individuals* differ from low cognitive complexity individuals not only in knowledge processing abilities in general but in understanding, in particular, due to their limited background knowledge as well as to their limited ability to process knowledge needed for analysis and perception and evaluation. Sirisonthi [2004] refers to cognitive complexity as a factor to understand/misunderstand in e-mail communications. Future applications of Semantic Web agents may take this aspect of bias in analyses of increasing volume of e-mail communications.

As a thought disorder, *Rorschach effect* in understanding may result in different interpretations and hence different misunderstandings by different individuals.

Illusion is a misinterpretation of a true sensation. Schizophrenic understanding –as an aberration– leads to misinterpretations of reality and hallucinations in the absence of stimulus.

6.3.4 *Emotive Bias* in Misunderstanding

It is well known that certain types of emotions affect reasoning abilities to cause misunderstanding. For example, anger affects reasoning negatively; hence understanding ability. Effect of anger in misunderstanding leads to anger-induced misunderstanding [14].

6.3.5 *Personality Bias* in Misunderstanding

Some personality types are prone to anger; hence their understanding ability can easily be affected to lead misunderstanding [26].

6.3.6 Effects of *Dysrationalia* and *Irrationality* in Misunderstanding

Dysrationalia is the inability to think and behave rationally despite adequate intelligence [30, 31]. It affects ability to properly understand. Irrationality is common in cognition [2]. Irrationality may have two types of effects in misunderstanding; lack of ability to understand properly and ability to distort understanding of others to cause distorted understanding.

6.4 Role of Fallacies in Misunderstanding

Fallacy is misconception resulting from incorrect reasoning. A logical fallacy is an element of an argument that is flawed, essentially rendering the line of reasoning, if not the entire argument, invalid. Fallacies in *information distortion* [10] as well as *deliberate misperception* and *misinterpretation* are sources of misunderstanding. Fallacies can be grouped in two categories: *Paralogism* is unintentional invalid argument in reasoning. *Sophism* is a deliberately invalid argument displaying ingenuity in reasoning in the hope of deceiving someone. Some recent techniques in lie detection in text analysis can also be used to detect sources of attempt to misguide in understanding [8].

6.4.1 *Deliberate* Misunderstanding

Another type of misunderstanding is deliberate misunderstanding which giving the illusion of not understanding. This may be functionally equivalent to misunderstanding. However, the deliberate aspect is important and–if detected–may need to be proven.

6.4.2 *Induced Misunderstanding*

To induce misunderstanding, data and evidences may be tempered or doctored. The individuals (or their representatives, such as software agents) need to notice that their understanding is being tempered. Hence, recognizing why a reality is presented in a certain way is helpful not to be trapped in misunderstanding. Some types of induced misunderstanding are: Socratically induced misunderstanding, and speaker-induced misunderstanding.

6.4.3 *Mutual Misunderstanding*

Avoiding mutual misunderstanding is very important to find reconciliatory solutions at different levels of relationships. For example Shi [28] explores ways to avoid mutual misunderstanding between China and the USA.

7 Conclusions and Future Research

Ability to understand is a pivotal feature of intelligent systems in general and in software agents and especially in semantic agents in particular. Levels of understanding are understanding truth, understanding metaphors, and understanding pataphors. Currently, we focus on understanding truth. In this chapter, spectrum of machine understanding is presented and an important aspect of machine understanding, namely misunderstanding is elaborated systematically for the first time in the literature. Hence, 90 concepts and terms related with machine understanding and almost 60 concepts and terms related with machine misunderstanding are presented; along with rationale why discriminating and identifying types and factors influencing misunderstanding has pragmatic importance.

Our machine understanding studies started in understanding software and especially understanding simulation languages and simulation programs [20, 21]. Last decade, we have concentrated on understanding in general and understanding emotions especially by agents [13, 25]. Our studies on agents able to understand and process emotions will continue [14]. Our research on development of training technology (based on agent simulation) for international conflict management continues. Advances in Semantic Web agents in analyzing text written in natural language may be very valuable to find reconciliations by analyzing several types of misunderstandings.

References

1. Allemang, D., Hendler, J.: Semantic Web for the Working Ontologist: Effective Modeling in RDFS and OWL. Morgan Kaufmann, Burlington (2008)
2. Ariely, D.: Predictably Irrational – The Hidden Forces That Shape Our Decisions. Harper/HarperCollins Publishers, New York (2008)
3. Brown, S.F.: Revolutionary Rail: High-Speed Rail Plan Will Bring Fast Trains to the U.S. Scientific American 302(5), 54–59 (2010)
4. Business dictionary,
   http://www.businessdictionary.com/definition/
   cognitive-bias.html

5. Changing Minds–on pataphor,
   `http://changingminds.org/techniques/language/metaphor/pataphor.htm`
6. Crane, T.: The Problem of Perception. Stanford Encyclopedia of Philosophy (2005),
   `http://plato.stanford.edu/entries/perception-problem/`
7. Culture Smart Guides, `http://www.kuperard.co.uk/culturesmart/`
8. Delen, D.: Business Intelligence and Advanced Analytics for Real World Problems. In: Güneş, M. (ed.) Proceedings of the 1st International Symposium on Computing in Science and Engineering, June 3-5. Gediz University, Turkey (2010) (2010-Invited Paper)
9. Dewey, J.: (Originally published by: D.C. Heath, Lexington, MA)How we Think. Prometheus Books, Buffalo, NY; Brown, S.F.: (original publication 1910) Revolutionary Rail – High-speed trains are coming to the U.S. Scientific American 302(5), 54–57 (2010)
10. Güvenen, O.: Information Systems Security and Information Distortion Impacts on the Socio-Economic Structures. In: Güneş, M. (ed.) Proceedings of the 1st International Symposium on Computing in Science and Engineering, June 3-5. Gediz University, Turkey (2010) (2010-Invited Paper)
11. Hofstede, G., Hofstede, G.J.: Cultures and Organizations – Cultures of the Mind. McGraw-Hill, New York (2005)
12. Itim International. Geert HofstedeTM Cultural Dimensions,
    `http://www.geert-hofstede.com/`
13. Kazemifard, M., Ghasem-Aghaee, N., Ören, T.I.: Agents with Ability to Understand Emotions. In: Proceedings of the Summer Computer Simulation Conference, Istanbul, Turkey, July 13-16. Simulation Series, vol. 41(3), pp. 254–260. SCS, Dan Diego (2009)
14. Kazemifard, M., Ghasem-Aghaee, N., Ören, T.I.: Design and Implementation of GEmA: A Generic Emotional Agent. In: Expert Systems with Applications (In Press, 2011)
15. Kruger, J., Dunning, D.: Unskilled and Unaware of It: How Difficulties in Recognizing One's Own Incompetence Lead to Inflated Self-Assessments. Psychology 1, 30–46 (2009)
16. Lewis, R.D.: When Cultures Collide – Managing Successfully Across Cultures. Nicholas Brealey, London (2000)
17. Minsky, M.: The Emotion Machine: Commonsense Thinking, Artificial Intelligence, and the Future of the Human Mind. Simon & Schuster, New York (2006)
18. Ören, T.I.: Understanding: A Taxonomy and Performance Factors. In: Thiel, D. (ed.) Proc. of FOODSIM 2000, Nantes, France, June 26-27, pp. 3–10. SCS, San Diego (2000) (2000 – Invited Opening Paper)
19. Ören, T.I.: Synergies of Simulation, Agents, and Systems Engineering. In: Güneş, M. (ed.) Proceedings of the 1st International Symposium on Computing in Science and Engineering, June 3-5. Gediz University, Turkey (2010) (2010–Invited Paper)
20. Ören, T.I., Abou-Rabia, O., King, D.G., Birta, L.G., Wendt, R.N.: Reverse Engineering in Simulation Program Understanding. In: Jávor, A. (ed.) Problem Solving by Simulation, Proceedings of IMACS European Simulation Meeting, Esztergom, Hungary, August 28-30, pp. 35–41 (1990/1990a) (Plenary Paper)
21. Ören, T.I., Birta, L.G., Abou-Rabia, O., King, D.G., Wendt, R.N.: E/Slam: A Software Understanding Environment for SLAM II Programs. In: Proceedings of European Simulation Multiconference, Erlangen-Nuremberg, Germany, June 10-13, pp. 235–240. SCS International, San Diego (1990/1990b)

22. Ören, T.I., Ghasem-Aghaee, N., Yilmaz, L.: An Ontology-Based Dictionary of Under-standing as a Basis for Software Agents with Understanding Abilities. In: Proceedings of the Spring Simulation Multiconference (SpringSim 2007), Norfolk, VA, March 25-29, pp. 19–27 (2007) ISBN: 1-56555-313-6

23. Ören, T.I., Yilmaz, L.: Failure Avoidance in Agent-Directed Simulation: Beyond Conventional V&V and QA. In: Yilmaz, L., Ören, T.I. (eds.) Agent-Directed Simulation and Systems Engineering. Systems Engineering Series, pp. 189–217. Wiley, Berlin (2009a)

24. Ören, T.I., Yilmaz, L.: On the Synergy of Simulation and Agents: An Innovation Paradigm Perspective. Special Issue on Agent-Directed Simulation. The International Journal of Intelligent Control and Systems (IJICS) 14(1), 4–19 (2009/2009b)

25. Ören, T.I., Yilmaz, L., Kazemifard, M., Ghasem-Aghaee, N.: Multi-understanding: A Basis for Switchable Understanding for Agents. In: Proceedings of the Summer Computer Simulation Conference, Istanbul, Turkey, July 13-16. Simulation Series, vol. 41(3), pp. 395–402. SCS, Dan Diego (2009)

26. PoorMohamadBagher, L., Kaedi, M., Ghasem-Aghaee, N., Ören, T.I.: Anger Evaluation for Fuzzy Agents with Dynamic Personality. In: Mathematical and Computer Modelling of Dynamical Systems (MCMDS), December 6, vol. 15(6), pp. 535–553 (2009)

27. XBRL Glossary, http://www.altova.com/xbrl-glossary.html

28. Shi, T.: Avoiding Mutual Misunderstanding: Sino-U.S. Relations and the New Administration. In: Carnegie Endowment for International Peace, Washington, D.C (2009)

29. Srisonthi, T.: A Cognitive Complexity as a Factor to Understanding/ Misunderstanding in e-Mail Communication. Academic Review 3(1) (2004)

30. Stanovich, K.E.: Dysrationalia: A New Specific Learning Disability. Journal of Learning Disabilities 26(8), 501–515 (1993)

31. Stanovich, K.E.: Rational and Irrational Thought: The Thinking That IQ Tests Miss. Scientific American Mind (November 2009)

32. Yilmaz, L., Ören, T.I. (eds.): (All Chapters by Invited Contributors). Agent-Directed Simulation and Systems Engineering. Wiley Series in Systems Engineering and Management, p. 520. Wiley, Berlin (2009)

33. Zeigler, B.P.: Systems Knowledge: A Definition and its Implications. In: Elzas, M.S., Ören, T.I., Zeigler, B.P. (eds.) Modelling and Simulation Methodology in the Artificial Intelligence Era, pp. 15–17. North-Holland, Amsterdam (1986)

Appendix A

Concepts and Terms Related with Machine Understanding

adaptive understanding	generalized understanding
analogical understanding	group understanding
apprehension	incorrupt understanding
associative understanding	in-depth understanding
autonomous understanding	indirect understanding
blackboard understanding	individual understanding
bottom up understanding	instantiated understanding
brittle understanding	internal understanding

broad understanding	invalid understanding
broadcasted understanding	learning understanding
clear understanding	legacy understanding
coarse understanding	lexical understanding
comprehension	local understanding
computational understanding	logical understanding
computer understanding	machine understanding
computerized understanding	mediated understanding
context-sensitive understanding	mode of understanding
context understanding	morphological understanding
cooperative understanding	multiaspect understanding
cumulative understanding	multimodal understanding
deep understanding	multivision understanding
delegated understanding	mutual understanding
detailed understanding	objective understanding
direct understanding	parallel understanding
distributed understanding	partially repetitive understanding
entire understanding	pragmatic understanding
evolving understanding	pre-understanding
external understanding	product of multi-understanding
focused understanding	product of single understanding
product of understanding	tailored understanding
re-initialized understanding	top-down understanding
reliable understanding	type of understanding
remote understanding	understander
product of understanding	understanding
re-initialized understanding	understanding context
repetitive understanding	understanding for subscribers
robust understanding	understanding initiator system
self understanding	understanding metaphor
semantic understanding	understanding pataphor
sequential understanding	understanding per command
shallow understanding	understanding process
single understanding	understanding reality
single vision understanding	understanding system

specialized understanding	understood knowledge
subjective understanding	unimodal understanding
switchable understanding	valid understanding
syntactic understanding	verified understanding
tabula rasa understanding	

Appendix B

Concepts and Terms Related with Machine Misunderstanding

anger-induced misunderstanding	fallacy in understanding
cognitive bias in understanding	incomplete meta-model in understanding
context-dependent misunderstanding	inconsistent meta-model in understanding
context-insensitive understanding	irrationality in understanding
corrupt understanding	irrelevant meta-model in understanding
culture-induced misunderstanding	paralogism in understanding
deliberate misinterpretation	personality in misunderstanding
deliberate misperception	sophism in understanding
deliberate misunderstanding	emotive bias in understanding
deliberate use of unfit meta-model	erroneous understanding
distorted understanding	hallucination
doctored (or tempered) data	illusion
doctored (or tempered) evidence	inability to understand
dogmatic understanding	incomplete understanding
double standards in understanding	inconsistent understanding
Dunning-Kruger effect in understanding	induced misunderstanding
effect of:	fallacy in understanding
-cognitive complexity in understanding	incomplete meta-model in understanding
-confusion in misunderstanding	inconsistent meta-model in understanding
-corrupt meta-model in understanding	irrationality in understanding
dysrationalia in understanding	irrelevant meta-model in understanding
erroneous interpretation in understanding	paralogism in understanding
erroneous meta-model in understanding	personality in misunderstanding
erroneous perception in understanding	sophism in understanding

emotive bias in understanding	mutual misunderstanding
erroneous understanding	perceptual confusion
hallucination	personal bias in understanding
illusion	Rorschach effect in understanding
inability to understand	schizophrenic understanding
incomplete understanding	Socratically induced misunderstanding
inconsistent understanding	speaker-induced misunderstanding
induced misunderstanding	stress-induced misunderstanding
information distortion	tempered (or doctored) data
irrelevant understanding	tempered (or doctored) evidence
miscognition	tunnel-vision understanding
misinterpretation	unfit meta-model in understanding
misperception	unreliable understanding
misunderstanding	unverified understanding

Author Index